THE UNSUPERVISED LEARNING WORKSHOP

Get started with unsupervised learning algorithms and simplify your unorganized data to help make future predictions

Aaron Jones, Christopher Kruger, and Benjamin Johnston

THE UNSUPERVISED LEARNING WORKSHOP

Copyright © 2020 Packt Publishing

All rights reserved. No part of this course may be reproduced, stored in a retrieval system, or transmitted in any form or by any means, without the prior written permission of the publisher, except in the case of brief quotations embedded in critical articles or reviews.

Every effort has been made in the preparation of this course to ensure the accuracy of the information presented. However, the information contained in this course is sold without warranty, either express or implied. Neither the authors, nor Packt Publishing, and its dealers and distributors will be held liable for any damages caused or alleged to be caused directly or indirectly by this course.

Packt Publishing has endeavored to provide trademark information about all of the companies and products mentioned in this course by the appropriate use of capitals. However, Packt Publishing cannot guarantee the accuracy of this information.

Authors: Aaron Jones, Christopher Kruger, and Benjamin Johnston

Reviewers: Richard Brooker, John Wesley Doyle, Priyanjit Ghosh, Sani Kamal, Ashish Pratik Patil, Geetank Raipuria, and Ratan Singh

Managing Editor: Rutuja Yerunkar

Acquisitions Editors: Manuraj Nair, Royluis Rodrigues, Anindya Sil, and Karan Wadekar

Production Editor: Salma Patel

Editorial Board: Megan Carlisle, Samuel Christa, Mahesh Dhyani, Heather Gopsill, Manasa Kumar, Alex Mazonowicz, Monesh Mirpuri, Bridget Neale, Dominic Pereira, Shiny Poojary, Abhishek Rane, Brendan Rodrigues, Erol Staveley, Ankita Thakur, Nitesh Thakur, and Jonathan Wray

First published: July 2020

Production reference: 2240221

ISBN: 978-1-80020-070-8

Published by Packt Publishing Ltd.

Livery Place, 35 Livery Street

Birmingham B3 2PB, UK

WHY LEARN WITH A PACKT WORKSHOP?

LEARN BY DOING

Packt Workshops are built around the idea that the best way to learn something new is by getting hands-on experience. We know that learning a language or technology isn't just an academic pursuit. It's a journey towards the effective use of a new tool—whether that's to kickstart your career, automate repetitive tasks, or just build some cool stuff.

That's why Workshops are designed to get you writing code from the very beginning. You'll start fairly small—learning how to implement some basic functionality—but once you've completed that, you'll have the confidence and understanding to move onto something slightly more advanced.

As you work through each chapter, you'll build your understanding in a coherent, logical way, adding new skills to your toolkit and working on increasingly complex and challenging problems.

CONTEXT IS KEY

All new concepts are introduced in the context of realistic use-cases, and then demonstrated practically with guided exercises. At the end of each chapter, you'll find an activity that challenges you to draw together what you've learned and apply your new skills to solve a problem or build something new.

We believe this is the most effective way of building your understanding and confidence. Experiencing real applications of the code will help you get used to the syntax and see how the tools and techniques are applied in real projects.

BUILD REAL-WORLD UNDERSTANDING

Of course, you do need some theory. But unlike many tutorials, which force you to wade through pages and pages of dry technical explanations and assume too much prior knowledge, Workshops only tell you what you actually need to know to be able to get started making things. Explanations are clear, simple, and to-the-point. So you don't need to worry about how everything works under the hood; you can just get on and use it.

Written by industry professionals, you'll see how concepts are relevant to real-world work, helping to get you beyond "Hello, world!" and build relevant, productive skills. Whether you're studying web development, data science, or a core programming language, you'll start to think like a problem solver and build your understanding and confidence through contextual, targeted practice.

ENJOY THE JOURNEY

Learning something new is a journey from where you are now to where you want to be, and this Workshop is just a vehicle to get you there. We hope that you find it to be a productive and enjoyable learning experience.

Packt has a wide range of different Workshops available, covering the following topic areas:

- Programming languages
- Web development
- Data science, machine learning, and artificial intelligence
- Containers

Once you've worked your way through this Workshop, why not continue your journey with another? You can find the full range online at http://packt.live/2MNkuyl.

If you could leave us a review while you're there, that would be great. We value all feedback. It helps us to continually improve and make better books for our readers, and also helps prospective customers make an informed decision about their purchase.

Thank you,
The Packt Workshop Team

Table of Contents

Preface   i

Chapter 1: Introduction to Clustering   1

Introduction 2

Unsupervised Learning versus Supervised Learning 2

Clustering 4

    Identifying Clusters 5

    Two-Dimensional Data 6

    Exercise 1.01: Identifying Clusters in Data 7

Introduction to k-means Clustering 11

    No-Math k-means Walkthrough 11

    K-means Clustering In-Depth Walkthrough 13

    Alternative Distance Metric – Manhattan Distance 14

    Deeper Dimensions 15

    Exercise 1.02: Calculating Euclidean Distance in Python 16

    Exercise 1.03: Forming Clusters with the Notion of Distance 18

    Exercise 1.04: K-means from Scratch – Part 1: Data Generation 20

    Exercise 1.05: K-means from Scratch – Part 2:
    Implementing k-means 24

    Clustering Performance – Silhouette Score 29

    Exercise 1.06: Calculating the Silhouette Score 31

    Activity 1.01: Implementing k-means Clustering 33

Summary 35

Chapter 2: Hierarchical Clustering — 37

Introduction 38

Clustering Refresher 38

    The k-means Refresher 39

The Organization of the Hierarchy 39

Introduction to Hierarchical Clustering 41

    Steps to Perform Hierarchical Clustering 43

    An Example Walkthrough of Hierarchical Clustering 43

    Exercise 2.01: Building a Hierarchy 47

Linkage 52

    Exercise 2.02: Applying Linkage Criteria 53

Agglomerative versus Divisive Clustering 58

    Exercise 2.03: Implementing Agglomerative Clustering
    with scikit-learn 60

    Activity 2.01: Comparing k-means with Hierarchical Clustering 64

k-means versus Hierarchical Clustering 68

Summary 69

Chapter 3: Neighborhood Approaches and DBSCAN — 71

Introduction 72

Clusters as Neighborhoods 73

Introduction to DBSCAN 75

    DBSCAN in Detail 76

    Walkthrough of the DBSCAN Algorithm 77

    Exercise 3.01: Evaluating the Impact
    of Neighborhood Radius Size 80

    DBSCAN Attributes – Neighborhood Radius 84

    Activity 3.01: Implementing DBSCAN from Scratch 86

    DBSCAN Attributes – Minimum Points 88

    Exercise 3.02: Evaluating the Impact of
    the Minimum Points Threshold 89

    Activity 3.02: Comparing DBSCAN with k-means
    and Hierarchical Clustering 93

DBSCAN versus k-means and Hierarchical Clustering 95

Summary 96

Chapter 4: Dimensionality Reduction Techniques and PCA   99

Introduction 100

What Is Dimensionality Reduction? 100

    Applications of Dimensionality Reduction 102

    The Curse of Dimensionality 104

Overview of Dimensionality Reduction Techniques 106

    Dimensionality Reduction 108

Principal Component Analysis 109

    Mean 109

    Standard Deviation 109

    Covariance 110

    Covariance Matrix 110

    Exercise 4.01: Computing Mean, Standard Deviation,
    and Variance Using the pandas Library 111

    Eigenvalues and Eigenvectors 116

    Exercise 4.02: Computing Eigenvalues and Eigenvectors 117

    The Process of PCA 121

    Exercise 4.03: Manually Executing PCA 123

Exercise 4.04: scikit-learn PCA 128

Activity 4.01: Manual PCA versus scikit-learn 133

Restoring the Compressed Dataset 136

Exercise 4.05: Visualizing Variance Reduction with Manual PCA 136

Exercise 4.06: Visualizing Variance Reduction with scikit-learn 143

Exercise 4.07: Plotting 3D Plots in Matplotlib 147

Activity 4.02: PCA Using the Expanded Seeds Dataset 150

Summary 153

Chapter 5: Autoencoders — 155

Introduction 156

Fundamentals of Artificial Neural Networks 157

    The Neuron 159

    The Sigmoid Function 160

    Rectified Linear Unit (ReLU) 161

    Exercise 5.01: Modeling the Neurons of
    an Artificial Neural Network 161

    Exercise 5.02: Modeling Neurons with
    the ReLU Activation Function 165

    Neural Networks: Architecture Definition 169

    Exercise 5.03: Defining a Keras Model 171

    Neural Networks: Training 173

    Exercise 5.04: Training a Keras Neural Network Model 175

    Activity 5.01: The MNIST Neural Network 185

Autoencoders 187

    Exercise 5.05: Simple Autoencoder 188

    Activity 5.02: Simple MNIST Autoencoder 193

Exercise 5.06: Multi-Layer Autoencoder	194
Convolutional Neural Networks	199
Exercise 5.07: Convolutional Autoencoder	200
Activity 5.03: MNIST Convolutional Autoencoder	205
Summary	207

Chapter 6: t-Distributed Stochastic Neighbor Embedding   209

Introduction	210
The MNIST Dataset	210
Stochastic Neighbor Embedding (SNE)	212
t-Distributed SNE	213
Exercise 6.01: t-SNE MNIST	214
Activity 6.01: Wine t-SNE	227
Interpreting t-SNE Plots	229
Perplexity	230
Exercise 6.02: t-SNE MNIST and Perplexity	230
Activity 6.02: t-SNE Wine and Perplexity	235
Iterations	236
Exercise 6.03: t-SNE MNIST and Iterations	237
Activity 6.03: t-SNE Wine and Iterations	242
Final Thoughts on Visualizations	243
Summary	243

Chapter 7: Topic Modeling   245

| Introduction | 246 |
| Topic Models | 247 |

Exercise 7.01: Setting up the Environment	249
A High-Level Overview of Topic Models	250
Business Applications	254
Exercise 7.02: Data Loading	256

Cleaning Text Data 259

Data Cleaning Techniques	260
Exercise 7.03: Cleaning Data Step by Step	261
Exercise 7.04: Complete Data Cleaning	266
Activity 7.01: Loading and Cleaning Twitter Data	268

Latent Dirichlet Allocation 270

Variational Inference	272
Bag of Words	275
Exercise 7.05: Creating a Bag-of-Words Model Using the Count Vectorizer	276
Perplexity	277
Exercise 7.06: Selecting the Number of Topics	279
Exercise 7.07: Running LDA	281
Visualization	286
Exercise 7.08: Visualizing LDA	287
Exercise 7.09: Trying Four Topics	291
Activity 7.02: LDA and Health Tweets	296
Exercise 7.10: Creating a Bag-of-Words Model Using TF-IDF	298

Non-Negative Matrix Factorization 299

The Frobenius Norm	301
The Multiplicative Update Algorithm	301
Exercise 7.11: Non-negative Matrix Factorization	302

Exercise 7.12: Visualizing NMF 306

Activity 7.03: Non-negative Matrix Factorization 309

Summary 310

Chapter 8: Market Basket Analysis 313

Introduction 314

Market Basket Analysis 314

    Use Cases 317

    Important Probabilistic Metrics 318

    Exercise 8.01: Creating Sample Transaction Data 319

    Support 321

    Confidence 322

    Lift and Leverage 323

    Conviction 324

    Exercise 8.02: Computing Metrics 325

Characteristics of Transaction Data 328

    Exercise 8.03: Loading Data 329

    Data Cleaning and Formatting 333

    Exercise 8.04: Data Cleaning and Formatting 334

    Data Encoding 339

    Exercise 8.05: Data Encoding 341

    Activity 8.01: Loading and Preparing Full Online Retail Data 343

The Apriori Algorithm 344

    Computational Fixes 347

    Exercise 8.06: Executing the Apriori Algorithm 348

    Activity 8.02: Running the Apriori Algorithm on
the Complete Online Retail Dataset 354

Association Rules 356
    Exercise 8.07: Deriving Association Rules 358
    Activity 8.03: Finding the Association Rules on
    the Complete Online Retail Dataset 365
Summary 367

Chapter 9: Hotspot Analysis   369

Introduction 370
Spatial Statistics 371
    Probability Density Functions 372
    Using Hotspot Analysis in Business 374
Kernel Density Estimation 375
    The Bandwidth Value 376
    Exercise 9.01: The Effect of the Bandwidth Value 376
    Selecting the Optimal Bandwidth 380
    Exercise 9.02: Selecting the Optimal Bandwidth
    Using Grid Search 381
    Kernel Functions 384
    Exercise 9.03: The Effect of the Kernel Function 387
    Kernel Density Estimation Derivation 389
    Exercise 9.04: Simulating the Derivation
    of Kernel Density Estimation 389
    Activity 9.01: Estimating Density in One Dimension 393
Hotspot Analysis 394
    Exercise 9.05: Loading Data and Modeling with Seaborn 396
    Exercise 9.06: Working with Basemaps 404
    Activity 9.02: Analyzing Crime in London 411

Summary	414
Appendix	**417**
Index	**521**

PREFACE

ABOUT THE BOOK

Do you find it difficult to understand how popular companies like WhatsApp and Amazon find valuable insights from large amounts of unorganized data? *The Unsupervised Learning Workshop* will give you the confidence to deal with cluttered and unlabeled datasets, using unsupervised algorithms in an easy and interactive manner.

The book starts by introducing the most popular clustering algorithms of unsupervised learning. You'll find out how hierarchical clustering differs from k-means, along with understanding how to apply DBSCAN to highly complex and noisy data. Moving ahead, you'll use autoencoders for efficient data encoding.

As you progress, you'll use t-SNE models to extract high-dimensional information into a lower dimension for better visualization, in addition to working with topic modeling for implementing Natural Language Processing. In later chapters, you'll find key relationships between customers and businesses using Market Basket Analysis, before going on to use Hotspot Analysis for estimating the population density of an area.

By the end of this book, you'll be equipped with the skills you need to apply unsupervised algorithms on cluttered datasets to find useful patterns and insights.

AUDIENCE

If you are a data scientist who is just getting started and want to learn how to implement machine learning algorithms to build predictive models, then this book is for you. To expedite the learning process, a solid understanding of the Python programming language is recommended, as you'll be editing classes and functions instead of creating them from scratch.

ABOUT THE CHAPTERS

Chapter 1, Introduction to Clustering, introduces clustering (the most well-known family of unsupervised learning algorithms), before digging into the simplest and most popular clustering algorithm—k-means.

Chapter 2, Hierarchical Clustering, covers another clustering technique, hierarchical clustering, and explains how it differs from k-means. The chapter teaches you two main approaches to this type of clustering: agglomerative and divisive.

Chapter 3, Neighborhood Approaches and DBSCAN, explores clustering approaches that involve neighbors. Unlike the two other clustering approaches, the neighborhood approaches allow outlier points that are not assigned to any particular cluster.

Chapter 4, Dimensionality Reduction and PCA, teaches you how to navigate large feature spaces by leveraging principal component analysis to reduce the number of features while maintaining the explanatory power of the whole feature space.

Chapter 5, Autoencoders, shows you how neural networks can be leveraged to find data encodings. Data encodings are like combinations of features that reduce the dimensionality of the feature space. Autoencoders also decode the data and put it back into its original form.

Chapter 6, t-Distributed Stochastic Neighbor Embedding, discusses the process of reducing high-dimensional datasets down to two or three dimensions for the purpose of visualization. Unlike PCA, t-SNE is a non-linear, probabilistic model.

Chapter 7, Topic Modeling, explores the fundamental methodology of natural language processing. You will learn how to work with text data and fit Latent Dirichlet Allocation and Non-negative Matrix Factorization models to tag topics relevant to the text.

Chapter 8, Market Basket Analysis, explores a classic analytical technique used in retail businesses. You will, in a scalable way, build association rules that explain the relationships between groups of items.

Chapter 9, Hotspot Analysis, teaches you to estimate the true population density of some random variable using sample data. This technique is applicable to many fields, including epidemiology, weather, crime, and demography.

CONVENTIONS

Code words in text, database table names, folder names, filenames, file extensions, pathnames, dummy URLs, user input, and Twitter handles are shown as follows:

"Plot the coordinate points using the scatterplot functionality we imported from `matplotlib.pyplot`."

Words that you see on the screen (for example, in menus or dialog boxes) appear in the same format.

A block of code is set as follows:

```
import pandas as pd
import numpy as np
import matplotlib.pyplot as plt
from sklearn.metrics import silhouette_score
from scipy.spatial.distance import cdist
seeds = pd.read_csv('Seed_Data.csv')
```

New terms and important words are shown like this:

"**Unsupervised learning** is the field of practice that helps find patterns in cluttered data and is one of the most exciting areas of development in machine learning today."

Long code snippets are truncated and the corresponding names of the code files on GitHub are placed at the top of the truncated code. The permalinks to the entire code are placed below the code snippet. It should look as follows:

Exercise1.04-Exercise1.05.ipynb

```
def k_means(X, K):
    # Keep track of history so you can see K-Means in action
    centroids_history = []
    labels_history = []
    rand_index = np.random.choice(X.shape[0], K)
    centroids = X[rand_index]
    centroids_history.append(centroids)
```

The complete code for this step can be found at https://packt.live/2JM8Q1S.

CODE PRESENTATION

Lines of code that span multiple lines are split using a backslash (\ ). When the code is executed, Python will ignore the backslash, and treat the code on the next line as a direct continuation of the current line.

For example:

```
history = model.fit(X, y, epochs=100, batch_size=5, verbose=1, \
                    validation_split=0.2, shuffle=False)
```

Comments are added into code to help explain specific bits of logic. Single-line comments are denoted using the **#** symbol, as follows:

```
# Print the sizes of the dataset
print("Number of Examples in the Dataset = ", X.shape[0])
print("Number of Features for each example = ", X.shape[1])
```

Multi-line comments are enclosed by triple quotes, as shown below:

```
"""
Define a seed for the random number generator to ensure the
result will be reproducible
"""
seed = 1
np.random.seed(seed)
random.set_seed(seed)
```

SETTING UP YOUR ENVIRONMENT

Before we explore the book in detail, we need to set up specific software and tools. In the following section, we shall see how to do that.

HARDWARE REQUIREMENTS

For the optimal user experience, we recommend 8 GB RAM.

INSTALLING PYTHON

The following section will help you to install Python in Windows, macOS, and Linux systems.

INSTALLING PYTHON ON WINDOWS

1. Find your desired version of Python on the official installation page at https://www.python.org/downloads/windows/.

2. Ensure that you install the correct "-bit" version depending on your computer system, either 32-bit or 64-bit. You can find out this information in the **System Properties** window of your OS.

3. After you download the installer, simply double-click the file and follow the user-friendly prompts on the screen.

INSTALLING PYTHON ON LINUX

1. Open a Terminal and verify Python 3 is not already installed by running **python3 --version**.

2. To install Python 3, run the following:

   ```
   sudo apt-get update
   sudo apt-get install python3.7
   ```

3. If you encounter problems, there are numerous sources online that can help you troubleshoot the issue.

INSTALLING PYTHON ON MACOS

Here are the steps to install Python on macOS:

1. Open the Terminal by holding *Cmd + Space*, typing **terminal** in the open search box, and hitting *Enter*.

2. Install Xcode through the command line by running **xcode-select --install**.

3. The easiest way to install Python 3 is with Homebrew, which is installed through the command line by running **ruby -e "$(curl -fsSL https://raw.githubusercontent.com/Homebrew/install/master/install)"**

4. Add Homebrew to your **PATH** environment variable. Open your profile in the command line by running **sudo nano ~/.profile** and inserting **export PATH="/usr/local/opt/python/libexec/bin:$PATH"** at the bottom.

5. The final step is to install Python. In the command line, run **brew install python**.

6. Note that if you install Anaconda, the latest version of Python will be installed automatically.

INSTALLING PIP

Python does not come with **pip** (the package manager for Python) pre-installed, so we need to install it manually. Once **pip** is installed, the remaining libraries can be installed as mentioned in the *Installing Libraries* section. The steps to install **pip** are as follows:

1. Go to https://bootstrap.pypa.io/get-pip.py and save the file as **get-pip.py**.

2. Go to the folder where you have saved **get-pip.py**. Open the command line in that folder (Bash for Linux users and Terminal for Mac users).

3. Execute following command in the command line:

    ```
    python get-pip.py
    ```

    Please note that you should have Python installed before executing this command.

4. Once **pip** is installed, you can install the desired libraries. To install pandas, you can simply execute **pip install pandas**. To install a specific version of a library, for example, version 0.24.2 of **pandas**, you can execute **pip install pandas=0.24.2**.

INSTALLING ANACONDA

Anaconda is a Python package manager that easily allows you to install and use the libraries needed for this course.

INSTALLING ANACONDA ON WINDOWS

1. Anaconda installation for Windows is very user-friendly. Visit the download page to get the installation executable at https://www.anaconda.com/distribution/#download-section.

2. Double-click the installer on your computer.

3. Follow the prompts on screen to complete the installation of Anaconda.

4. After installation, you can access Anaconda Navigator, which will be available alongside the rest of your applications as normal.

INSTALLING ANACONDA ON LINUX

1. Visit the Anaconda download page to get the installation shell script, at https://www.anaconda.com/distribution/#download-section.

2. To download the shell script directly to your Linux instance you can use the **curl** or **wget** retrieval libraries. The example here shows how to use **curl** to retrieve the file located at the URL you found on the Anaconda download page:

   ```
   curl -O https://repo.anaconda.com/archive/Anaconda3-2019.03-Linux-x86_64.sh
   ```

3. After downloading the shell script, you can run it with the following command:

   ```
   bash Anaconda3-2019.03-Linux-x86_64.sh
   ```

   Running the preceding command will move you to a very user-friendly installation process. You will be prompted on where you want to install Anaconda and how you wish Anaconda to work. In this case, you should just keep all the standard settings.

INSTALLING ANACONDA ON MACOS X

1. Anaconda installation for macOS is very user-friendly. Visit the download page to get the installation executable, at https://www.anaconda.com/distribution/#download-section.

2. Make sure macOS is selected and double-click the **Download** button for the Anaconda installer.

3. Follow the prompts on screen to complete the installation of Anaconda.

4. After installation, you can access Anaconda Navigator, which will be available alongside the rest of your applications as normal.

SETTING UP A VIRTUAL ENVIRONMENT

1. After Anaconda is installed, you must create environments where you will install packages you wish to use. The great thing about Anaconda environments is that you can build individual environments for specific projects you're working on. To create a new environment, use the following command:

   ```
   conda create --name my_packt_env python=3.7
   ```

INSTALLING PIP

Python does not come with **pip** (the package manager for Python) pre-installed, so we need to install it manually. Once **pip** is installed, the remaining libraries can be installed as mentioned in the *Installing Libraries* section. The steps to install **pip** are as follows:

1. Go to https://bootstrap.pypa.io/get-pip.py and save the file as **get-pip.py**.

2. Go to the folder where you have saved **get-pip.py**. Open the command line in that folder (Bash for Linux users and Terminal for Mac users).

3. Execute following command in the command line:

   ```
   python get-pip.py
   ```

   Please note that you should have Python installed before executing this command.

4. Once **pip** is installed, you can install the desired libraries. To install pandas, you can simply execute **pip install pandas**. To install a specific version of a library, for example, version 0.24.2 of **pandas**, you can execute **pip install pandas=0.24.2**.

INSTALLING ANACONDA

Anaconda is a Python package manager that easily allows you to install and use the libraries needed for this course.

INSTALLING ANACONDA ON WINDOWS

1. Anaconda installation for Windows is very user-friendly. Visit the download page to get the installation executable at https://www.anaconda.com/distribution/#download-section.

2. Double-click the installer on your computer.

3. Follow the prompts on screen to complete the installation of Anaconda.

4. After installation, you can access Anaconda Navigator, which will be available alongside the rest of your applications as normal.

INSTALLING ANACONDA ON LINUX

1. Visit the Anaconda download page to get the installation shell script, at https://www.anaconda.com/distribution/#download-section.

2. To download the shell script directly to your Linux instance you can use the `curl` or `wget` retrieval libraries. The example here shows how to use `curl` to retrieve the file located at the URL you found on the Anaconda download page:

   ```
   curl -O https://repo.anaconda.com/archive/Anaconda3-2019.03-Linux-x86_64.sh
   ```

3. After downloading the shell script, you can run it with the following command:

   ```
   bash Anaconda3-2019.03-Linux-x86_64.sh
   ```

   Running the preceding command will move you to a very user-friendly installation process. You will be prompted on where you want to install Anaconda and how you wish Anaconda to work. In this case, you should just keep all the standard settings.

INSTALLING ANACONDA ON MACOS X

1. Anaconda installation for macOS is very user-friendly. Visit the download page to get the installation executable, at https://www.anaconda.com/distribution/#download-section.

2. Make sure macOS is selected and double-click the **Download** button for the Anaconda installer.

3. Follow the prompts on screen to complete the installation of Anaconda.

4. After installation, you can access Anaconda Navigator, which will be available alongside the rest of your applications as normal.

SETTING UP A VIRTUAL ENVIRONMENT

1. After Anaconda is installed, you must create environments where you will install packages you wish to use. The great thing about Anaconda environments is that you can build individual environments for specific projects you're working on. To create a new environment, use the following command:

   ```
   conda create --name my_packt_env python=3.7
   ```

Here, we are naming our environment **my_packt_env** and specifying the version of Python to be 3.7. Thus you can have multiple versions of Python installed in the environment that will be virtually separate.

2. Once the environment is created, you can activate it using the well-named **activate** command:

```
conda activate my_packt_env
```

That's it. You are now in your own customized environment that will allow you to install packages as needed for your projects. To exit your environment, you can simply use the **conda deactivate** command.

INSTALLING LIBRARIES

pip comes pre-installed with Anaconda. Once Anaconda is installed on your machine, all the required libraries can be installed using **pip**, for example, **pip install numpy**. Alternatively, you can install all the required libraries using **pip install -r requirements.txt**. You can find the **requirements.txt** file at https://packt.live/2CnpCEp.

The exercises and activities will be executed in Jupyter Notebooks. Jupyter is a Python library and can be installed in the same way as the other Python libraries – that is, with **pip install jupyter**, but fortunately, it comes pre-installed with Anaconda. To open a notebook, simply run the command **jupyter notebook** in the Terminal or Command Prompt.

In *Chapter 9, Hotspot Analysis*, the **basemap** module from **mpl_toolkits** is used to generate maps. This library can be difficult to install. The easiest way is to install Anaconda, which includes **mpl_toolkits**. Once Anaconda is installed, **basemap** can be installed using **conda install basemap**. If you want to avoid installing libraries repeatedly, and instead want to install them all at once, you can follow the instructions in the next section.

SETTING UP THE MACHINE

It might be that if you are installing dependencies chapter by chapter, the version of the libraries could be different. In order to sync the system, we provide a **requirements.txt** file that contains the versions of the libraries used. Once you have installed the libraries using this, you don't have to install any other libraries throughout the book. Assuming you have installed Anaconda by now, you can follow these steps:

1. Download the **requirements.txt** file from GitHub.
2. Go to the folder where **requirements.txt** is placed and open Command Prompt (Bash for Linux and Terminal for Mac).
3. Execute the following command on it:

   ```
   conda install --yes --file requirements.txt --channel conda-forge
   ```

   It should install all the packages necessary for the coding activities in the book.

ACCESSING THE CODE FILES

You can find the complete code files of this book at https://packt.live/34kXeMw. You can also run many activities and exercises directly in your web browser by using the interactive lab environment at https://packt.live/2ZMUWW0.

We've tried to support interactive versions of all activities and exercises, but we recommend a local installation as well for instances where this support isn't available.

If you have any issues or questions about installation, please email us at **workshops@packt.com**.

1
INTRODUCTION TO CLUSTERING

OVERVIEW

Finding insights and value in data is the ambitious promise that has been seen in the rise of machine learning. Within machine learning, there are predictive approaches to understanding dense information in deeper ways, as well as approaches to predicting outcomes based on changing inputs. In this chapter, we will learn what supervised learning and unsupervised learning are, and how they are applied to different use cases. Once you have a deeper understanding of where unsupervised learning is useful, we will walk through some foundational techniques that provide value quickly.

By the end of this chapter, you will be able to implement k-means clustering algorithms using built-in Python packages and calculate the silhouette score.

INTRODUCTION

Have you ever been asked to take a look at some data and came up empty handed? Maybe you weren't familiar with the dataset, or maybe you didn't even know where to start. This may have been extremely frustrating, and even embarrassing, depending on who asked you to take care of the task.

You are not alone, and, interestingly enough, there are many times the data itself is simply too confusing to be made sense of. As you try and figure out what all those numbers in your spreadsheet mean, you're most likely mimicking what many unsupervised algorithms do when they try to find meaning in data. The reality is that many unprocessed real-world datasets may not have any useful insights. One example to consider is the fact that these days, individuals generate massive amounts of granular data on a daily basis – whether it's their actions on a website, their purchase history, or what apps they use on their phone. If you were to look at this information on the surface, it would be a big, unorganized mess with no hope of clarity. Don't fret, however; this book will prepare you for such tall tasks so that you'll never be frustrated again when dealing with data exploration tasks, no matter how large.

For this book, we have developed some best-in-class content to help you understand how unsupervised algorithms work and where to use them. We'll cover some of the foundations of finding clusters in your data, how to reduce the size of your data so it's easier to understand, and how each of these sides of unsupervised learning can be applied in the real world. We hope you will come away from this book with a strong real-world understanding of unsupervised learning, the problems that it can solve, and those it cannot.

UNSUPERVISED LEARNING VERSUS SUPERVISED LEARNING

Unsupervised learning is the field of practice that helps find patterns in cluttered data and is one of the most exciting areas of development in machine learning today. If you have explored machine learning bookwork before, you are probably familiar with the common breakout of problems in either supervised or unsupervised learning. **Supervised learning** encompasses the problem set of having a labeled dataset that can be used to either classify data (for example, predicting smokers and non-smokers, if you're looking at a lung health dataset) or finding a pattern in clearly defined data (for example, predicting the sale price of a home based on how many bedrooms it has). This model most closely mirrors an intuitive human approach to learning.

For example, if you wanted to learn how to not burn your food with a basic understanding of cooking, you could build a dataset by putting your food on the burner and seeing how long it takes (input) for your food to burn (output). Eventually, as you continue to burn your food, you will build a mental model of when burning will occur and how to avoid it in the future. Development in supervised learning was once fast paced and valuable, but it has simmered down in recent years. Many of the obstacles around getting to know your data have already been tackled and are listed in the following image:

Figure 1.1: Differences between unsupervised and supervised learning

Conversely, unsupervised learning encompasses the problem set of having a tremendous amount of data that is unlabeled. Labeled data, in this case, would be data that has a supplied "target" outcome that you are trying to find the correlation to with supplied data. For instance, in the preceding example, you know that your "target outcome" is whether your food was burned; this is an example of labeled data. Unlabeled data is when you do not know what the "target" outcome is, and you have only supplied input data.

Building upon the previous example, imagine you were just dropped on planet Earth with zero knowledge of how cooking works. You are given 100 days, a stove, and a fridge full of food without any instructions on what to do. Your initial exploration of a kitchen could go in infinite directions. On day 10, you may finally learn how to open the fridge; on day 30, you may learn that food can go on the stove; and after many more days, you may unwittingly make an edible meal. As you can see, trying to find meaning in a kitchen devoid of adequate informational structure leads to very noisy data that is completely irrelevant to actually preparing a meal.

Unsupervised learning can be an answer to this problem. Looking back at your 100 days of data, you can use **clustering** to find patterns of similar attributes across days and deduce which foods are similar and may lead to a "good" meal. However, unsupervised learning isn't a magical answer. Simply finding clusters can be just as likely to help you find pockets of similar, yet ultimately useless, data. Expanding on the cooking example, we can illustrate this shortcoming with the concept of the "third variable". Just because you have a cluster of really great recipes doesn't mean they are infallible. During your research, you may have found a unifying factor that all good meals were cooked on a stove. This does not mean that every meal cooked on a stove will be good, and you cannot easily jump to that conclusion for all future scenarios.

This challenge is what makes unsupervised learning so exciting. How can we find smarter techniques to speed up the process of finding clusters of information that are beneficial to our end goals? The following sections would help us answer this question.

CLUSTERING

Clustering is the overarching process that involves finding groups of similar data that exist in your dataset, which can be extremely valuable if you are trying to find its underlying meaning. If you were a store owner and you wanted to understand which customers are more valuable without a set idea of what valuable is, clustering would be a great place to start to find patterns in your data. You may have a few high-level ideas of what denotes a valuable customer, but you aren't entirely sure in the face of a large mountain of available data. Through clustering, you can find commonalities among similar groups in your data. For example, if you look more deeply at a cluster of similar people, you may learn that everyone in that group visits your website for longer periods of time than others. This can show you what the value is and also provide a clean sample size for future supervised learning experiments.

IDENTIFYING CLUSTERS

The following image shows two scatterplots:

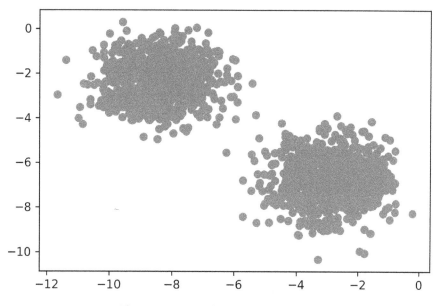

Figure 1.2: Two distinct scatterplots

The following image separates the two scatterplots into two distinct clusters:

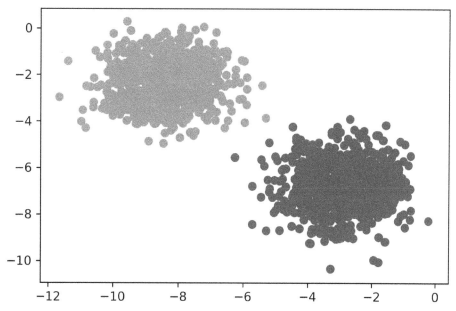

Figure 1.3: Scatterplots clearly showing clusters that exist in a provided dataset

Figure 1.2 and *Figure 1.3* display randomly generated number pairs (x and y coordinates) pulled from two distinct Gaussian distributions centered at different locations. Simply by glancing at the first image, it should be obvious where the clusters exist in your data; in real life, it will never be this easy. Now that you know that the data can be clearly separated into two clusters, you can start to understand what differences exist between the two groups.

Rewinding a bit from where unsupervised learning fits into the larger machine learning environment, let's begin by understanding the building blocks of clustering. The most basic definition finds clusters simply as groupings of similar data as subsets of a larger dataset. As an example, imagine that you had a room with 10 people in it and each person had a job either in finance or as a scientist. If you told all the financial workers to stand together and all the scientists to do the same, you would have effectively formed two clusters based on job types. Finding clusters can be immensely valuable in identifying items that are more similar and, on the other end of the scale, quite different from one another.

TWO-DIMENSIONAL DATA

To understand this, imagine that you were given a simple 1,000-row dataset by your employer that had two columns of numerical data, as follows:

```
array([[-0.72690901,  2.76012303],
       [-1.38504876,  2.16558784],
       [-1.12519969,  0.78279526],
       ...,
       [-0.92272983, -0.44782031],
       [ 8.26124228, -0.37099837],
       [-1.01204517,  0.3228703 ]])
```

Figure 1.4: Two-dimensional raw data in an array

At first glance, this dataset provides no real structure or understanding.

A **dimension** in a dataset is another way of simply counting the number of features available. In most organized data tables, you can view the number of features as the number of columns. So, using the 1,000-row dataset example of size (1,000 x 2), you will have 1,000 observations across two dimensions. Please note that dimensions of dataset should not be confused with the dimensions of an array.

Clustering | 7

You begin by plotting the first column against the second column to get a better idea of what the data structure looks like. There will be plenty of times where the cause of differences between groups will prove to be underwhelming; however, the cases that have differences that you can take action on are extremely rewarding.

EXERCISE 1.01: IDENTIFYING CLUSTERS IN DATA

You are given two-dimensional plots of data that you suspect have clusters of similar data. Please look at the two-dimensional graphs provided in the exercise and identify the groups of data points to drive the point home that machine learning is important. Without using any algorithmic approaches, identify where these clusters exist in the data.

This exercise will help you start building your intuition of how we can identify clusters using our own eyes and thought processes. As you complete this exercise, think of the rationale of why a group of data points should be considered a cluster versus a group that should not be considered a cluster. Follow these steps to complete this exercise:

1. Identify the clusters in the following scatterplot:

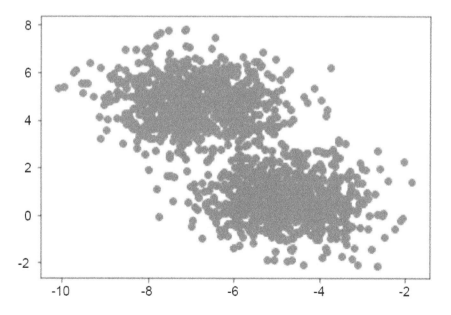

Figure 1.5: Two-dimensional scatterplot

8 | Introduction to Clustering

The clusters are as follows:

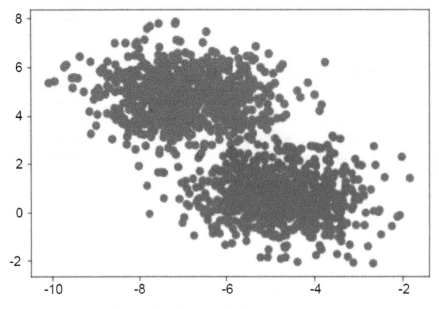

Figure 1.6: Clusters in the scatterplot

2. Identify the clusters in the following scatterplot:

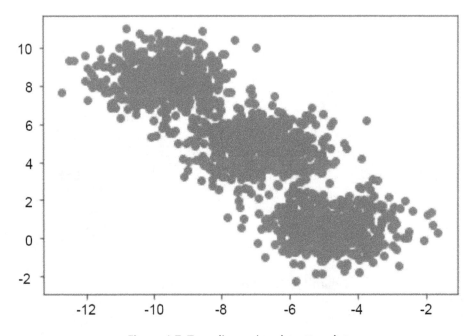

Figure 1.7: Two-dimensional scatterplot

The clusters are as follows:

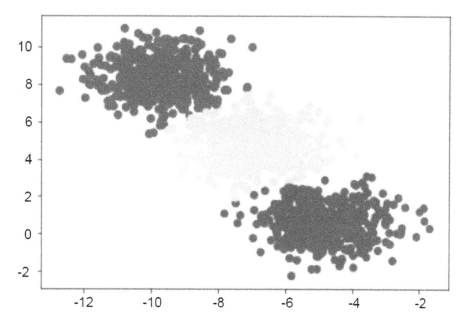

Figure 1.8: Clusters in the scatterplot

3. Identify the clusters in the following scatterplot:

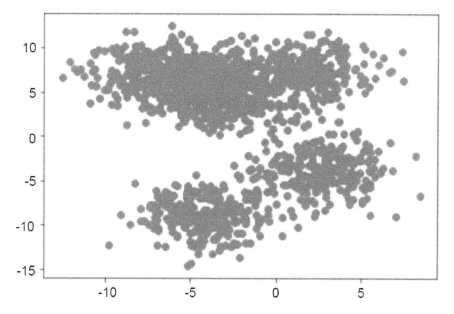

Figure 1.9: Two-dimensional scatterplot

The clusters are as follows:

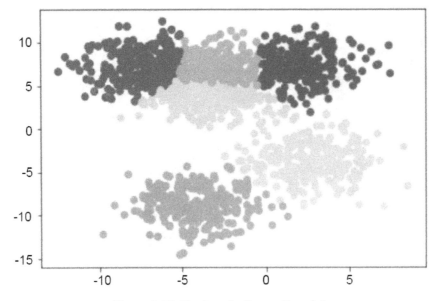

Figure 1.10: Clusters in the scatterplot

Most of these examples were likely quite easy for you to understand, and that's the point. The human brain and eyes are incredible at finding patterns in the real world. Within milliseconds of viewing each plot, you could tell what fitted together and what didn't. While it is easy for you, a computer does not have the ability to see and process plots in the same manner that we do.

However, this is not always a bad thing. Look back at the preceding scatterplot. Were you able to find the six discrete clusters in the data just by looking at the plot? You probably found only three to four clusters in this scatterplot, while a computer would be able to see all six. The human brain is magnificent, but it also lacks the nuances that come with a strictly logic-based approach. Through algorithmic clustering, you will learn how to build a model that works even better than a human at these tasks.

We'll look at the clustering algorithm in the next section.

INTRODUCTION TO K-MEANS CLUSTERING

Hopefully, by now, you can see that finding clusters is extremely valuable in a machine learning workflow. But, how can you actually find these clusters? One of the most basic yet popular approaches is to use a cluster analysis technique called **k-means clustering**. The k-means clustering works by searching for k clusters in your data and the workflow is actually quite intuitive. We will start with the no-math introduction to k-means, followed by an implementation in Python. **Cluster membership** refers to where the points go as the algorithm processes the data. Consider it like choosing players for a sports team, where all the players are in a pool but, for each successive run, the player is assigned to a team (in this case, a cluster).

NO-MATH K-MEANS WALKTHROUGH

The no-math algorithm for k-means clustering is pretty simple:

1. First, we'll pick "k" centroids, where "k" would be the expected distinct number of clusters. The value of k will be chosen by us and determines the type of clustering we obtain.

2. Then, we will place the "k" centroids at random places among the existing training data.

3. Next, the distance from each centroid to all the points in the training data will be calculated. We will go into detail about distance functions shortly, but for now, let's just consider it as how far points are from each other.

4. Now, all the training points will be grouped with their nearest centroid.

5. Isolating the grouped training points along with their respective centroid, calculate the mean data point in the group and move the previous centroid to the mean location.

6. This process is to be repeated until convergence or until maximum iteration limit has been achieved.

And that's it. The following image represents original raw data:

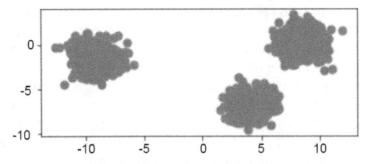

Figure 1.11: Original raw data charted on x and y coordinates

Provided with the original data in the preceding image, we can visualize the iterative process of k-means by showing the predicted clusters in each step:

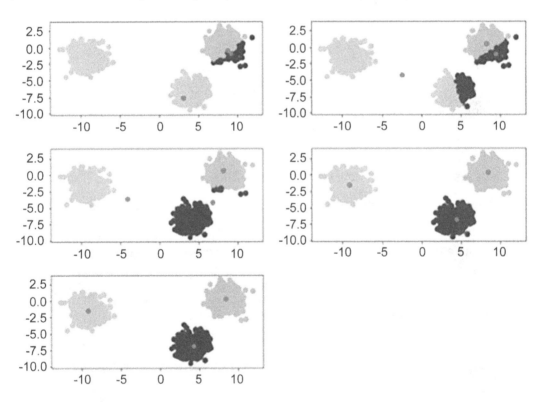

Figure 1.12: Reading from left to right, red points are randomly initialized centroids, and the closest data points are assigned to groupings of each centroid

K-MEANS CLUSTERING IN-DEPTH WALKTHROUGH

To understand k-means at a deeper level, let's walk through the example that was provided in the introduction again with some of the math that supports k-means. The most important math that underpins this algorithm is the distance function. A distance function is basically any formula that allows you to quantitatively understand how far one object is from another, with the most popular one being the Euclidean distance formula. This formula works by subtracting the respective components of each point and squaring to remove negatives, followed by adding the resulting distances and square rooting them:

$$d((x,y),(a,b)) = \sqrt{(x-a)^2 + (y-b)^2}$$

Figure 1.13: Euclidean distance formula

If you notice, the preceding formula holds true for data points having only two dimensions (the number of co-ordinates). A generic way of representing the preceding equation for higher-dimensional points is as follows:

$$\text{Euclidean Distance } (p,q) = \sqrt{\sum_{i=1}^{n}(p_i - q_i)^2}$$

Figure 1.14: Euclidean distance formula for higher dimensional points

Let's see the terms involved in calculation of Euclidean distance between two points p and q in a higher dimensional space. Here, n is the number of dimensions of the two points. We compute the difference between the respective components of points p and q (p_i and q_i are known as the ith component of point p and q respectively) and square each of them. This squared value of the difference is summed up for all n components, and then square root of this sum is obtained. This value represents the Euclidean distance between point p and q. If you substitute n = 2 in the preceding equation, it will decompose to the equation represented in *Figure 1.13*.

Now coming back again to our discussion on k-means. Centroids are randomly set at the beginning as points in your n-dimensional space. Each of these centers is fed into the preceding formula as (*a, b*), and a point in your space is fed in as (*x, y*). Distances are calculated between each point and the coordinates of every centroid, with the centroid the shortest distance away chosen as the point's group.

As an example, let's pick three random centroids, an arbitrary point, and, using the Euclidean distance formula, calculate the distance from each point to the centroid:

- Random centroids: [(2,5), (8,3), (4,5)].
- Arbitrary point x: (0, 8).
- Distance from point to each centroid: [3.61, 9.43, 5.00].

Since the arbitrary point x is closest to the first centroid, it will be assigned to the first centroid.

ALTERNATIVE DISTANCE METRIC – MANHATTAN DISTANCE

Euclidean distance is the most common distance metric for many machine learning applications and is often known colloquially as the distance metric; however, it is not the only, or even the best, distance metric for every situation. Another popular distance metric that can be used for clustering is **Manhattan distance**.

Manhattan distance is called as such because it mirrors the concept of traveling through a metropolis (such as New York City) that has many square blocks. Euclidean distance relies on diagonals due to its basis in Pythagorean theorem, while Manhattan distance constrains distance to only right angles. The formula for Manhattan distance is as follows:

$$\text{Manhattan Distance } (p,q) = \sum_{i=1}^{n} |p_i - q_i|$$

Figure 1.15: Manhattan distance formula

Here p_i and q_i are the ith component of points p and q, respectively. Building upon our examples of Euclidean distance, where we want to find the distance between two points, if our two points were (1,2) and (2,3), then the Manhattan distance would equal |1-2| + |2-3| = 1 + 1 = 2. This functionality scales to any number of dimensions. In practice, Manhattan distance may outperform Euclidean distance when it comes to high dimensional data.

DEEPER DIMENSIONS

The preceding examples can be clearly visualized when your data is only two-dimensional. This is for convenience, to help drive the point home of how k-means works and could lead you into a false understanding of how easy clustering is. In many of your own applications, your data will likely be orders of magnitude larger to the point that it cannot be perceived by visualization (anything beyond three dimensions will be unperceivable to humans). In the previous examples, you could mentally work out a few two-dimensional lines to separate the data into its own groups. At higher dimensions, you will need to be aided by a computer to find an n-dimensional hyperplane that adequately separates the dataset. In practice, this is where clustering methods such as k-means provide significant value. The following image shows the two-dimensional, three-dimensional, and n-dimensional plots:

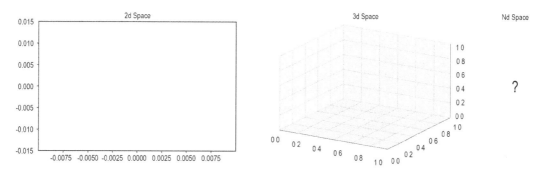

Figure 1.16: Two-dimensional, three-dimensional, and n-dimensional plots

In the next exercise, we will calculate Euclidean distance. We'll build our set of tools by using the **NumPy** and **Math** Python packages. **NumPy** is a scientific computing package for Python that pre-packages common mathematical functions in highly optimized formats.

As the name implies, the **Math** package is a basic library that makes implementing foundational math building blocks, such as exponentials and square roots, much easier. By using a package such as **NumPy** or **Math**, we help cut down the time spent creating custom math functions from scratch and instead focus on developing our solutions. You will see how each of these packages is used in practice in the following exercise.

EXERCISE 1.02: CALCULATING EUCLIDEAN DISTANCE IN PYTHON

In this exercise, we will create an example point along with three sample centroids to help illustrate how Euclidean distance works. Understanding this distance formula is the basis for the rest of our work in clustering.

Perform the following steps to complete this exercise:

1. Open a Jupyter notebook and create a naïve formula that captures the direct math of Euclidean distance, as follows:

```
import math
import numpy as np
def dist(a, b):
    return math.sqrt(math.pow(a[0]-b[0],2) \
                    + math.pow(a[1]-b[1],2))
```

> **NOTE**
>
> The code snippet shown here uses a backslash (\ ) to split the logic across multiple lines. When the code is executed, Python will ignore the backslash, and treat the code on the next line as a direct continuation of the current line.

This approach is considered naïve because it performs element-wise calculations on your data points (slow) compared to a more real-world implementation using vectors and matrix math to achieve significant performance increases.

2. Create the data points in Python as follows:

```
centroids = [ (2, 5), (8, 3), (4,5) ]
x = (0, 8)
```

3. Use the formula you created to calculate the Euclidean distance in *Step 1*:

```
# Calculating Euclidean Distance between x and centroid
centroid_distances =[]
for centroid in centroids:
    print("Euclidean Distance between x {} and centroid {} is {}"\
        .format(x ,centroid, dist(x,centroid)))
    centroid_distances.append(dist(x,centroid))
```

> **NOTE**
>
> The # symbol in the code snippet above denotes a code comment. Comments are added into code to help explain specific bits of logic.

The output is as follows:

```
Euclidean Distance between x (0, 8) and centroid (2, 5)
is 3.605551275463989
Euclidean Distance between x (0, 8) and centroid (8, 3)
is 9.433981132056603
Euclidean Distance between x (0, 8) and centroid (4, 5) is 5.0
```

The shortest distance between our point, **x**, and the centroids is `3.61`, which is equivalent to the distance between `(0, 8)` and `(2, 5)`. Since this is the minimum distance, our example point, **x**, will be assigned as a member of the first centroid's group.

In this example, our formula was used on a single point, x (0, 8). Beyond this single point, the same process will be repeated for every remaining point in your dataset until each point is assigned to a cluster. After each point is assigned, the mean point is calculated among all of the points within each cluster. The calculation of the mean among these points is the same as calculating the mean between single integers.

18 | Introduction to Clustering

While there was only one point in this example, by completing this process, you have effectively assigned a point to its first cluster using Euclidean distance. We'll build upon this approach with more than one point in the following exercise.

> **NOTE**
>
> To access the source code for this specific section, please refer to https://packt.live/2VUvCuz.
>
> You can also run this example online at https://packt.live/3ebDwpZ.

EXERCISE 1.03: FORMING CLUSTERS WITH THE NOTION OF DISTANCE

It is very intuitive for our human minds to see groups of dots on a plot and determine which dots belong to discrete clusters. However, how do we ask a computer to repeat this same task? In this exercise, you'll help teach a computer an approach to forming clusters of its own with the notion of distance. We will build upon how we use these distance metrics in the next exercise:

1. Create a list of points, [(0,8), (3,8), (3,4)], that are assigned to cluster one:

    ```
    cluster_1_points =[ (0,8), (3,8), (3,4) ]
    ```

2. To find the new centroid among your list of points, calculate the mean point between all of the points. Calculation of the mean scales to infinite points, as you simply add the integers at each position and divide by the total number of points. For example, if your two points are (0,1,2) and (3,4,5), the mean calculation would be [(0+3)/2, (1+4)/2, (2+5)/2]:

    ```
    mean =[ (0+3+3)/3,  (8+8+4)/3 ]
    print(mean)
    ```

    The output is as follows:

    ```
    [2.0, 6.666666666666667]
    ```

After a new centroid is calculated, repeat the cluster membership calculation we looked at in *Exercise 1.02, Calculating Euclidean Distance in Python*, and then repeat the previous two steps to find the new cluster centroid. Eventually, the new cluster centroid will be the same as the centroid before the cluster membership calculation and the exercise will be complete. How many times this repeats depends on the data you are clustering.

Once you have moved the centroid location to the new mean point of (2, 6.67), you can compare it to the initial list of centroids you entered the problem with. If the new mean point is different than the centroid that is currently in your list, you will have to go through another iteration of the preceding two exercises. Once the new mean point you calculate is the same as the centroid you started the problem with, you have completed a run of k-means and reached a point called **convergence**. However, in practice, sometimes the number of iterations required to reach convergence is very large and such large computations may not be practically feasible. In such cases, we need to set a maximum limit to the number of iterations. Once this iteration limit is reached, we stop further processing.

> **NOTE**
>
> To access the source code for this specific section, please refer to https://packt.live/3iJ3jiT.
>
> You can also run this example online at https://packt.live/38CCpOG.

In the next exercise, we will implement k-means from scratch. To do this, we will start employing common packages from the Python ecosystem that will serve as building blocks for the rest of your career. One of the most popular machine learning libraries is called scikit-learn (https://scikit-learn.org/stable/user_guide.html), which has many built-in algorithms and functions to support your understanding of how the algorithms work. We will also be using functions from SciPy (https://docs.scipy.org/doc/scipy/reference/), which is a package much like NumPy and abstracts away basic scientific math functions that allow for more efficient deployment. Finally, the next exercise will introduce `matplotlib` (https://matplotlib.org/3.1.1/contents.html), which is a plotting library that creates graphical representations of the data you are working with.

EXERCISE 1.04: K-MEANS FROM SCRATCH – PART 1: DATA GENERATION

The next two exercises focus on the creation of exercise data and the implementation of k-means from scratch on your training data. This exercise relies on scikit-learn, an open source Python package that enables the fast prototyping of popular machine learning models. Within scikit-learn, we will be using the **datasets** functionality to create a synthetic blob dataset. In addition to harnessing the power of scikit-learn, we will also rely on Matplotlib, a popular plotting library for Python that makes it easy for us to visualize our data. To do this, perform the following steps:

1. Import the necessary libraries:

```
from sklearn.datasets import make_blobs
from sklearn.cluster import KMeans
import matplotlib.pyplot as plt
import numpy as np
import math
np.random.seed(0)
%matplotlib inline
```

> **NOTE**
>
> You can find more details on the **KMeans** library at https://scikit-learn.org/stable/modules/clustering.html#k-means.

2. Generate a random cluster dataset to experiment on X = coordinate points, y = cluster labels, and define random centroids. We will achieve this with the **make_blobs** function that we imported from **sklearn.datasets**, which, as the name implies, generates blobs of data points.

```
X, y = make_blobs(n_samples=1500, centers=3, \
                  n_features=2, random_state=800)
centroids = [[-6,2],[3,-4],[-5,10]]
```

Here the **n_samples** parameter determines the total number of data points generated by the blobs. The **centers** parameter determines the number of centroids for the blob. The **n_feature** attribute defines the number of dimensions generated by the dataset. Here, the data will be two dimensional.

In order to generate the same data points in all the iterations (which in turn are generated randomly) for reproducibility of results, we set the **random_state** parameter to **800**. Different values of the **random_state** parameter would yield different results. If we do not set the **random_state** parameter, each time on execution we will obtain different results.

3. Print the data:

    ```
    X
    ```

    The output is as follows:

    ```
    array([[-3.83458347,  6.09210705],
           [-4.62571831,  5.54296865],
           [-2.87807159, -7.48754592],
           ...,
           [-3.709726  , -7.77993633],
           [-8.44553266, -1.83519866],
           [-4.68308431,  6.91780744]])
    ```

4. Plot the coordinate points using the scatterplot functionality we imported from **matplotlib.pyplot**. This function takes input lists of points and presents them graphically for ease of understanding. Please review the **matplotlib** documentation if you want to explore the parameters provided at a deeper level:

    ```
    plt.scatter(X[:, 0], X[:, 1], s=50, cmap='tab20b')
    plt.show()
    ```

The plot appears as follows:

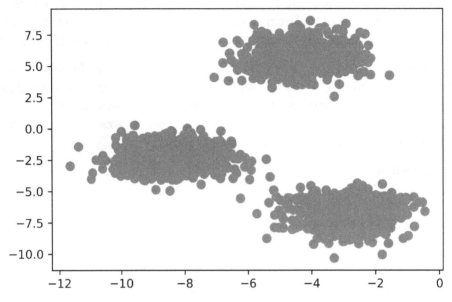

Figure 1.17: Plot of the coordinates

5. Print the array of **y**, which is the labels provided by scikit-learn and serves as the ground truth for comparison.

> **NOTE**
>
> These labels will not be known to us in practice. This is just for us to cross verify our clustering in later stages.

Use the following code to print the array:

```
y
```

The output is as follows:

```
array([2, 2, 1, ..., 1, 0, 2])
```

6. Plot the coordinate points with the correct cluster labels:

```
plt.scatter(X[:, 0], X[:, 1], c=y,s=50, cmap='tab20b')
plt.show()
```

The plot appears as follows:

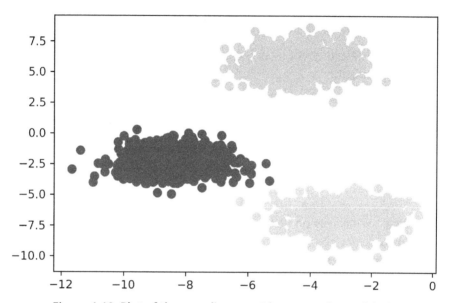

Figure 1.18: Plot of the coordinates with correct cluster labels

By completing the preceding steps, you have generated the data and visually explored how it is put together. By visualizing the ground truth, you have established a baseline that provides a relative metric for algorithm accuracy.

> **NOTE**
>
> To access the source code for this specific section, please refer to https://packt.live/2JM8Q1S.
>
> You can also run this example online at https://packt.live/3ecjKdT.

With data in hand, in the next exercise, we'll continue by building your unsupervised learning toolset with an optimized version of the Euclidean distance function from the `SciPy` package, `cdist`. You will compare a non-vectorized, clearly understandable version of the approach with `cdist`, which has been specially tweaked for maximum performance.

EXERCISE 1.05: K-MEANS FROM SCRATCH – PART 2: IMPLEMENTING K-MEANS

Let's recreate these results on our own. We will go over an example implementing this with some optimizations.

> **NOTE**
>
> This exercise is a continuation of the previous exercise and should be performed in the same Jupyter notebook.

For this exercise, we will rely on SciPy, a Python package that allows easy access to highly optimized versions of scientific calculations. In particular, we will be implementing Euclidean distance with `cdist`, the functionally of which replicates the barebones implementation of our distance metric in a much more efficient manner. Follow these steps to complete this exercise:

1. The basis of this exercise will be comparing a basic implementation of Euclidean distance with an optimized version provided in SciPy. First, import the optimized Euclidean distance reference:

   ```
   from scipy.spatial.distance import cdist
   ```

2. Identify a subset of **X** you want to explore. For this example, we are only selecting five points to make the lesson clearer; however, this approach scales to any number of points. We chose points 105-109, inclusive:

   ```
   X[105:110]
   ```

   The output is as follows:

   ```
   array([[-3.09897933,  4.79407445],
          [-3.37295914, -7.36901393],
          [-3.372895  ,  5.10433846],
          [-5.90267987, -3.28352194],
          [-3.52067739,  7.7841276 ]])
   ```

3. Calculate the distances and choose the index of the shortest distance as a cluster:

```
"""
Finds distances from each of 5 sampled points to all of the centroids
"""
for x in X[105:110]:
    calcs = cdist(x.reshape([1,-1]),centroids).squeeze()
    print(calcs, "Cluster Membership: ", np.argmin(calcs))
```

NOTE

The triple-quotes (""") shown in the code snippet above are used to denote the start and end points of a multi-line code comment. Comments are added into code to help explain specific bits of logic.

The preceding code will result in the following output:

```
[4.027750355981394, 10.70202290628413, 5.542160268055164]
 Cluster Membership:  0
[9.73035280174993, 7.208665829113462, 17.44505393393603]
 Cluster Membership:  1
[4.066767506545852, 11.113179986633003, 5.1589701124301515]
 Cluster Membership:  0
[5.284418164665783, 8.931464028407861, 13.314157359115697]
 Cluster Membership:  0
[6.293105164930943, 13.467921029846712, 2.664298385076878]
 Cluster Membership:  2
```

4. Define the **k_means** function as follows and initialize the k-centroids randomly. Repeat this process until the difference between the new/old **centroids** equals **0**, using the **while** loop:

Exercise1.04-Exercise1.05.ipynb

```
def k_means(X, K):
    # Keep track of history so you can see K-Means in action
    centroids_history = []
    labels_history = []
    rand_index = np.random.choice(X.shape[0], K)
    centroids = X[rand_index]
    centroids_history.append(centroids)
```

The complete code for this step can be found at https://packt.live/2JM8Q1S.

26 | Introduction to Clustering

> **NOTE**
> Do not break this code, as it might lead to an error.

5. Zip together the historical steps of centers and their labels:

```
history = zip(centers_hist, labels_hist)
for x, y in history:
    plt.figure(figsize=(4,3))
    plt.scatter(X[:, 0], X[:, 1], c=y, s=50, cmap='tab20b');
    plt.scatter(x[:, 0], x[:, 1], c='red')
    plt.show()
```

The following plots may differ from what you can see if we haven't set the random seed. The first plot looks as follows:

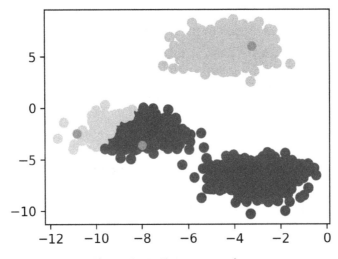

Figure 1.19: First scatterplot

The second plot appears as follows:

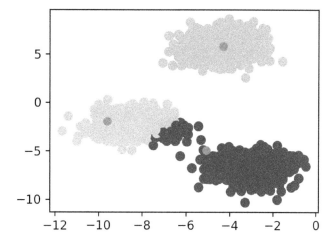

Figure 1.20: Second scatterplot

The third plot appears as follows:

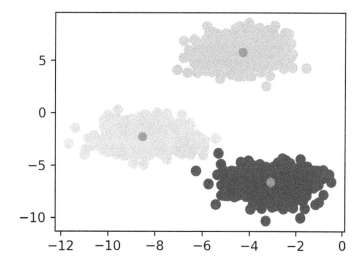

Figure 1.21: Third scatterplot

The fourth plot appears as follows:

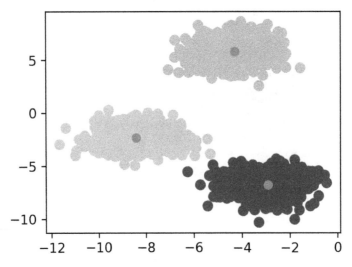

Figure 1.22: Fourth scatterplot

The fifth plot looks as follows:

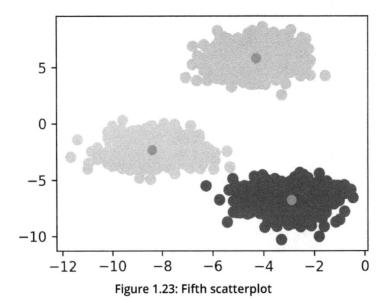

Figure 1.23: Fifth scatterplot

As shown by the preceding images, k-means takes an iterative approach to refine optimal clusters based on distance. The algorithm starts with random initialization of centroids and, depending on the complexity of the data, quickly finds the separations that make the most sense.

> **NOTE**
>
> To access the source code for this specific section, please refer to https://packt.live/2JM8Q1S.
>
> You can also run this example online at https://packt.live/3ecjKdT.

CLUSTERING PERFORMANCE – SILHOUETTE SCORE

Understanding the performance of unsupervised learning methods is inherently much more difficult than supervised learning methods because there is no ground truth available. For supervised learning, there are many robust performance metrics—the most straightforward of these being accuracy in the form of comparing model-predicted labels to actual labels and seeing how many the model got correct. Unfortunately, for clustering, we do not have labels to rely on and need to build an understanding of how "different" our clusters are. We achieve this with the silhouette score metric. We can also use silhouette scores to find the optimal "K" numbers of clusters for our unsupervised learning methods.

The silhouette metric works by analyzing how well a point fits within its cluster. The metric ranges from -1 to 1. If the average silhouette score across your clustering is one, then you will have achieved perfect clusters and there will be minimal confusion about which point belongs where. For the plots in the previous exercise, the silhouette score will be much closer to one since the blobs are tightly condensed and there is a fair amount of distance between each blob. This is very rare, though; the silhouette score should be treated as an attempt at doing the best you can, since hitting one is highly unlikely. If the silhouette score is positive, it means that a point is closer to the assigned cluster than it is to the neighboring clusters. If the silhouette score is 0, then a point lies on the boundary between the assigned cluster and the next closest cluster. If the silhouette score is negative, then it indicates that a given point is assigned to an incorrect cluster, and the given point in fact likely belongs to a neighboring cluster.

Mathematically, the silhouette score calculation is quite straightforward and is obtained using the **Simplified Silhouette Index (SSI)**:

$$SSI_i = b_i - a_i / \max(a_i, b_i)$$

Here a_i is the distance from point *i* to its own cluster centroid, and b_i is the distance from point *i* to the nearest cluster centroid.

The intuition captured here is that a_i represents how cohesive the cluster of point *i'* is as a clear cluster, and b_i represents how far apart the clusters lie. We will use the optimized implementation of `silhouette_score` in scikit-learn in *Activity 1.01, Implementing k-means Clustering*. Using it is simple and only requires that you pass in the feature array and the predicted cluster labels from your k-means clustering method.

In the next exercise, we will use the **pandas** library (https://pandas.pydata.org/pandas-docs/stable/) to read a CSV file. Pandas is a Python library that makes data wrangling easier through the use of DataFrames. If you look back at the arrays you built with NumPy, you probably noticed that the resulting data structures are quite unwieldly. To extract subsets from the data, you had to index using brackets and specific numbers of rows. Instead of this approach, pandas allows an easier-to-understand approach to moving data around and getting it into the format necessary for unsupervised learning and other machine learning techniques.

> **NOTE**
>
> To read data in Python, you will use `variable_name = pd.read_csv('file_name.csv', header=None)`
>
> Here, the parameter `header = None` explicitly mentions that there is no presence of column names. If your file contains column names, then retain those default values. Also, if you specify `header = None` for a file which contains column names, Pandas will treat the row containing names of column as the row containing data only.

EXERCISE 1.06: CALCULATING THE SILHOUETTE SCORE

In this exercise, we will calculate the silhouette score of a dataset with a fixed number of clusters. For this, we will use the seeds dataset, which is available at https://packt.live/2UQA79z. The following note outlines more information regarding this dataset, in addition to further exploration in the next activity. For the purpose of this exercise, please disregard the specific details of what this dataset is comprised of as it is of greater importance to learn about the silhouette score. As we go into the next activity, you will gain more context as needed to create a smart machine learning system. Follow these steps to complete this exercise:

> **NOTE**
>
> This dataset is sourced from https://archive.ics.uci.edu/ml/datasets/seeds. It can be accessed at https://packt.live/2UQA79z
>
> Citation: Contributors gratefully acknowledge support of their work by the Institute of Agrophysics of the Polish Academy of Sciences in Lublin.

1. Load the seeds data file using pandas, a package that makes data wrangling much easier through the use of DataFrames:

```
import pandas as pd
import numpy as np
import matplotlib.pyplot as plt
from sklearn.metrics import silhouette_score
from scipy.spatial.distance import cdist
np.random.seed(0)

seeds = pd.read_csv('Seed_Data.csv')
```

2. Separate the **X** features, since we want to treat this as an unsupervised learning problem:

```
X = seeds[['A','P','C','LK','WK','A_Coef','LKG']]
```

3. Bring back the **k_means** function we made earlier for reference:

Exercise 1.06.ipynb

```
def k_means(X, K):
    # Keep track of history so you can see K-Means in action
    centroids_history = []
    labels_history = []
    rand_index = np.random.choice(X.shape[0], K)
    centroids = X[rand_index]
    centroids_history.append(centroids)
```

The complete code for this step can be found at https://packt.live/2UOqW9H.

4. Convert our seeds **X** feature DataFrame into a **NumPy** matrix:

```
X_mat = X.values
```

5. Run our **k_means** function on the seeds matrix:

```
centroids, labels, centroids_history, labels_history = \
k_means(X_mat, 3)
```

6. Calculate the silhouette score for the **Area ('A')** and **Length of Kernel ('LK')** columns:

```
silhouette_score(X[['A','LK']], labels)
```

The output should be similar to the following:

```
0.5875704550892767
```

In this exercise, we calculated the silhouette score for the **Area ('A')** and **Length of Kernel ('LK')** columns of the seeds dataset. We will use this technique in the next activity to determine the performance of our k-means clustering algorithm.

> **NOTE**
>
> To access the source code for this specific section, please refer to https://packt.live/2UOqW9H.
>
> You can also run this example online at https://packt.live/3fbtJ4y.

ACTIVITY 1.01: IMPLEMENTING K-MEANS CLUSTERING

You are implementing a k-means clustering algorithm from scratch to prove that you understand how it works. You will be using the seeds dataset provided by the UCI ML repository. The seeds dataset is a classic in the data science world and contains features of wheat kernels that are used to predict three different types of wheat species. The download location can be found later in this activity.

For this activity, you should use Matplotlib, NumPy, scikit-learn metrics, and pandas.

By loading and reshaping data easily, you can focus more on learning k-means instead of writing data loader functionality.

The following seeds data features are provided for reference:

1. area (A),
2. perimeter (P)
3. compactness (C)
4. length of kernel (LK)
5. width of kernel (WK)
6. asymmetry coefficient (A_Coef)
7. length of kernel groove (LKG)

The aim here is to truly understand how k-means works. To do so, you need to take what you have learned in the previous sections and implement k-means from scratch in Python.

Please open your favorite editing platform and try the following steps:

1. Using **NumPy** or the **math** package and the Euclidean distance formula, write a function that calculates the distance between two coordinates.

2. Write a function that calculates the distance from the centroids to each of the points in your dataset and returns the cluster membership.

3. Write a k-means function that takes in a dataset and the number of clusters (K) and returns the final cluster centroids, as well as the data points that make up that cluster's membership. After implementing k-means from scratch, apply your custom algorithm to the seeds dataset, which is located here: https://packt.live/2Xh2FdS.

34 | Introduction to Clustering

> **NOTE**
>
> This dataset is sourced from https://archive.ics.uci.edu/ml/datasets/seeds.
> It can be accessed at https://packt.live/2Xh2FdS.
>
> UCI Machine Learning Repository [http://archive.ics.uci.edu/ml]. Irvine, CA: University of California, School of Information and Computer Science.
>
> Citation: Contributors gratefully acknowledge support of their work by the Institute of Agrophysics of the Polish Academy of Sciences in Lublin.

4. Remove the classes supplied in this dataset and see whether your k-means algorithm can group the different wheat species into their proper groups just based on plant characteristics!

5. Calculate the silhouette score using the scikit-learn implementation.

In completing this exercise, you have gained hands-on experience of tuning a k-means clustering algorithm for a real-world dataset. The seeds dataset is seen as a classic "hello world"-type problem in the data science space and is helpful for testing foundational techniques. Your final clustering algorithm should do a decent job of finding the three clusters of wheat species types that exist in the data, as follows:

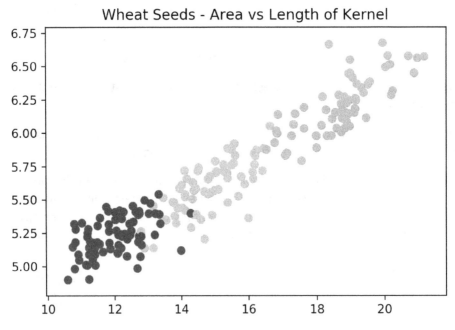

Figure 1.24: Expected plot of three clusters of wheat species

> **NOTE**
>
> The solution to this activity can be found on page 418.

SUMMARY

In this chapter, we have explored what clustering is and why it is important in a variety of data challenges. Building upon this foundation of clustering knowledge, you implemented k-means, which is one of the simplest, yet most popular, methods of unsupervised learning. If you have reached this summary and can repeat what k-means does step by step to a friend, then you're ready to move on to more complex forms of clustering.

From here, we will be moving on to hierarchical clustering, which, in one configuration, reuses the centroid learning approach that we used in k-means. We will build upon this approach by outlining additional clustering methodologies and approaches in the next chapter.

2

HIERARCHICAL CLUSTERING

OVERVIEW

In this chapter, we will implement the hierarchical clustering algorithm from scratch using common Python packages and perform agglomerative clustering. We will also compare k-means with hierarchical clustering. We will use hierarchical clustering to build stronger groupings that make more logical sense. By the end of this chapter, we will be able to use hierarchical clustering to build stronger groupings that make more logical sense.

INTRODUCTION

In this chapter, we will expand on the basic ideas that we built in Chap*ter 1*, *Introduction to Clustering*, by surrounding clustering with the concept of similarity. Once again, we will be implementing forms of the Euclidean distance to capture the notion of similarity. It is important to bear in mind that the Euclidean distance just happens to be one of the most popular distance metrics; it's not the only one. Through these distance metrics, we will expand on the simple neighbor calculations that we explored in the previous chapter by introducing the concept of hierarchy. By using hierarchy to convey clustering information, we can build stronger groupings that make more logical sense. Similar to k-means, hierarchical clustering can be helpful for cases such as customer segmentation or identifying similar product types. However, there is a slight benefit in being able to explain things in a clearer fashion with hierarchical clustering. In this chapter, we will outline some cases where hierarchical clustering can be the solution you're looking for.

CLUSTERING REFRESHER

Chapter 1, Introduction to Clustering, covered both the high-level concepts and in-depth details of one of the most basic clustering algorithms: k-means. While it is indeed a simple approach, do not discredit it; it will be a valuable addition to your toolkit as you continue your exploration of the unsupervised learning world. In many real-world use cases, companies experience valuable discoveries through the simplest methods, such as k-means or linear regression (for supervised learning). An example of this is evaluating a large selection of customer data – if you were to evaluate it directly in a table, it would be unlikely that you'd find anything helpful. However, even a simple clustering algorithm can identify where groups within the data are similar and dissimilar. As a refresher, let's quickly walk through what clusters are and how k-means works to find them:

Unsupervised	Supervised
- No labels provided - Finds structure in unlabeled data - Uses techniques such as clustering and dimensionality reduction	- Labels provided - Finds patterns in existing structure - Uses techniques such as regression and classification

Figure 2.1: The attributes that separate supervised and unsupervised problems

If you were given a random collection of data without any guidance, you would probably start your exploration using basic statistics – for example, the mean, median, and mode values for each of the features. Given a dataset, choosing supervised or unsupervised learning as an approach to derive insights is dependent on the data goals that you have set for yourself. If you were to determine that one of the features was actually a label and you wanted to see how the remaining features in the dataset influence it, this would become a supervised learning problem. However, if, after initial exploration, you realized that the data you have is simply a collection of features without a target in mind (such as a collection of health metrics, purchase invoices from a web store, and so on), then you could analyze it through unsupervised methods.

A classic example of unsupervised learning is finding clusters of similar customers in a collection of invoices from a web store. Your hypothesis is that by finding out which people are the most similar, you can create more granular marketing campaigns that appeal to each cluster's interests. One way to achieve these clusters of similar users is through k-means.

THE K-MEANS REFRESHER

The k-means clustering works by finding "k" number of clusters in your data through certain distance calculations such as Euclidean, Manhattan, Hamming, Minkowski, and so on. "K" points (also called centroids) are randomly initialized in your data and the distance is calculated from each data point to each of the centroids. The minimum of these distances designates which cluster a data point belongs to. Once every point has been assigned to a cluster, the mean intra-cluster data point is calculated as the new centroid. This process is repeated until the newly calculated cluster centroid no longer changes position or until the maximum limit of iterations is reached.

THE ORGANIZATION OF THE HIERARCHY

Both the natural and human-made world contain many examples of organizing systems into hierarchies and why, for the most part, it makes a lot of sense. A common representation that is developed from these hierarchies can be seen in tree-based data structures. Imagine that you have a parent node with any number of child nodes that can subsequently be parent nodes themselves. By organizing information into a tree structure, you can build an information-dense diagram that clearly shows how things are related to their peers and their larger abstract concepts.

An example from the natural world to help illustrate this concept can be seen in how we view the hierarchy of animals, which goes from parent classes to individual species:

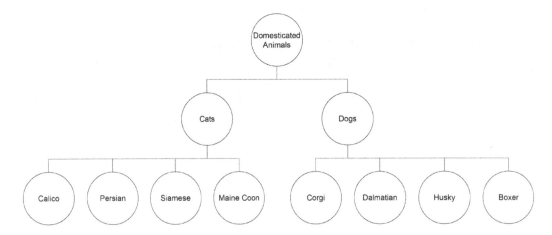

Figure 2.2: The relationships of animal species in a hierarchical tree structure

In the preceding diagram, you can see an example of how relational information between varieties of animals can be easily mapped out in a way that both saves space and still transmits a large amount of information. This example can be seen as both a tree of its own (showing how cats and dogs are different, but both are domesticated animals) and as a potential piece of a larger tree that shows a breakdown of domesticated versus non-domesticated animals.

As a business-facing example, let's go back to the concept of a web store selling products. If you sold a large variety of products, then you would probably want to create a hierarchical system of navigation for your customers. By preventing all of the information in your product catalog from being presented at once, customers will only be exposed to the path down the tree that matches their interests. An example of the hierarchical system of navigation can be seen in the following diagram:

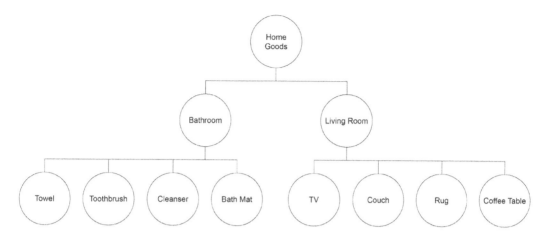

Figure 2.3: Product categories in a hierarchical tree structure

Clearly, the benefits of a hierarchical system of navigation cannot be overstated in terms of improving your customer experience. By organizing information into a hierarchical structure, you can build an intuitive structure into your data that demonstrates explicit nested relationships. If this sounds like another approach to finding clusters in your data, then you're definitely on the right track. Through the use of similar distance metrics, such as the Euclidean distance from k-means, we can develop a tree that shows the many cuts of data that allow a user to subjectively create clusters at their discretion.

INTRODUCTION TO HIERARCHICAL CLUSTERING

So far, we have shown you that hierarchies can be excellent structures to organize information that clearly shows nested relationships among data points. While this helps us gain an understanding of the parent/child relationships between items, it can also be very handy when forming clusters. Expanding on the animal example in the previous section, imagine that you were simply presented with two features of animals: their height (measured from the tip of the nose to the end of the tail) and their weight. Using this information, you then have to recreate a hierarchical structure in order to identify which records in your dataset correspond to dogs and cats, as well as their relative subspecies.

42 | Hierarchical Clustering

Since you are only given animal heights and weights, you won't be able to deduce the specific names of each species. However, by analyzing the features that you have been provided with, you can develop a structure within the data that serves as an approximation of what animal species exist in your data. This perfectly sets the stage for an unsupervised learning problem that is well solved with hierarchical clustering. In the following plot, you can see the two features that we created on the left, with animal height in the left-hand column and animal weight in the right-hand column. This is then charted on a two-axis plot with the height on the X-axis and the weight on the Y-axis:

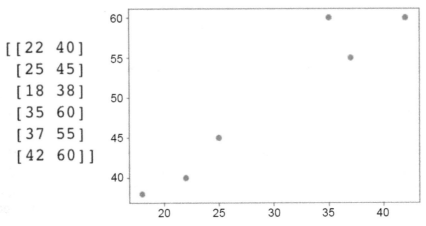

Figure 2.4: An example of a two-feature dataset comprising animal height and animal weight

One way to approach hierarchical clustering is by starting with each data point, serving as its own cluster, and recursively joining the similar points together to form clusters – this is known as **agglomerative** hierarchical clustering. We will go into more detail about the different ways of approaching hierarchical clustering in the *Agglomerative versus Divisive Clustering* section.

In the agglomerative hierarchical clustering approach, the concept of data point similarity can be thought of in the paradigm that we saw during k-means. In k-means, we used the Euclidean distance to calculate the distance from the individual points to the centroids of the expected "k" clusters. In this approach to hierarchical clustering, we will reuse the same distance metric to determine the similarity between the records in our dataset.

Eventually, by grouping individual records from the data with their most similar records recursively, you end up building a hierarchy from the bottom up. The individual single-member clusters join into one single cluster at the top of our hierarchy.

STEPS TO PERFORM HIERARCHICAL CLUSTERING

To understand how agglomerative hierarchical clustering works, we can trace the path of a simple toy program as it merges to form a hierarchy:

1. Given n sample data points, view each point as an individual "cluster" with just that one point as a member (the centroid).

2. Calculate the pairwise Euclidean distance between the centroids of all the clusters in your data. (Here, minimum distance between clusters, maximum distance between clusters, average distance between clusters, or distance between two centroids can also be considered. In this example, we are considering the distance between two cluster centroids).

3. Group the closest clusters/points together.

4. Repeat *Step 2* and *Step 3* until you get a single cluster containing all the data in your set.

5. Plot a dendrogram to show how your data has come together in a hierarchical structure. A dendrogram is simply a diagram that is used to represent a tree structure, showing an arrangement of clusters from top to bottom. We will go into the details of how this may be helpful in the following walkthrough.

6. Decide what level you want to create the clusters at.

AN EXAMPLE WALKTHROUGH OF HIERARCHICAL CLUSTERING

While slightly more complex than k-means, hierarchical clustering is, in fact, quite similar to it from a logistical perspective. Here is a simple example that walks through the preceding steps in slightly more detail:

1. Given a list of four sample data points, view each point as a centroid that is also its own cluster with the point indices from 0 to 3:

   ```
   Clusters (4): [ (1,7) ], [ (-5,9) ], [ (-9,4) ] , [ (4, -2) ]
   Centroids (4): [ (1,7) ], [ (-5,9) ], [ (-9,4) ] , [ (4, -2) ]
   ```

2. Calculate the pairwise Euclidean distance between the centroids of all the clusters.

   > **NOTE**
   >
   > Refer to the *K-means Clustering In-Depth Walkthrough* section in *Chapter 1, Introduction to Clustering* for a refresher on Euclidean distance.

44 | Hierarchical Clustering

In the matrix displayed in *Figure 2.5*, the point indices are between 0 and 3 both horizontally and vertically, showing the distance between the respective points. Notice that the values are mirrored across the diagonal – this happens because you are comparing each point against all the other points, so you only need to worry about the set of numbers on one side of the diagonal:

```
              (1,7)           (-5,9)          (-9,4)          (4,-2)
(1,7)  [  0.            6.32455532  10.44030651   9.48683298]
(-5,9) [  6.32455532    0.           6.40312424  14.2126704 ]
(-9,4) [ 10.44030651    6.40312424   0.          14.31782106]
(4,-2) [  9.48683298   14.2126704   14.31782106   0.        ]
```

Figure 2.5: An array of distances

3. Group the closest point pairs together.

   In this case, points [1,7] and [-5,9] join into a cluster since they are the closest, with the remaining two points left as single-member clusters:

```
              (1,7)           (-5,9)          (-9,4)          (4,-2)
(1,7)  [  0.            6.32455532  10.44030651   9.48683298]
(-5,9) [  6.32455532    0.           6.40312424  14.2126704 ]
(-9,4) [ 10.44030651    6.40312424   0.          14.31782106]
(4,-2) [  9.48683298   14.2126704   14.31782106   0.        ]
```

Figure 2.6: An array of distances

Here are the resulting three clusters:

```
[ [1,7], [-5,9] ]
[-9,4]
[4,-2]
```

4. Calculate the mean point between the points of the two-member cluster to find the new centroid:

```
mean([ [1,7], [-5,9] ]) = [-2,8]
```

Introduction to Hierarchical Clustering | 45

5. Add the centroid to the two single-member centroids and recalculate the distances:

```
Clusters (3):
[ [1,7], [-5,9] ]
[-9,4]
[4,-2]
```

Centroids (3):

```
[-2,8]
[-9,4]
[4,-2]
```

Once again, we'll calculate the Euclidean distance between the points and the centroid:

```
                (-2,8)          (-9,4)          (4,-2)
(-2,8)  [ 0.            8.06225775  11.66190379]
(-9,4)  [ 8.06225775    0.          14.31782106]
(4,-2)  [11.66190379   14.31782106   0.        ]
```

Figure 2.7: An array of distances

6. As shown in the preceding image, point [-9,4] is the shortest distance from the centroid and thus it is added to cluster 1. Now, the cluster list changes to the following:

```
Clusters (2):
[ [1,7], [-5,9], [-9,4] ]
[4,-2]
```

7. With only point [4,-2] left as the furthest distance away from its neighbors, you can just add it to cluster 1 to unify all the clusters:

```
Clusters (1):
[ [ [1,7], [-5,9], [-9,4], [4,-2] ] ]
```

8. Plot a dendrogram to show the relationship between the points and the clusters:

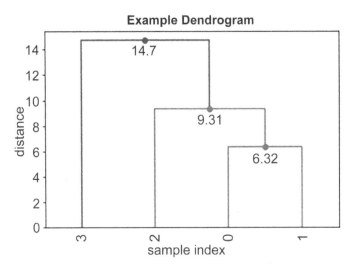

Figure 2.8: A dendrogram showing the relationship between the points and the clusters

Dendrograms show how data points are similar and will look familiar to the hierarchical tree structures that we discussed earlier. There is some loss of information, as with any visualization technique; however, dendrograms can be very helpful when determining how many clusters you want to form. In the preceding example, you can see four potential clusters across the X-axis, if each point was its own cluster. As you travel vertically, you can see which points are closest together and can potentially be clubbed into their own cluster. For example, in the preceding dendrogram, the points at indices 0 and 1 are the closest and can form their own cluster, while index 2 remains a single-point cluster.

Revisiting the previous animal taxonomy example that involved dog and cat species, imagine that you were presented with the following dendrogram:

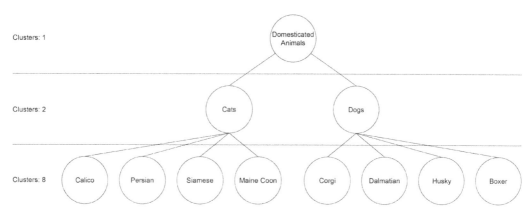

Figure 2.9: An animal taxonomy dendrogram

If you were just interested in grouping your species dataset into dogs and cats, you could stop clustering at the first level of the grouping. However, if you wanted to group all species into domesticated or non-domesticated animals, you could stop clustering at level two. The great thing about hierarchical clustering and dendrograms is that you can see the entire breakdown of potential clusters to choose from.

EXERCISE 2.01: BUILDING A HIERARCHY

Let's implement the preceding hierarchical clustering approach in Python. With the framework for the intuition laid out, we can now explore the process of building a hierarchical cluster with some helper functions provided in **sciPy**. SciPy (https://www.scipy.org/docs.html) is an open source library that packages functions that are helpful in scientific and technical computing. Examples of this include easy implementations of linear algebra and calculus-related methods. In this exercise, we will specifically be using helpful functions from the **cluster** subsection of SciPy. In addition to **scipy**, we will be using **matplotlib** to complete this exercise. Follow these steps to complete this exercise:

1. Generate some dummy data, as follows:

   ```
   from scipy.cluster.hierarchy import linkage, dendrogram, fcluster
   from sklearn.datasets import make_blobs
   import matplotlib.pyplot as plt
   %matplotlib inline
   ```

2. Generate a random cluster dataset to experiment with. **X** = coordinate points, **y** = cluster labels (not needed):

   ```
   X, y = make_blobs(n_samples=1000, centers=8, \
                     n_features=2, random_state=800)
   ```

3. Visualize the data, as follows:

   ```
   plt.scatter(X[:,0], X[:,1])
   plt.show()
   ```

   The output is as follows:

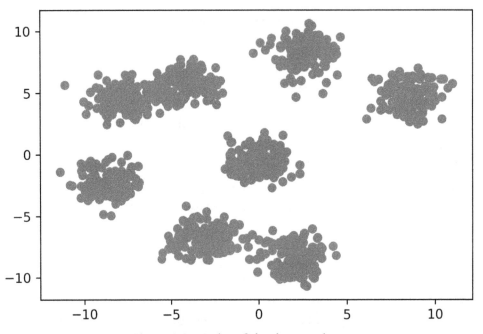

Figure 2.10: A plot of the dummy data

After plotting this simple toy example, it should be pretty clear that our dummy data comprises eight clusters.

4. We can easily generate the distance matrix using the built-in **SciPy** package, **linkage**. We will go further into what's happening with the linkage function shortly; however, for now it's good to know that there are pre-built tools that calculate distances between points:

   ```
   # Generate distance matrix with 'linkage' function
   distances = linkage(X, method="centroid", metric="euclidean")
   print(distances)
   ```

The output is as follows:

```
distances = linkage(X, method="centroid", metric="euclidean")
```

```
print(distances)
```

```
[[5.720e+02 7.620e+02 7.694e-03 2.000e+00]
 [3.000e+01 1.960e+02 8.879e-03 2.000e+00]
 [5.910e+02 8.700e+02 1.075e-02 2.000e+00]
 ...
 [1.989e+03 1.992e+03 7.812e+00 3.750e+02]
 [1.995e+03 1.996e+03 1.024e+01 7.500e+02]
 [1.994e+03 1.997e+03 1.200e+01 1.000e+03]]
```

Figure 2.11: A matrix of the distances

If you experiment with different methods by trying to autofill the **method** hyperparameter of the `linkage` function, you will see how they affect overall performance. Linkage works by simply calculating the distances between each of the data points. We will go into specifically what it is calculating in the *Linkage* topic. In the `linkage` function, we have the option to select both the metric and the method (we will cover this in more detail later).

After we determine the linkage matrix, we can easily pass it through the **dendrogram** function provided by **SciPy**. As the name suggests, the **dendrogram** function uses the distances calculated in *Step 4* to generate a visually clean way of parsing grouped information.

5. We will be using a custom function to clean up the styling of the original output (note that the function provided in the following snippet is using the base SciPy implementation of the dendrogram, and the only custom code is for cleaning up the visual output):

```
# Take normal dendrogram output and stylize in cleaner way

def annotated_dendrogram(*args, **kwargs):
    # Standard dendrogram from SciPy
    scipy_dendro = dendrogram(*args, truncate_mode='lastp', \
                              show_contracted=True,\
                              leaf_rotation=90.)
    
    plt.title('Blob Data Dendrogram')
```

50 | Hierarchical Clustering

```
            plt.xlabel('cluster size')
            plt.ylabel('distance')
            for i, d, c in zip(scipy_dendro['icoord'], \
                            scipy_dendro['dcoord'], \
                            scipy_dendro['color_list']):
                x = 0.5 * sum(i[1:3])
                y = d[1]
                if y > 10:
                    plt.plot(x, y, 'o', c=c)
                    plt.annotate("%.3g" % y, (x, y), xytext=(0, -5), \
                            textcoords='offset points', \
                            va='top', ha='center')
            return scipy_dendro

dn = annotated_dendrogram(distances)
plt.show()
```

The output is as follows:

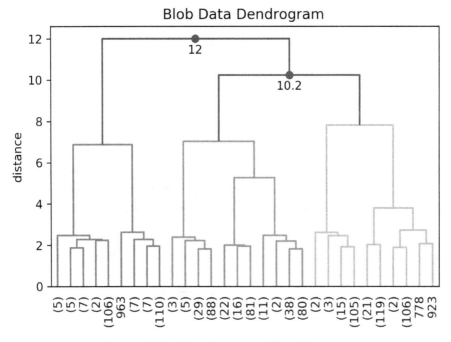

Figure 2.12: A dendrogram of the distances

Introduction to Hierarchical Clustering | 51

This plot will give us some perspective on the potential breakouts of our data. Based on the distances calculated in prior steps, it shows a potential path that we can use to create three separate groups around the distance of seven that are distinctly different enough to stand on their own.

6. Using this information, we can wrap up our exercise on hierarchical clustering by using the **fcluster** function from **SciPy**:

```
scipy_clusters = fcluster(distances, 3, criterion="distance")
plt.scatter(X[:,0], X[:,1], c=scipy_clusters)
plt.show()
```

The **fcluster** function uses the distances and information from the dendrogram to cluster our data into a number of groups based on a stated threshold. The number **3** in the preceding example represents the maximum inter-cluster distance threshold hyperparameter that you can set. This hyperparameter can be tuned based on the dataset that you are looking at; however, it is supplied to you as **3** for this exercise. The final output is as follows:

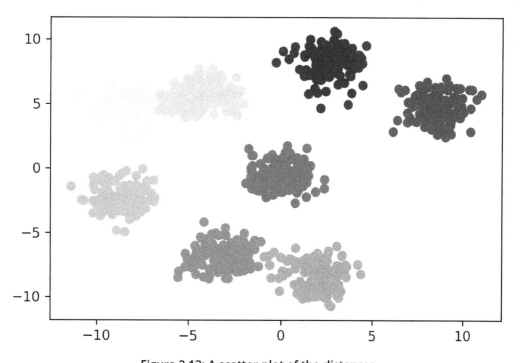

Figure 2.13: A scatter plot of the distances

In the preceding plot, you can see that by using our threshold hyperparameter, we've identified eight distinct clusters. By simply calling a few helper functions provided by `SciPy`, you can easily implement agglomerative clustering in just a few lines of code. While SciPy does help with many of the intermediate steps, this is still an example that is a bit more verbose than what you will probably see in your regular work. We will cover more streamlined implementations later.

> **NOTE**
>
> To access the source code for this specific section, please refer to https://packt.live/2VTRp5K.
>
> You can also run this example online at https://packt.live/2Cdyiww.

LINKAGE

In *Exercise 2.01, Building a Hierarchy*, you implemented hierarchical clustering using what is known as **Centroid Linkage**. Linkage is the concept of determining how you can calculate the distances between clusters and is dependent on the type of problem you are facing. Centroid linkage was chosen for *Exercise 2.02, Applying Linkage Criteria*, as it essentially mirrors the new centroid search that we used in k-means. However, this is not the only option when it comes to clustering data points. Two other popular choices for determining distances between clusters are single linkage and complete linkage.

Single Linkage works by finding the minimum distance between a pair of points between two clusters as its criteria for linkage. Simply put, it essentially works by combining clusters based on the closest points between the two clusters. This is expressed mathematically as follows:

```
dist(a,b) = min( dist( a[i]), b[j] ) )
```

In the preceding code, `a[i]` is the ith point within first cluster where `b[j]` is j^{th} point of second cluster.

Complete Linkage is the opposite of single linkage and it works by finding the maximum distance between a pair of points between two clusters as its criteria for linkage. Simply put, it works by combining clusters based on the furthest points between the two clusters. This is mathematically expressed as follows:

```
dist(a,b) = max( dist( a[i]), b[j] ) )
```

In the preceding code, **a[i]** and **b[j]** are i^{th} and j^{th} point of first and second cluster respectively. Determining what linkage criteria is best for your problem is as much art as it is science, and it is heavily dependent on your particular dataset. One reason to choose single linkage is if your data is similar in a nearest-neighbor sense; therefore, when there are differences, the data is extremely dissimilar. Since single linkage works by finding the closest points, it will not be affected by these distant outliers. However, as single linkage works by finding the smallest distance between a pair of points, it is quite prone to the noise distributed between the clusters. Conversely, complete linkage may be a better option if your data is distant in terms of inter-cluster state; complete linkage causes incorrect splitting when the spatial distribution of cluster is fairly imbalanced. Centroid linkage has similar benefits but falls apart if the data is very noisy and there are less clearly defined "centers" of clusters. Typically, the best approach is to try a few different linkage criteria options and see which fits your data in a way that's the most relevant to your goals.

EXERCISE 2.02: APPLYING LINKAGE CRITERIA

Recall the dummy data of the eight clusters that we generated in the previous exercise. In the real world, you may be given real data that resembles discrete Gaussian blobs in the same way. Imagine that the dummy data represents different groups of shoppers in a particular store. The store manager has asked you to analyze the shopper data in order to classify the customers into different groups so that they can tailor marketing materials to each group.

Using the data we generated in the previous exercise, or by generating new data, you are going to analyze which linkage types do the best job of grouping the customers into distinct clusters.

Once you have generated the data, view the documents supplied using SciPy to understand what linkage types are available in the **linkage** function. Then, evaluate the linkage types by applying them to your data. The linkage types you should test are shown in the following list:

```
['centroid', 'single', 'complete', 'average', 'weighted']
```

We haven't covered all of the previously mentioned linkage types yet – a key part of this activity is to learn how to parse the docstrings that are provided using packages to explore all of their capabilities. Follow these steps to complete this exercise:

1. Visualize the **x** dataset that we created in *Exercise 2.01, Building a Hierarchy*:

   ```
   from scipy.cluster.hierarchy import linkage, dendrogram, fcluster
   from sklearn.datasets import make_blobs
   import matplotlib.pyplot as plt
   %matplotlib inline
   ```

2. Generate a random cluster dataset to experiment on. **X** = coordinate points, **y** = cluster labels (not needed):

   ```
   X, y = make_blobs(n_samples=1000, centers=8, \
                     n_features=2, random_state=800)
   ```

3. Visualize the data, as follows:

   ```
   plt.scatter(X[:,0], X[:,1])
   plt.show()
   ```

   The output is as follows:

   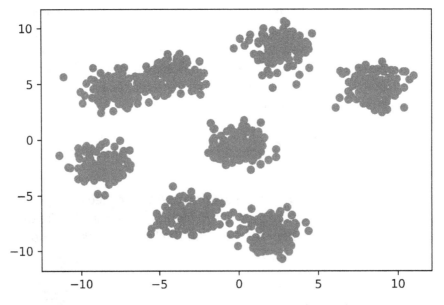

   Figure 2.14: A scatter plot of the generated cluster dataset

4. Create a list with all the possible linkage method hyperparameters:

```
methods = ['centroid', 'single', 'complete', \
           'average', 'weighted']
```

5. Loop through each of the methods in the list that you just created and display the effect that they have on the same dataset:

```
for method in methods:
    distances = linkage(X, method=method, metric="euclidean")
    clusters = fcluster(distances, 3, criterion="distance")
    plt.title('linkage: ' + method)
    plt.scatter(X[:,0], X[:,1], c=clusters, cmap='tab20b')
    plt.show()
```

The plot for centroid linkage is as follows:

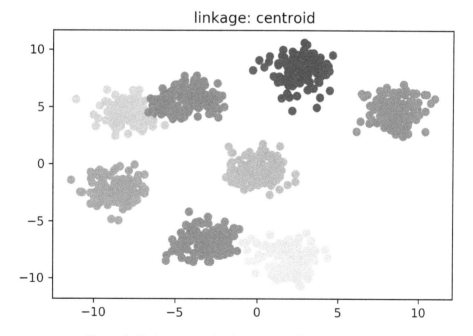

Figure 2.15: A scatter plot for centroid linkage method

56 | Hierarchical Clustering

The plot for single linkage is as follows:

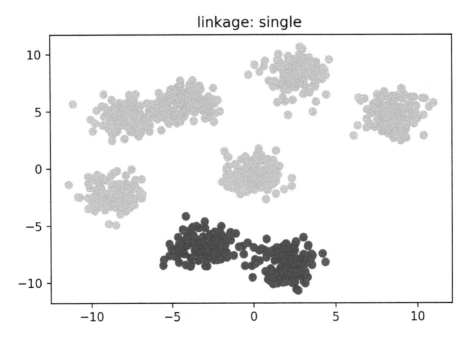

Figure 2.16: A scatter plot for single linkage method

The plot for complete linkage is as follows:

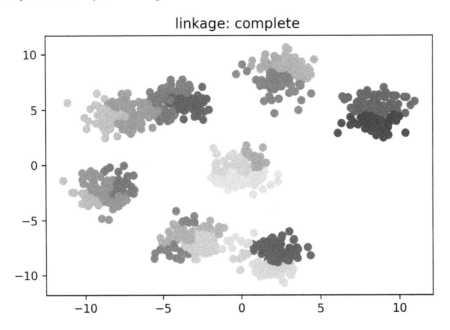

Figure 2.17: A scatter plot for complete linkage method

The plot for average linkage is as follows:

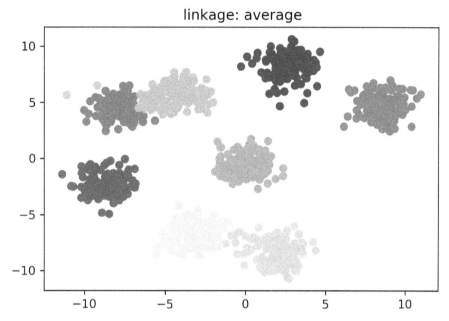

Figure 2.18: A scatter plot for average linkage method

The plot for weighted linkage is as follows:

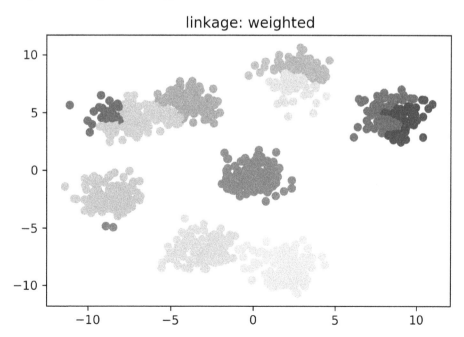

Figure 2.19: A scatter plot for weighted linkage method

As shown in the preceding plots, by simply changing the linkage criteria, you can dramatically change the efficacy of your clustering. In this dataset, centroid and average linkage work best at finding discrete clusters that make sense. This is clear from the fact that we generated a dataset of eight clusters, and centroid and average linkage are the only ones that show the clusters that are represented using eight different colors. The other linkage types fall short – most noticeably, single linkage. Single linkage falls short because it operates on the assumption that the data is in a thin "chain" format versus the clusters. The other linkage methods are superior due to their assumption that the data is coming in as clustered groups.

> **NOTE**
>
> To access the source code for this specific section, please refer to https://packt.live/2VWwbEv.
>
> You can also run this example online at https://packt.live/2Zb4zgN.

AGGLOMERATIVE VERSUS DIVISIVE CLUSTERING

So far, our instances of hierarchical clustering have all been agglomerative – that is, they have been built from the bottom up. While this is typically the most common approach for this type of clustering, it is important to know that it is not the only way a hierarchy can be created. The opposite hierarchical approach, that is, built from the top up, can also be used to create your taxonomy. This approach is called **divisive** hierarchical clustering and works by having all the data points in your dataset in one massive cluster. Many of the internal mechanics of the divisive approach will prove to be quite similar to the agglomerative approach:

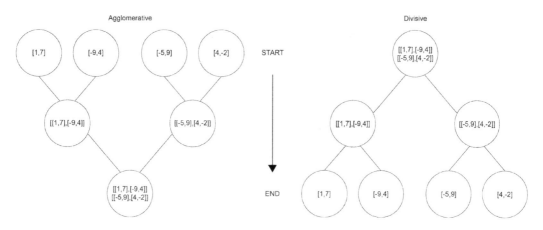

Figure 2.20: Agglomerative versus divisive hierarchical clustering

As with most problems in unsupervised learning, deciding on the best approach is often highly dependent on the problem you are faced with solving.

Imagine that you are an entrepreneur who has just bought a new grocery store and needs to stock it with goods. You receive a large shipment of food and drink in a container, but you've lost track of all the shipment information. In order to effectively sell your products, you must group similar products together (your store will be a huge mess if you just put everything on the shelves in a random order). Setting out on this organizational goal, you can take either a bottom-up or top-down approach. On the bottom-up side, you will go through the shipping container and think of everything as disorganized – you will then pick up a random object and find its most similar product. For example, you may pick up apple juice and realize that it makes sense to group it together with orange juice. With the top-down approach, you will view everything as organized in one large group. Then, you will move through your inventory and split the groups based on the largest differences in similarity. For example, if you were organizing a grocery store, you may originally think that apples and apple juice go together, but on second thoughts, they are quite different. Therefore, you will break them into smaller, dissimilar groups.

Hierarchical Clustering

In general, it helps to think of agglomerative as the bottom-up approach and divisive as the top-down approach – but how do they trade off in terms of performance? This behavior of immediately grabbing the closest thing is known as "greedy learning;" it has the potential to be fooled by local neighbors and not see the larger implications of the clusters it forms at any given time. On the flip side, the divisive approach has the benefit of seeing the entire data distribution as one from the beginning and choosing the best way to break down clusters. This insight into what the entire dataset looks like is helpful for potentially creating more accurate clusters and should not be overlooked. Unfortunately, a top-down approach typically trades off greater accuracy for deeper complexity. In practice, an agglomerative approach works most of the time and should be the preferred starting point when it comes to hierarchical clustering. If, after reviewing the hierarchies, you are unhappy with the results, it may help to take a divisive approach.

EXERCISE 2.03: IMPLEMENTING AGGLOMERATIVE CLUSTERING WITH SCIKIT-LEARN

In most business use cases, you will likely find yourself implementing hierarchical clustering with a package that abstracts everything away, such as scikit-learn. Scikit-learn is a free package that is indispensable when it comes to machine learning in Python. It conveniently provides highly optimized forms of the most popular algorithms, such as regression, classification, and clustering. By using an optimized package such as scikit-learn, your work becomes much easier. However, you should only use it when you fully understand how hierarchical clustering works, as we discussed in the previous sections. This exercise will compare two potential routes that you can take when forming clusters – using SciPy and scikit-learn. By completing this exercise, you will learn what the pros and cons are of each, and which suits you best from a user perspective. Follow these steps to complete this exercise:

1. Scikit-learn makes implementation as easy as just a few lines of code. First, import the necessary packages and assign the model to the **ac** variable. Then, create the blob data as shown in the previous exercises:

   ```
   from sklearn.cluster import AgglomerativeClustering
   from sklearn.datasets import make_blobs
   import matplotlib.pyplot as plt
   from scipy.cluster.hierarchy import linkage, dendrogram, fcluster
   ```

```
ac = AgglomerativeClustering(n_clusters = 8, \
                             affinity="euclidean", \
                             linkage="average")
X, y = make_blobs(n_samples=1000, centers=8, \
                  n_features=2, random_state=800)
```

First, we assign the model to the **ac** variable by passing in parameters that we are familiar with, such as **affinity** (the distance function) and **linkage**.

2. Then reuse the **linkage** function and **fcluster** objects we used in prior exercises:

```
distances = linkage(X, method="centroid", metric="euclidean")
sklearn_clusters = ac.fit_predict(X)
scipy_clusters = fcluster(distances, 3, criterion="distance")
```

After instantiating our model into a variable, we can simply fit the dataset to the desired model using **.fit_predict()** and assign it to an additional variable. This will give us information on the ideal clusters as part of the model fitting process.

3. Then, we can compare how each of the approaches work by comparing the final cluster results through plotting. Let's take a look at the clusters from the scikit-learn approach:

```
plt.figure(figsize=(6,4))
plt.title("Clusters from Sci-Kit Learn Approach")
plt.scatter(X[:, 0], X[:, 1], c = sklearn_clusters ,\
            s=50, cmap='tab20b')
plt.show()
```

Here is the output for the clusters from the scikit-learn approach:

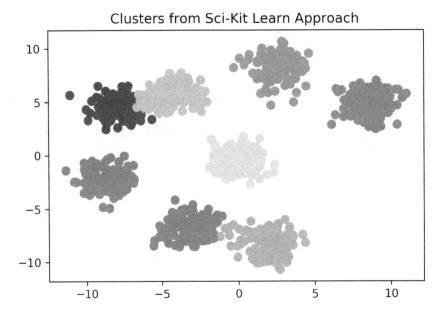

Figure 2.21: A plot of the scikit-learn approach

Take a look at the clusters from the SciPy approach:

```
plt.figure(figsize=(6,4))
plt.title("Clusters from SciPy Approach")
plt.scatter(X[:, 0], X[:, 1], c = scipy_clusters ,\
            s=50, cmap='tab20b')
plt.show()
```

The output is as follows:

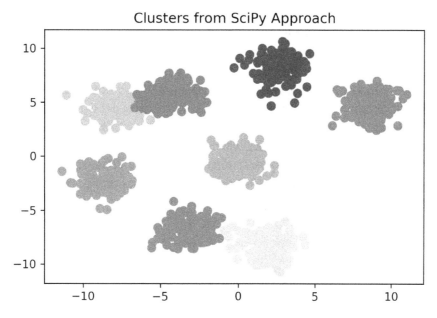

Figure 2.22: A plot of the SciPy approach

As you can see, the two converge to basically the same clusters.

> **NOTE**
>
> To access the source code for this specific section, please refer to https://packt.live/2DngJuz.
>
> You can also run this example online at https://packt.live/3f5PRgy.

While this is great from a toy problem perspective, in the next activity, you will learn that small changes to the input parameters can lead to wildly different results.

ACTIVITY 2.01: COMPARING K-MEANS WITH HIERARCHICAL CLUSTERING

You are managing a store's inventory and receive a large shipment of wine, but the brand labels fell off the bottles in transit. Fortunately, your supplier has provided you with the chemical readings for each bottle, along with their respective serial numbers. Unfortunately, you aren't able to open each bottle of wine and taste test the difference – you must find a way to group the unlabeled bottles back together according to their chemical readings. You know from the order list that you ordered three different types of wine and are given only two wine attributes to group the wine types back together. In this activity, we will be using the wine dataset. This dataset comprises chemical readings from three different types of wine, and as per the source on the UCI Machine Learning Repository, it contains these features:

- Alcohol
- Malic acid
- Ash
- Alkalinity of ash
- Magnesium
- Total phenols
- Flavanoids
- Nonflavanoid phenols
- Proanthocyanins
- Color intensity
- Hue
- OD280/OD315 of diluted wines
- Proline

> **NOTE**
>
> The wine dataset is sourced from https://archive.ics.uci.edu/ml/machine-learning-databases/wine/.[UCI Machine Learning Repository [http://archive.ics.uci.edu/ml]. Irvine, CA: University of California, School of Information and Computer Science.] It can also be accessed at https://packt.live/3aP8Tpv.

The aim of this activity is to implement k-means and hierarchical clustering on the wine dataset and to determine which of these approaches is more accurate in forming three separate clusters for each wine type. You can try different combinations of scikit-learn implementations and use helper functions in SciPy and NumPy. You can also use the silhouette score to compare the different clustering methods and visualize the clusters on a graph.

After completing this activity, you will see first-hand how two different clustering algorithms perform on the same dataset, allowing easy comparison when it comes to hyperparameter tuning and overall performance evaluation. You will probably notice that one method performs better than the other, depending on how the data is shaped. Another key outcome from this activity is gaining an understanding of how important hyperparameters are in any given use case.

Here are the steps to complete this activity:

1. Import the necessary packages from scikit-learn (**KMeans**, **AgglomerativeClustering**, and **silhouette_score**).
2. Read the wine dataset into a pandas DataFrame and print a small sample.
3. Visualize some features from the dataset by plotting the OD Reading feature against the proline feature.
4. Use the **sklearn** implementation of k-means on the wine dataset, knowing that there are three wine types.
5. Use the **sklearn** implementation of hierarchical clustering on the wine dataset.
6. Plot the predicted clusters from k-means.
7. Plot the predicted clusters from hierarchical clustering.
8. Compare the silhouette score of each clustering method.

66 | Hierarchical Clustering

Upon completing this activity, you should have plotted the predicted clusters you obtained from k-means as follows:

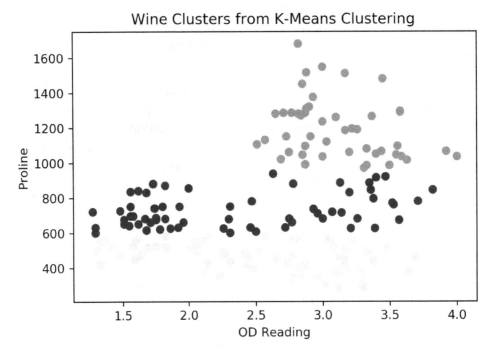

Figure 2.23: The expected clusters from the k-means method

A similar plot should also be obtained for the cluster that was predicted by hierarchical clustering, as shown here:

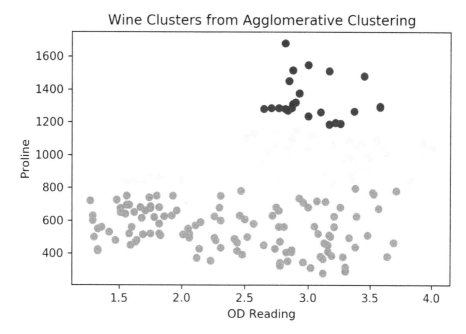

Figure 2.24: The expected clusters from the agglomerative method

NOTE

The solution to this activity can be found on page 423.

K-MEANS VERSUS HIERARCHICAL CLUSTERING

In the previous chapter, we explored the merits of k-means clustering. Now, it is important to explore where hierarchical clustering fits into the picture. As we mentioned in the *Linkage* section, there is some potential direct overlap when it comes to grouping data points together using centroids. Universal to all of the approaches we've mentioned so far is the use of a distance function to determine similarity. Due to our in-depth exploration in the previous chapter, we used the Euclidean distance here, but we understand that any distance function can be used to determine similarities.

In practice, here are some quick highlights for choosing one clustering method over another:

- Hierarchical clustering benefits from not needing to pass in an explicit "k" number of clusters a priori. This means that you can find all the potential clusters and decide which clusters make the most sense after the algorithm has completed.

- The k-means clustering benefits from a simplicity perspective – oftentimes, in business use cases, there is a challenge when it comes to finding methods that can be explained to non-technical audiences but are still accurate enough to generate quality results. k-means can easily fill this niche.

- Hierarchical clustering has more parameters to tweak than k-means clustering when it comes to dealing with abnormally shaped data. While k-means is great at finding discrete clusters, it can falter when it comes to mixed clusters. By tweaking the parameters in hierarchical clustering, you may find better results.

- Vanilla k-means clustering works by instantiating random centroids and finding the closest points to those centroids. If they are randomly instantiated in areas of the feature space that are far away from your data, then it can end up taking quite some time to converge, or it may never even get to that point. Hierarchical clustering is less prone to falling prey to this weakness.

SUMMARY

In this chapter, we discussed how hierarchical clustering works and where it may be best employed. In particular, we discussed various aspects of how clusters can be subjectively chosen through the evaluation of a dendrogram plot. This is a huge advantage over k-means clustering if you have absolutely no idea of what you're looking for in the data. Two key parameters that drive the success of hierarchical clustering were also discussed: the agglomerative versus divisive approach and linkage criteria. Agglomerative clustering takes a bottom-up approach by recursively grouping nearby data together until it results in one large cluster. Divisive clustering takes a top-down approach by starting with the one large cluster and recursively breaking it down until each data point falls into its own cluster. Divisive clustering has the potential to be more accurate since it has a complete view of the data from the start; however, it adds a layer of complexity that can decrease the stability and increase the runtime.

Linkage criteria grapples with the concept of how distance is calculated between candidate clusters. We have explored how centroids can make an appearance again beyond k-means clustering, as well as single and complete linkage criteria. Single linkage finds cluster distances by comparing the closest points in each cluster, while complete linkage finds cluster distances by comparing more distant points in each cluster. With the knowledge that you have gained in this chapter, you are now able to evaluate how both k-means and hierarchical clustering can best fit the challenge that you are working on.

While hierarchical clustering can result in better performance than k-means due to its increased complexity, please remember that more complexity is not always good. Your duty as a practitioner of unsupervised learning is to explore all the options and identify the solution that is both resource-efficient and performant. In the next chapter, we will cover a clustering approach that will serve us best when it comes to highly complex and noisy data: **Density-Based Spatial Clustering of Applications with Noise**.

3
NEIGHBORHOOD APPROACHES AND DBSCAN

OVERVIEW

In this chapter, we will see how neighborhood approaches to clustering work from start to end and implement the **Density-Based Spatial Clustering of Applications with Noise** (**DBSCAN**) algorithm from scratch by using packages. We will also identify the most suitable algorithm to solve your problem from k-means, hierarchical clustering, and DBSCAN. By the end of this chapter, we will see how the DBSCAN clustering approach will serve us best in the sphere of highly complex data.

INTRODUCTION

In previous chapters, we evaluated a number of different approaches to data clustering, including k-means and hierarchical clustering. While k-means is the simplest form of clustering, it is still extremely powerful in the right scenarios. In situations where k-means can't capture the complexity of the dataset, hierarchical clustering proves to be a strong alternative.

One of the key challenges in unsupervised learning is that you will be presented with a collection of feature data but no complementary labels telling you what a target state will be. While you may not get a discrete view of what the target labels are, you can get some semblance of structure out of the data by clustering similar groups together and seeing what is similar within groups. The first approach we covered to achieve this goal of clustering similar data points is k-means. K-means clustering works best for simple data challenges where speed is paramount. Simply looking at the closest data point (cluster centroid) does not require a lot of computational overhead; however, there is also a greater challenge posed when it comes to higher-dimensional datasets. K-means clustering is also not ideal if you are unaware of the potential number of clusters you are looking for. An example we worked with in *Chapter 2, Hierarchical Clustering*, entailed looking at chemical profiles to determine which wines belonged together in a disorganized shipment. This exercise only worked well because we knew that three wine types were ordered; however, k-means would have been less successful if you had no idea regarding what the original order constituted.

The second clustering approach we explored was hierarchical clustering. This method can work in two ways – either agglomerative or divisive. Agglomerative clustering works with a bottom-up approach, treating each data point as its own cluster and recursively grouping them together with linkage criteria. Divisive clustering works in the opposite way by treating all data points as one large class and recursively breaking them down into smaller clusters. This approach has the benefit of fully understanding the entire data distribution, as it calculates splitting potential; however, it is typically not implemented in practice due to its greater complexity. Hierarchical clustering is a strong contender for your clustering needs when it comes to not knowing anything about the data. Using a dendrogram, you can visualize all the splits in your data and consider what number of clusters makes sense after the fact. This can be really helpful in your specific use case; however, it also comes at a higher computational cost than is associated with k-means.

In this chapter, we will cover a clustering approach that will serve us best in the sphere of highly complex data: **Density-Based Spatial Clustering of Applications with Noise** (**DBSCAN**). Canonically, this method has always been seen as a high performer in datasets that have a lot of densely interspersed data. Let's walk through why it does so well in these use cases.

CLUSTERS AS NEIGHBORHOODS

Until now, we have explored the concept of likeness being described as a function of Euclidean distance – data points that are closer to any one point can be seen as similar, while those that are further away in Euclidean space can be seen as dissimilar. This notion is seen once again in the DBSCAN algorithm. As alluded to by the lengthy name, the DBSCAN approach expands upon basic distance metric evaluation by also incorporating the notion of density. If there are clumps of data points that all exist in the same area as one another, they can be seen as members of the same cluster:

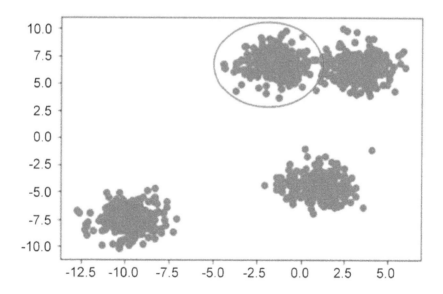

Figure 3.1: Neighbors have a direct connection to clusters

In the preceding figure, we can see four neighborhoods. The density-based approach has a number of benefits when compared to the past approaches we've covered that focus exclusively on distance. If you were just focusing on distance as a clustering threshold, then you may find your clustering makes little sense if faced with a sparse feature space with outliers. Both k-means and hierarchical clustering will automatically group together all data points in the space until no points are left.

While hierarchical clustering does provide a path around this issue somewhat, since you can dictate where clusters are formed using a dendrogram post-clustering run, k-means is the most susceptible to failure as it is the simplest approach to clustering. These pitfalls are less evident when we begin evaluating neighborhood approaches to clustering. In the following dendrogram, you can see an example of the pitfall where all data points are grouped together. Clearly, as you travel down the dendrogram, there is a lot of potential variation that gets grouped together since every point needs to be a member of a cluster. This is less of an issue with neighborhood-based clustering:

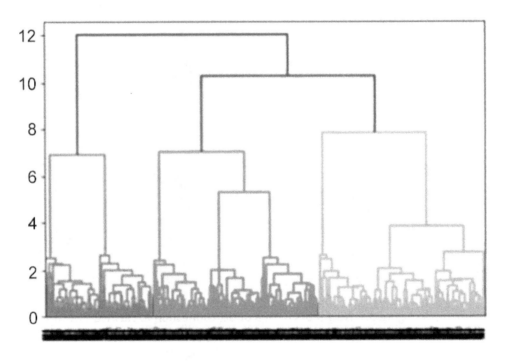

Figure 3.2: Example dendrogram

By incorporating the notion of neighbor density in DBSCAN, we can leave outliers out of clusters if we choose to, based on the hyperparameters we choose at runtime. Only the data points that have close neighbors will be seen as members within the same cluster, and those that are farther away can be left as unclustered outliers.

INTRODUCTION TO DBSCAN

In DBSCAN, density is evaluated as a combination of neighborhood radius and minimum points found in a neighborhood deemed a cluster. This concept can be driven home if we reconsider the scenario where you are tasked with organizing an unlabeled shipment of wine for your store. In the previous example, it was made clear that we can find similar wines based on their features, such as chemical traits. Knowing this information, we can more easily group together similar wines and efficiently have our products organized for sale in no time. In the real world, however, the products that you order to stock your store will reflect real-world purchase patterns. To promote variety in your inventory, but still have sufficient stock of the most popular wines, there is a highly uneven distribution of product types that you have available. Most people love the classic wines, such as white and red; however, you may still carry more exotic wines for your customers who love expensive varieties. This makes clustering more difficult, since there are uneven class distributions (you don't order 10 bottles of every wine available, for example).

DBSCAN differs from k-means and hierarchical clustering because you can build this intuition into how we evaluate the clusters of customers we are interested in forming. It can cut through the noise in an easier fashion and only point out customers who have the highest potential for remarketing in a campaign.

By clustering through the concept of a neighborhood, we can separate out the one-off customers who can be seen as random noise, relative to the more valuable customers who come back to our store time and time again. This approach calls into question how we establish the best numbers when it comes to neighborhood radius and minimum points per neighborhood.

As a high-level heuristic, we want our neighborhood radius to be small, but not too small. At one end of the extreme, you can have the neighborhood radius quite high – this can max out at treating all points in the feature space as one massive cluster. At the opposite end of the extreme, you can have a very small neighborhood radius. Overly small neighborhood radii can result in no points being clustered together and having a large collection of single-member clusters.

Similar logic applies when it comes to the minimum number of points that can make up a cluster. Minimum points can be seen as a secondary threshold that tunes the neighborhood radius a bit, depending on what data you have available in your space. If all of the data in your feature space is extremely sparse, minimum points become extremely valuable, in tandem with the neighborhood radius, to make sure you don't just have a large number of uncorrelated data points. When you have very dense data, the minimum points threshold becomes less of a driving factor than neighborhood radius.

As you can see from these two hyperparameter rules, the best options are, as usual, dependent on what your dataset looks like. Oftentimes, you will want to find the perfect "goldilocks" zone of not being too small in your hyperparameters, but also not too large.

DBSCAN IN DETAIL

To see how DBSCAN works, we can trace the path of a simple toy program as it merges together to form a variety of clusters and noise-labeled data points:

1. Out of *n* unvisited sample data points, we'll first move through each point in a loop and mark each one as visited.

2. From each point, we'll look at the distance to every other point in the dataset.

3. All points that fall within the neighborhood radius hyperparameter should be considered as neighbors.

4. The number of neighbors should be at least as many as the minimum points required.

5. If the minimum point threshold is reached, the points should be grouped together as a cluster, or else marked as noise.

6. This process should be repeated until all data points are categorized in clusters or as noise.

DBSCAN is fairly straightforward in some senses – while there are the new concepts of density through neighborhood radius and minimum points, at its core, it is still just evaluating using a distance metric.

WALKTHROUGH OF THE DBSCAN ALGORITHM

The following steps will walk you through this path in slightly more detail:

1. Given six sample data points, view each point as its own cluster [(1,3)], [(-8,6)], [(-6,4)], [(4,-2)],] (2,5)], [(-2,0)]:

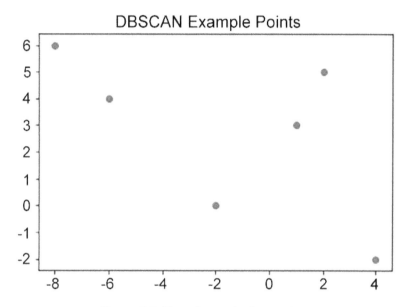

Figure 3.3: Plot of sample data points

2. Calculate the pairwise Euclidean distance between each of the points:

POINTS	(1,3)	(-8,6)	(-6,4)	(4,-2)	(2,5)	(-2,0)
(1,3) [	0.	9.48683298	7.07106781	5.83095189	2.23606798	4.24264069]
(-8,6) [	9.48683298	0.	2.82842712	14.4222051	10.04987562	8.48528137]
(-6,4) [	7.07106781	2.82842712	0.	11.66190379	8.06225775	5.65685425]
(4,-2) [	5.83095189	14.4222051	11.66190379	0.	7.28010989	6.32455532]
(2,5) [	2.23606798	10.04987562	8.06225775	7.28010989	0.	6.40312424]
(-2,0) [	4.24264069	8.48528137	5.65685425	6.32455532	6.40312424	0.]

Figure 3.4: Point distances

3. From each point, expand a neighborhood size outward and form clusters. For the purpose of this example, imagine you pass through a neighborhood radius of five. This means that any two points will be neighbors if the distance between them is less than five units. For example, point (1,3) has points (2,5) and (-2,0) as neighbors.

   Depending on the number of points in the neighborhood of a given point, the point can be classified into the following three categories:

   Core Point: If the point under observation has data points greater than the minimum number of points in its neighborhood that make up a cluster, then that point is called a core point of the cluster. All core points within the neighborhood of other core points are part of the same cluster. However, all the core points that are not in same neighborhood are part of another cluster.

   Boundary Point: If the point under observation does not have sufficient neighbors (data points) of its own, but it has at least one core point (in its neighborhood), then that point represents the boundary point of the cluster. Boundary points belong to the same cluster of their nearest core point.

   Noise Point: A data point is treated as a noise point if it does not have the required minimum number of data points in its neighborhood and is not associated with a core point. This point is treated as pure noise and is excluded from clustering.

4. Points that have neighbors are then evaluated to see whether they pass the minimum points threshold. In this example, if we had passed through a minimum points threshold of two, then points (1,3), (2,5), and (-2,0) could formally be grouped together as a cluster. If we had a minimum points threshold of four, then these three data points would be considered superfluous noise.

5. Points that have fewer neighbors than the minimum number of neighboring points required and whose neighborhood does not contain a core point are marked as noise and remain unclustered. Thus, points (-6,4), (4,-2), and (-8,6) fall under this category. However, points such as (2,5) and (2,0), though don't satisfy the criteria of the minimum number of points in neighborhood, do contain a core point as their neighbor, and are therefore marked as boundary points.

6. The following table summarizes the neighbors of a particular point and classifies them as core, boundary, and noise data points (mentioned in the preceding step) for a neighborhood radius of 5 and a minimum-neighbor criterion of 2.

Point of observation	Neighbor	Has min number of neighbors	Does contain a core point as a neighbor	Classification
(1,3)	(2,5) , (-2,0)	Yes	-	Core Point
(-8,6)	(-6,4)	No	No	Noise
(-6,4)	(-8,6)	No	No	Noise
(4,-2)		No	No	Noise
(2,5)	(1,3)	No	Yes	Boundary Point
(-2,0)	(1,3)	No	Yes	Boundary Point

Figure 3.5: Table showing details of neighbors for given points

7. Repeat this process on any remaining unvisited data points.

At the end of this process, you will have sorted your entire dataset into either clusters or unrelated noise. DBSCAN performance is highly dependent on the threshold hyperparameters you choose. This means that you may have to run DBSCAN a couple of times with different hyperparameter options to get an understanding of how they influence overall performance.

Note that DBSCAN does not require the centroids that we saw in both k-means and centroid-focused implementation of hierarchical clustering. This feature allows DBSCAN to work better for complex datasets, since most data is not shaped like clean blobs. DBSCAN is also more effective against outliers and noise than k-means or hierarchical clustering.

Let's now see how the performance of DBSCAN changes with varying neighborhood radius sizes.

EXERCISE 3.01: EVALUATING THE IMPACT OF NEIGHBORHOOD RADIUS SIZE

For this exercise, we will work in reverse of what we have typically seen in previous examples by first seeing the packaged implementation of DBSCAN in scikit-learn, and then implementing it on our own. This is done on purpose to fully explore how different neighborhood radius sizes drastically impact DBSCAN performance.

By completing this exercise, you will become familiar with how tuning neighborhood radius size can change how well DBSCAN performs. It is important to understand these facets of DBSCAN, as they can save you time in the future by troubleshooting your clustering algorithms efficiently:

1. Import the packages from scikit-learn and matplotlib that are necessary for this exercise:

```
from sklearn.cluster import DBSCAN
from sklearn.datasets import make_blobs
import matplotlib.pyplot as plt
%matplotlib inline
```

2. Generate a random cluster dataset to experiment on; X = coordinate points, and y = cluster labels (not needed):

```
X, y = make_blobs(n_samples=1000, centers=8, \
                  n_features=2, random_state=800)
# Visualize the data
plt.scatter(X[:,0], X[:,1])
plt.show()
```

The output is as follows:

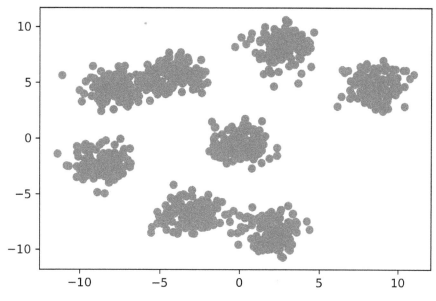

Figure 3.6: Visualized toy data example

3. After plotting the dummy data for this toy problem, you will see that the dataset has two features and approximately seven to eight clusters. To implement DBSCAN using scikit-learn, you will need to instantiate a new scikit-learn class:

```
db = DBSCAN(eps=0.5, min_samples=10, metric='euclidean')
```

Our example DBSCAN instance is stored in the **db** variable, and our hyperparameters are passed through on creation. For the sake of this example, you can see that the neighborhood radius (**eps**) is set to **0.5**, while the minimum number of points is set to **10**. To keep in line with our past chapters, we will once again be using Euclidean distance as our distance metric.

> **NOTE**
>
> **eps** stands for epsilon and is the radius of the neighborhood that your algorithm will look within when searching for neighbors.

4. Let's set up a loop that allows us to explore potential neighborhood radius size options interactively:

```
eps = [0.2,0.7,4]
for ep in eps:
    db = DBSCAN(eps=ep, min_samples=10, metric='euclidean')
    plt.scatter(X[:,0], X[:,1], c=db.fit_predict(X))
    plt.title('Toy Problem with eps: ' + str(ep))
    plt.show()
```

Introduction to DBSCAN | 83

The preceding code results in the following plots:

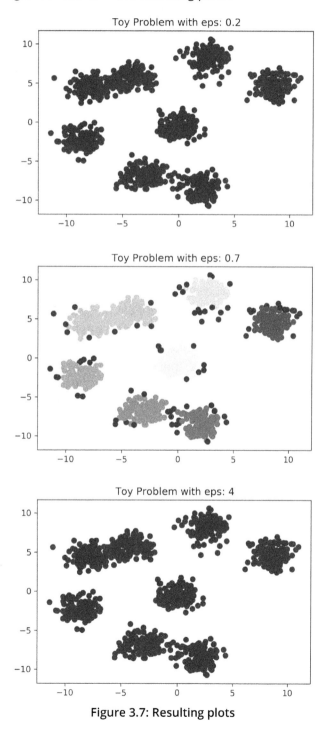

Figure 3.7: Resulting plots

As you can see from the plots, setting our neighborhood size too small will cause everything to be seen as random noise (purple points). Bumping our neighborhood size up a little bit allows us to form clusters that make more sense. A larger epsilon value would again convert the entire dataset into a single cluster (purple data points). Try recreating the preceding plots and experiment with varying **eps** sizes.

> **NOTE**
>
> To access the source code for this specific section, please refer to https://packt.live/3gEijGC.
>
> You can also run this example online at https://packt.live/2ZPBfeJ.

DBSCAN ATTRIBUTES – NEIGHBORHOOD RADIUS

In the preceding exercise, you saw how impactful setting the proper neighborhood radius is on the performance of your DBSCAN implementation. If your neighborhood is too small, then you will run into issues where all the data will be treated as noise and is left unclustered. If you set your neighborhood too large, then all of the data will similarly be grouped together into one cluster and not provide any value. If you explored the preceding exercise further with your own **eps** sizes, you may have noticed that it is very difficult to perform effective clustering using only the neighborhood size. This is where a minimum points threshold comes in handy. We will visit that topic later.

To go deeper into the neighborhood concept of DBSCAN, let's take a deeper look at the **eps** hyperparameter you pass at instantiation time. This epsilon value is converted to a radius that sweeps around any given data point in a circular manner to serve as a neighborhood:

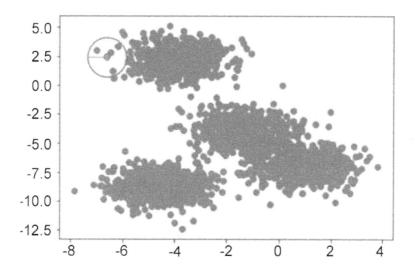

Figure 3.8: Visualization of the neighborhood radius; the red circle is the neighborhood

In this instance, there will be four neighbors of the center point, as can be seen in the preceding plot.

86 | Neighborhood Approaches and DBSCAN

One key aspect to observe here is that the shape formed by your neighborhood search is a circle in two dimensions, and a sphere in three dimensions. This may impact the performance of your model simply based on how the data is structured. Once again, blobs may seem like an intuitive structure to find – this may not always be the case. Fortunately, DBSCAN is well equipped to handle this dilemma of clusters that you may be interested in, but that do not fit the explicit blob structure:

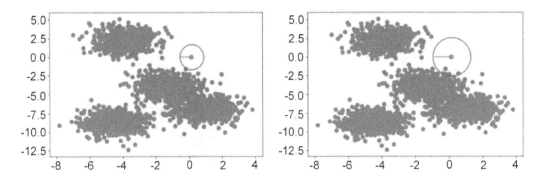

Figure 3.9: Impact of varying neighborhood radius size

On the left, the data point will be classified as random noise. On the right, the data point has multiple neighbors and could be its own cluster.

ACTIVITY 3.01: IMPLEMENTING DBSCAN FROM SCRATCH

During an interview, you are asked to create the DBSCAN algorithm from scratch using a generated two-dimensional dataset. To do this, you will need to convert the theory behind neighborhood searching into production code, with a recursive call that adds neighbors. As explained in the previous section, you will use a distance scan in space surrounding a specified point to add these neighbors.

Given what you've learned about DBSCAN and distance metrics from prior chapters, build an implementation of DBSCAN from scratch in Python. You are free to use NumPy and SciPy to evaluate distances here.

These steps will help you to complete the activity:

1. Generate a random cluster dataset.
2. Visualize the data.

3. Create functions from scratch that allow you to call DBSCAN on a dataset.

4. Use your created DBSCAN implementation to find clusters in the generated dataset. Feel free to use hyperparameters as you see fit, tuning them based on their performance.

5. Visualize the clustering performance of your DBSCAN implementation from scratch.

The desired outcome of this exercise is for you to implement how DBSCAN works from the ground up before you use the fully packaged implementation in scikit-learn. Taking this approach to any machine learning algorithm from scratch is important, as it helps you "earn" the ability to use easier implementations, while still being able to discuss DBSCAN in depth in the future:

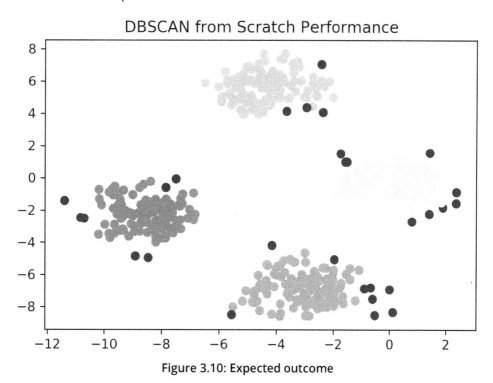

Figure 3.10: Expected outcome

> **NOTE**
>
> The solution to this activity can be found on page 428.

DBSCAN ATTRIBUTES – MINIMUM POINTS

The other core component to a successful implementation of DBSCAN beyond the neighborhood radius is the minimum number of points required to justify membership within a cluster. As mentioned earlier, it is more obvious that this lower bound benefits your algorithm when it comes to sparser datasets. That's not to say that it is a useless parameter when you have very dense data; however, while having single data points randomly interspersed through your feature space can be easily bucketed as noise, it becomes more of a gray area when we have random patches of two to three, for example. Should these data points be their own cluster, or should they also be categorized as noise? Minimum points thresholding helps to solve this problem.

In the scikit-learn implementation of DBSCAN, this hyperparameter is seen in the `min_samples` field passed on DBSCAN instance creation. This field is very valuable in tandem with the neighborhood radius size hyperparameter to fully round out your density-based clustering approach:

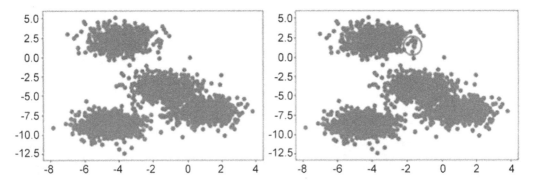

Figure 3.11: Minimum points threshold deciding whether a group of data points is noise or a cluster

On the right, if the minimum points threshold is 10 points, it will classify data in this neighborhood as noise.

In real-world scenarios, you can see minimum points being highly impactful when you have truly large amounts of data. Going back to the wine-clustering example, if your store was actually a large wine warehouse, you could have thousands of individual wines with only one or two bottles that could easily be viewed as their own cluster. This may be helpful depending on your use case; however, it is important to keep in mind the subjective magnitudes that come with your data. If you have millions of data points, then random noise can easily be seen as hundreds or even thousands of random one-off sales. However, if your data is on the scale of hundreds or thousands, single data points can be seen as random noise.

EXERCISE 3.02: EVALUATING THE IMPACT OF THE MINIMUM POINTS THRESHOLD

Similar to *Exercise 3.01, Evaluating the Impact of Neighborhood Radius Size*, where we explored the value of setting a proper neighborhood radius size, we will repeat the exercise, but instead will change the minimum points threshold on a variety of datasets.

Using our current implementation of DBSCAN, we can easily tune the minimum points threshold. Tune this hyperparameter and see how it performs on generated data.

By tuning the minimum points threshold for DBSCAN, you will understand how it can affect the quality of your clustering predictions.

Once again, let's start with randomly generated data:

1. Generate a random cluster dataset, as follows:

    ```
    from sklearn.cluster import DBSCAN
    from sklearn.datasets import make_blobs
    import matplotlib.pyplot as plt
    %matplotlib inline
    X, y = make_blobs(n_samples=1000, centers=8,\
                     n_features=2, random_state=800)
    ```

2. Visualize the data as follows:

```
# Visualize the data
plt.scatter(X[:,0], X[:,1])
plt.show()
```

The output is as follows:

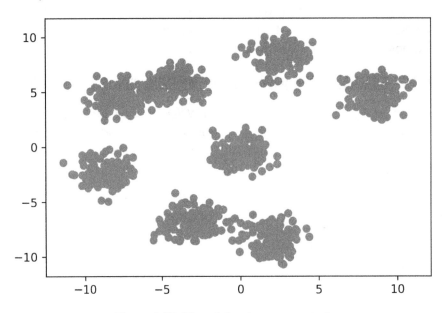

Figure 3.12: Plot of the data generated

3. With the same plotted data as before, let's grab one of the better-performing neighborhood radius sizes from *Exercise 3.01, Evaluating the Impact of Neighborhood Radius Size* – **eps = 0.7**:

```
db = DBSCAN(eps=0.7, min_samples=10, metric='euclidean')
```

> **NOTE**
>
> **eps** is a tunable hyperparameter. Earlier in *Step 3* of the previous exercise, we used a value of **0.5**. In this step, we are using **eps = 0.7** based on our experimentation with this parameter.

4. After instantiating the DBSCAN clustering algorithm, let's treat the **min_samples** hyperparameters as the variable we wish to tune. We can cycle through a loop to find which minimum number of points works best for our use case:

```
num_samples = [10,19,20]
for min_num in num_samples:
    db = DBSCAN(eps=0.7, min_samples=min_num, metric='euclidean')
    plt.scatter(X[:,0], X[:,1], c=db.fit_predict(X))
    plt.title('Toy Problem with Minimum Points: ' + str(min_num))
    plt.show()
```

Looking at the first plot generated, we can see where we ended if you followed *Exercise 3.01, Evaluating the Impact of Neighborhood Radius Size* exactly, using 10 minimum points to mark the threshold for cluster membership:

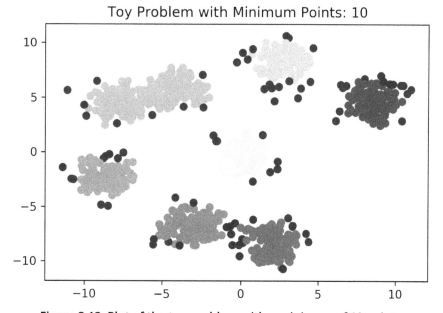

Figure 3.13: Plot of the toy problem with a minimum of 10 points

92 | Neighborhood Approaches and DBSCAN

The remaining two hyperparameter options can be seen to greatly impact the performance of your DBSCAN clustering algorithm, and show how a shift in one number can greatly influence performance:

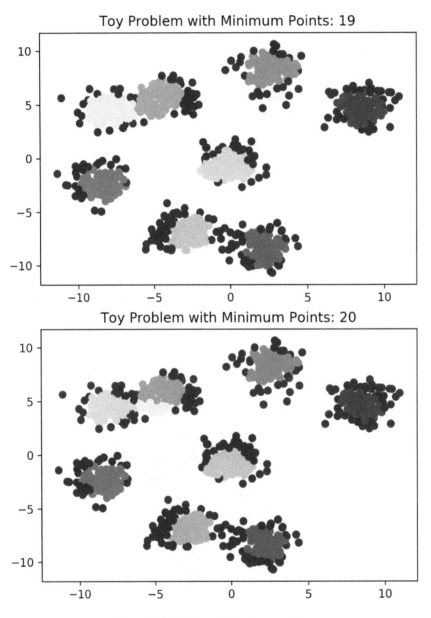

Figure 3.14: Plots of the toy problem

As you can see, simply changing the number of minimum points from 19 to 20 adds an additional (incorrect!) cluster to our feature space. Given what you've learned about minimum points through this exercise, you can now tweak both epsilon and minimum points thresholding in your scikit-learn implementation to achieve the optimal number of clusters.

> **NOTE**
>
> In our original generation of the data, we created eight clusters. These indicate that small changes in minimum points can add entire new clusters that we know shouldn't be there.
>
> To access the source code for this specific section, please refer to https://packt.live/3fa4L5F.
>
> You can also run this example online at https://packt.live/31XUeqi.

ACTIVITY 3.02: COMPARING DBSCAN WITH K-MEANS AND HIERARCHICAL CLUSTERING

In the preceding chapter, we attempted to group different wines together using hierarchical clustering. Let's attempt this approach again with DBSCAN and see whether a neighborhood search fares any better. As a reminder, you are managing store inventory and have received a large shipment of wine, but the brand labels fell off the bottles during transit. Fortunately, your supplier provided you with the chemical readings for each bottle along with their respective serial numbers. Unfortunately, you aren't able to open each bottle of wine and taste test the difference – you must find a way to group the unlabeled bottles back together according to their chemical readings! You know from the order list that you ordered three different types of wine and are given only two wine attributes to group the wine types back together.

In the previous sections, we were able to see how k-means and hierarchical clustering performed on the wine dataset. In our best-case scenario, we were able to achieve a silhouette score of 0.59. Using scikit-learn's implementation of DBSCAN, let's see whether we can get even better clustering.

These steps will help you to complete the activity:

1. Import the necessary packages.
2. Load the wine dataset and check what the data looks like.
3. Visualize the data.
4. Generate clusters using k-means, agglomerative clustering, and DSBSCAN.
5. Evaluate a few different options for DSBSCAN hyperparameters and their effect on the silhouette score.
6. Generate the final clusters based on the highest silhouette score.
7. Visualize clusters generated using each of the three methods.

> **NOTE**
>
> We have sourced this dataset from https://archive.ics.uci.edu/ml/datasets/wine. [Citation: Dua, D. and Graff, C. (2019). UCI Machine Learning Repository [http://archive.ics.uci.edu/ml]. Irvine, CA: University of California, School of Information and Computer Science]. You can also access it at https://packt.live/3bW8NME.

By completing this activity, you will be recreating a full workflow of a clustering problem. You have already made yourself familiar with the data in *Chapter 2, Hierarchical Clustering*, and, by the end of this activity, you will have performed model selection to find the best model and hyperparameters for your dataset. You will have silhouette scores of the wine dataset for each type of clustering.

> **NOTE**
>
> The solution to this activity can be found on page 431.

DBSCAN VERSUS K-MEANS AND HIERARCHICAL CLUSTERING

Now that you've reached an understanding of how DBSCAN is implemented and how many different hyperparameters you can tweak to drive performance, let's survey how it compares to the clustering methods we have covered previously – k-means clustering and hierarchical clustering.

You may have noticed in *Activity 3.02, Comparing DBSCAN with k-means and Hierarchical Clustering,* that DBSCAN can be a bit finicky when it comes to finding the optimal clusters via a silhouette score. This is a downside of the neighborhood approach – k-means and hierarchical clustering really excel when you have some idea regarding the number of clusters in your data. In most cases, this number is low enough that you can iteratively try a few different numbers and see how it performs. DBSCAN, instead, takes a more bottom-up approach by working with your hyperparameters and finding the clusters it views as important. In practice, it is helpful to consider DBSCAN when the first two options fail, simply because of the amount of tweaking needed to get it to work properly. That said, when your DBSCAN implementation is working correctly, it will often immensely outperform k-means and hierarchical clustering (in practice, this often happens with highly intertwined, yet still discrete, data, such as a feature space containing two half-moons).

Compared to k-means and hierarchical clustering, DBSCAN can be seen as being potentially more efficient, since it only has to look at each data point once. Instead of multiple iterations of finding new centroids and evaluating where their nearest neighbors are, once a point has been assigned to a cluster in DBSCAN, it does not change cluster membership. The other key feature that DBSCAN and hierarchical clustering both share, in comparison with k-means, is not needing to explicitly pass a number of clusters expected at the time of creation. This can be extremely helpful when you have no external guidance on how to break your dataset down.

SUMMARY

In this chapter, we discussed hierarchical clustering and DBSCAN, and in what type of situations they are best employed. While hierarchical clustering can, in some respects, be seen as an extension of the nearest-neighbor approach seen in k-means, DBSCAN approaches the problem of finding neighbors by applying a notion of density.
This can prove extremely beneficial when it comes to highly complex data that is intertwined in a complex fashion. While DBSCAN is very powerful, it is not infallible and can even be overkill, depending on what your original data looks like.

Combined with k-means and hierarchical clustering, however, DBSCAN completes a strong toolbox when it comes to the unsupervised learning task of clustering your data. When faced with any problem in this space, it is worthwhile comparing the performance of each method and seeing which performs best.

With clustering explored, we will now move onto another key piece of rounding out your skills in unsupervised learning: dimensionality reduction. Through the smart reduction of dimensions, we can make clustering easier to understand and communicate to stakeholders. Dimensionality reduction is also key to creating all types of machine learning models in the most efficient manner possible. In the next chapter, we will dive deeper into topic models and see how the aspects of clustering learned in these chapters apply to NLP-type problems.

4

DIMENSIONALITY REDUCTION TECHNIQUES AND PCA

OVERVIEW

In this chapter, we will apply dimension reduction techniques and describe the concepts behind principal components and dimensionality reduction. We will apply **Principal Component Analysis** (**PCA**) when solving problems using scikit-learn. We will also compare manual PCA versus scikit-learn. By the end of this chapter, you will be able to reduce the size of a dataset by extracting only the most important components of variance within the data.

INTRODUCTION

In the previous chapter, we discussed clustering algorithms and how they can be helpful to find underlying meaning in large volumes of data. This chapter investigates the use of different feature sets (or spaces) in our unsupervised learning algorithms, and we will start with a discussion regarding dimensionality reduction, specifically, **Principal Component Analysis** (**PCA**). We will then extend our understanding of the benefits of the different feature spaces through an exploration of two independently powerful machine learning architectures in neural network-based autoencoders. Neural networks certainly have a well-deserved reputation for being powerful models in supervised learning problems. Furthermore, through the use of an autoencoder stage, neural networks have been shown to be sufficiently flexible for their application to unsupervised learning problems. Finally, we will build on our neural network implementation and dimensionality reduction as we cover t-distributed nearest neighbors in *Chapter 6, t-Distributed Stochastic Neighbor Embedding*. These techniques will prove helpful when dealing with high-dimensional data, such as image processing or datasets with many features. One strong business benefit of some types of dimension reduction is that it helps to remove features that do not have much impact on final outputs. This creates opportunities to make your algorithms more efficient without any loss in performance.

WHAT IS DIMENSIONALITY REDUCTION?

Dimensionality reduction is an important tool in any data scientist's toolkit, and due to its wide variety of use cases, is essentially assumed knowledge within the field. So, before we can consider reducing the dimensionality and why we would want to reduce it, we must first have a good understanding of what dimensionality is. To put it simply, dimensionality is the number of dimensions, features, or variables associated with a sample of data. Often, this can be thought of as a number of columns in a spreadsheet, where each sample is on a new row, and each column describes an attribute of the sample. The following table is an example:

Pressure (hPa)	Temperature (°C)	Humidity (%)
1050	32.2	12
1026	27.8	80

Figure 4.1: Two samples of data with three different features

In the preceding table, we have two samples of data, each with three independent features or dimensions. Depending on the problem being solved, or the origin of this dataset, we may want to reduce the number of dimensions per sample without losing the provided information. This is where dimensionality reduction can be helpful. But how exactly can dimensionality reduction help us to solve problems? We will cover the applications in more detail in the following section. However, let's say that we had a very large dataset of time series data, such as echocardiogram or ECG (also known as an EKG in some countries) signals, as shown in the following diagram:

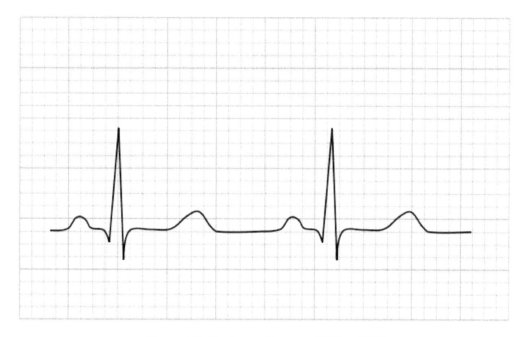

Figure 4.2: Electrocardiogram (ECG or EKG)

These signals were captured from your company's new model of watch, and we need to look for signs of a heart attack or stroke. After looking through the dataset, we can make a few observations:

- Most of the individual heartbeats are very similar.
- There is some noise in the data from the recording system or from the patient moving during the recording.
- Despite the noise, the heartbeat signals are still visible.
- There is a lot of data – too much to be able to process using the hardware available on the watch.

It is in such a situation that dimensionality reduction really shines. By using dimensionality reduction, we are able to remove much of the noise from the signal, which, in turn, will assist with the performance of the algorithms that are applied to the data as well as reduce the size of the dataset to allow for reduced hardware requirements. The techniques that we are going to discuss in this chapter, in particular, PCA and autoencoders, have been well applied in research and industry to effectively process, cluster, and classify such datasets.

APPLICATIONS OF DIMENSIONALITY REDUCTION

Before we start a detailed investigation of dimensionality reduction and PCA, we will discuss some of the common applications for these techniques:

- **Preprocessing/feature engineering**: One of the most common applications is in the preprocessing or feature engineering stages of developing a machine learning solution. The quality of the information provided during the algorithm development, as well as the correlation between the input data and the desired result, is critical in order for a high-performing solution to be designed. In this situation, PCA can provide assistance, as we are able to isolate the most important components of information from the data and provide this to the model so that only the most relevant information is being provided. This can also have a secondary benefit in that we have reduced the number of features being provided to the model, so there can be a corresponding reduction in the number of calculations to be completed. This can reduce the overall training time for the system. An example use case of this feature engineering would be predicting whether a transaction is at risk of credit card theft. In this scenario, you may be presented with millions of transactions that each have tens or hundreds of features. This would be resource-intensive or even impossible to run a predictive algorithm on in real time; however, by using feature preprocessing, we can distill the many features down to just the top 3-4 most important ones, thereby reducing runtime.

- **Noise reduction**: Dimensionality reduction can also be used as an effective noise reduction/filtering technique. It is expected that the noise within a signal or dataset does not comprise a large component of the variation within the data. Thus, we can remove some of the noise from the signal by removing the smaller components of variation and then restoring the data back to the original dataspace. In the following example, the image on the left has been filtered to the first 20 most significant sources of data, which gives us the image on the right. We can see that the quality of the image has been reduced, but the critical information is still there:

Figure 4.3: An image filtered with dimensionality reduction

> **NOTE**
>
> This photograph was taken by Arthur Brognoli from Pexels and is available for free use under https://www.pexels.com/photo-license/. In this case, we have the original image on the left and the filtered image on the right.

- **Generating plausible artificial datasets**: As PCA divides the dataset into the components of information (or variation), we can investigate the effects of each component or generate new dataset samples by adjusting the ratios between the eigenvalues. We will cover more on eigenvalues later on in this chapter. We can scale these components, which, in effect, increases or decreases the importance of that specific component. This is also referred to as **statistical shape modeling**, as one common method is to use it to create plausible variants of shapes. It is also used to detect facial landmarks in images in the process of **active shape modeling**.

- **Financial modeling/risk analysis**: Dimensionality reduction provides a useful toolkit for the finance industry, since being able to consolidate a large number of individual market metrics or signals into a smaller number of components allows for faster, and more efficient, computations. Similarly, the components can be used to highlight those higher-risk products/companies.

THE CURSE OF DIMENSIONALITY

Before we can understand the benefits of using dimensionality reduction techniques, we must first understand why the dimensionality of feature sets needs to be reduced at all. The **curse of dimensionality** is a phrase commonly used to describe issues that arise when working with data that has a high number of dimensions in the feature space; for example, the number of attributes that are collected for each sample. Consider a dataset of point locations within a game of *Pac-Man*. Your character, Pac-Man, occupies a position within the virtual world defined by two dimensions or coordinates (x, y). Let's say that we are creating a new computer enemy: an AI-driven ghost to play against, and that it requires some information regarding our character to make its own game logic decisions. For the bot to be effective, we require the player's position (x, y) and their velocity in each of the directions (vx, vy) in addition to the players last five (x, y) positions, the number of remaining hearts, and the number of remaining power pellets in the maze (power pellets temporarily allow Pac-Man to eat ghosts). Now, for each moment in time, our bot requires 16 individual features (or dimensions) to make its decisions. These 16 features correspond to 5 previous positions times the 2 x and y coordinates + the 2 x and y coordinates of the player's current position + the 2 x and y coordinates of player's velocity + 1 feature for the number of hearts + 1 feature for the power pellets = 16. This is clearly a lot more than just the two dimensions as provided by the position:

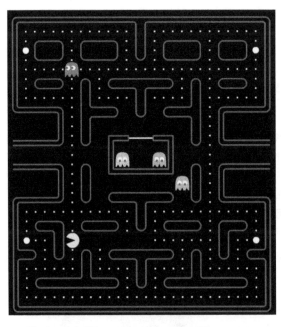

Figure 4.4: Dimensions in a PacMan game

What Is Dimensionality Reduction? | 105

To explain the concept of dimensionality reduction, we will consider a fictional dataset (see *Figure 4.5*) of *x* and *y* coordinates as features, giving two dimensions in the feature space. It should be noted that this example is by no means a mathematical proof but is rather intended to provide a means of visualizing the effect of increased dimensionality. In this dataset, we have six individual samples (or points), and we can visualize the currently occupied volume within the feature space of approximately (3 – 1) x (4 – 2) = 2 x 2 = 4 squared units:

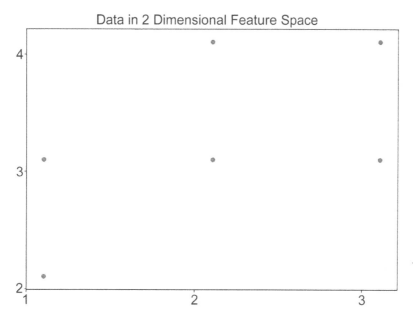

Figure 4.5: Data in a 2D feature space

Suppose the dataset comprises the same number of points, but with an additional feature (the *z* coordinate) to each sample. The occupied data volume is now approximately 2 x 2 x 2 = 8 cubed units. So, we now have the same number of samples, but the space enclosing the dataset is now larger. As such, the data takes up less relative volume in the available space and is now sparser. This is the curse of dimensionality; as we increase the number of available features, we increase the sparsity of the data, and, in turn, make statistically valid correlations more difficult. Looking back to our example of creating a video game bot to play against a human player, we have 16 features that are a mix of different feature types: positions, velocity, power-ups, and hearts. Depending on the range of possible values for each of these features and the variance to the dataset provided by each feature, the data could be extremely sparse. Even within the constrained world of Pac-Man, the potential variance of each of the features could be quite large, some much larger than others.

So, without dealing with the sparsity of the dataset, we have more information with the additional feature(s), but may not be able to improve the performance of our machine learning model, as the statistical correlations are more difficult. What we would like to do is to keep the useful information provided by the extra features but minimize the negative effect of sparsity. This is exactly what dimensionality reduction techniques are designed to do and these can be extremely powerful in increasing the performance of your machine learning model.

Throughout this chapter, we will discuss a number of different dimensionality reduction techniques and will cover one of the most important and useful methods, PCA, in greater detail with an example.

OVERVIEW OF DIMENSIONALITY REDUCTION TECHNIQUES

The goal of any dimensionality reduction technique is to manage the sparsity of the dataset while keeping any useful information that is provided. In our case of classification, dimensionality reduction is typically used as an important preprocessing step used before the actual classification. Most dimensionality reduction techniques aim to complete this task using a process of **feature projection**, which adjusts the data from the higher-dimensional space into a space with fewer dimensions to remove the sparsity from the data. Again, as a means of visualizing the projection process, consider a sphere in a 3D space. We can project the sphere into a lower 2D space into a circle with some information loss (the value for the z coordinate), but retaining much of the information that describes its original shape. We still know the origin, radius, and manifold (outline) of the shape, and it is still very clear that it is a circle. So, depending on the problem that we are trying to solve, we may have reduced the dimensionality while retaining the important information:

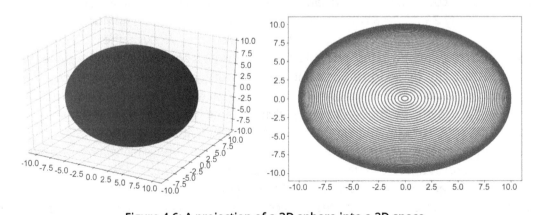

Figure 4.6: A projection of a 3D sphere into a 2D space

The secondary benefit that can be obtained by preprocessing the dataset with a dimensionality reduction stage is the improved computational performance that can be achieved. As the data has been projected into a lower-dimensional space, it will contain fewer, but potentially more powerful, features. The fact that there are fewer features means that, during later classification or regression stages, the size of the dataset being processed is significantly smaller. This will potentially reduce the required system resources and processing time for classification/regression, and, in some cases, the dimensionality reduction technique can also be used directly to complete the analysis.

This analogy also introduces one of the important considerations of dimensionality reduction. We are always trying to balance the information loss resulting from the projection into lower dimensional space by reducing the sparsity of the data. Depending on the nature of the problem and the dataset being used, the correct balance could present itself and be relatively straightforward. In some applications, this decision may rely on the outcome of additional validation methods, such as cross-validation (particularly in supervised learning problems) or the assessment of experts in your problem domain. In this scenario, cross-validation refers to the practice of partitioning a rolling section of the data to test on, with the inverse serving as the train set until all parts of the dataset are used. This approach allows for the reduction of bias within a machine learning problem.

One way we like to think about this trade-off in dimensionality reduction is to consider compressing a file or image on a computer for transfer. Dimensionality reduction techniques, such as PCA, are essentially methods of compressing information into a smaller size for transfer, and, in many compression methods, some losses occur as a result of the compression process. Sometimes, these losses are acceptable; if we are transferring a 50 MB image and need to shrink it to 5 MB for transfer, we can expect to still be able to see the main subject of the image, but perhaps some smaller background features will become too blurry to see. We would also not expect to be able to restore the original image to a pixel-perfect representation from the compressed version, but we could expect to restore it with some additional artifacts, such as blurring.

DIMENSIONALITY REDUCTION

Dimensionality reduction techniques have many uses in machine learning, as the ability to extract the useful information of a dataset can provide performance boosts in many machine learning problems. They can be particularly useful in unsupervised as opposed to supervised learning methods because the dataset does not contain any ground truth labels or targets to achieve. In unsupervised learning, the training environment is being used to organize the data in a way that is appropriate for the problem being solved (for example, classification via clustering), which is typically based on the most important information in the dataset. Dimensionality reduction provides an effective means of extracting important information, and, as there are a number of different methods that we could use, it is beneficial to review some of the available options:

- **Linear Discriminant Analysis (LDA):** This is a particularly handy technique that can be used for both classification as well as dimensionality reduction. LDA will be covered in more detail in *Chapter 7, Topic Modeling*.

- **Non-negative matrix factorization (NMF):** Like many of the dimensionality reduction techniques, this relies on the properties of linear algebra to reduce the number of features in the dataset. NMF will also be covered in more detail in *Chapter 7, Topic Modeling*.

- **Singular Value Decomposition (SVD):** This is somewhat related to PCA (which is covered in more detail in this chapter) and is also a matrix decomposition process not too dissimilar to NMF.

- **Independent Component Analysis (ICA):** This also shares some similarities to SVD and PCA, but relaxing the assumption of the data being a Gaussian distribution allows for non-Gaussian data to be separated.

Each of the methods described so far all use linear transformation to reduce the sparsity of the data in their original implementation. Some of these methods also have variants that use non-linear kernel functions in the separation process, providing the ability to reduce the sparsity in a non-linear fashion. Depending on the dataset being used, a non-linear kernel may be more effective at extracting the most useful information from the signal.

PRINCIPAL COMPONENT ANALYSIS

As described previously, PCA is a commonly used and very effective dimensionality reduction technique, which often forms a preprocessing stage for a number of machine learning models and techniques. For this reason, we will dedicate this section of the book to looking at PCA in more detail than any of the other methods. PCA reduces the sparsity in the dataset by separating the data into a series of components where each component represents a source of information within the data. As its name suggests, the first component produced in PCA, the **principal component**, comprises the majority of information or variance within the data. The principal component can often be thought of as contributing the most amount of interesting information in addition to the mean. With each subsequent component, less information, but more subtlety, is contributed to the compressed data. If we consider all of these components together, there will be no benefit of using PCA, as the original dataset will be returned. To clarify this process and the information returned by PCA, we will use a worked example, completing the PCA calculations by hand. But first, we must review some foundational statistical concepts, which are required to execute the PCA calculations.

MEAN

The mean, or the average value, is simply the addition of all values divided by the number of values in the set.

STANDARD DEVIATION

Often referred to as the spread of the data and related to the variance, the standard deviation is a measure of how much of the data lies within proximity to the mean. In a normally distributed dataset, approximately 68% of the dataset lies within one standard deviation of the mean, (that is, between (mean - 1*std) to (mean + 1*std), you can find 68% of the data if it is normally distributed.)

The relationship between the variance and standard deviation is quite a simple one – the variance is the standard deviation squared.

COVARIANCE

Where standard deviation or variance is the spread of the data calculated on a single dimension, the covariance is the variance of one dimension (or feature) against another. When the covariance of a dimension is computed against itself, the result is the same as simply calculating the variance for the dimension.

COVARIANCE MATRIX

A covariance matrix is a matrix representation of the possible covariance values that can be computed for a dataset. Other than being particularly useful in data exploration, covariance matrices are also required to execute the PCA of a dataset. To determine the variance of one feature with respect to another, we simply look up the corresponding value in the covariance matrix. In the following diagram, we can see that, in column 1, row 2, the value is the variance of feature or dataset Y with respect to X (cov(Y, X)). We can also see that there is a diagonal column of covariance values computed against the same feature or dataset; for example, cov(X, X). In this situation, the value is simply the variance of X:

$$\overline{cov} = \begin{bmatrix} cov(X,X) & cov(X,Y) & cov(X,Z) \\ cov(Y,X) & cov(Y,Y) & cov(Y,Z) \\ cov(Z,X) & cov(Z,Y) & cov(Z,Z) \end{bmatrix}$$

Figure 4.7: The covariance matrix

Typically, the exact values of each of the covariances are not as interesting as looking at the magnitude and relative size of each of the covariances within the matrix. A large value of the covariance of one feature against another would suggest that one feature changes significantly with respect to the other, while a value close to zero would signify very little change. The other interesting aspect of the covariance to look for is the sign associated with the covariance; a positive value indicates that as one feature increases or decreases, then so does the other, while a negative covariance indicates that the two features diverge from one another, with one increasing as the other decreases or vice versa.

Thankfully, **numpy** and **scipy** provide functions to efficiently perform these calculations for you. In the next exercise, we will compute these values in Python.

EXERCISE 4.01: COMPUTING MEAN, STANDARD DEVIATION, AND VARIANCE USING THE PANDAS LIBRARY

In this exercise, we will briefly review how to compute some of the foundational statistical concepts using both the **numpy** and **pandas** Python packages. In this exercise, we will use a dataset of the measurements of seeds from different varieties of wheat, created using X-ray imaging. The dataset, which can be found in the accompanying source code, comprises seven individual measurements (`area A`, `perimeter P`, `compactness C`, `length of kernel LK`, `width of kernel WK`, `asymmetry coefficient A_Coef`, and `length of kernel groove LKG`) of three different wheat varieties: Kama, Rosa, and Canadian.

> **NOTE**
>
> This dataset is sourced from https://archive.ics.uci.edu/ml/datasets/seed (UCI Machine Learning Repository [http://archive.ics.uci.edu/ml]. Irvine, CA: University of California, School of Information and Computer Science) Citation: Contributors gratefully acknowledge the support of their work by the Institute of Agrophysics of the Polish Academy of Sciences in Lublin. The dataset can also be downloaded from https://packt.live/2RjpDxk.

The steps to be performed are as follows:

1. Import the **pandas**, **numpy**, and **matplotlib** packages for use:

   ```
   import pandas as pd
   import numpy as np
   import matplotlib.pyplot as plt
   ```

2. Load the dataset and preview the first five lines of data:

   ```
   df = pd.read_csv('../Seed_Data.csv')
   df.head()
   ```

The output is as follows:

	A	P	C	LK	WK	A_Coef	LKG	target
0	15.26	14.84	0.8710	5.763	3.312	2.221	5.220	0
1	14.88	14.57	0.8811	5.554	3.333	1.018	4.956	0
2	14.29	14.09	0.9050	5.291	3.337	2.699	4.825	0
3	13.84	13.94	0.8955	5.324	3.379	2.259	4.805	0
4	16.14	14.99	0.9034	5.658	3.562	1.355	5.175	0

Figure 4.8: The head of the data

3. We only require the area, **A**, and the length of the kernel **LK** features, so remove the other columns:

```
df = df[['A', 'LK']]
df.head()
```

The output is as follows:

	A	LK
0	15.26	5.763
1	14.88	5.554
2	14.29	5.291
3	13.84	5.324
4	16.14	5.658

Figure 4.9: The head after cleaning the data

4. Visualize the dataset by plotting the **A** versus **LK** values:

```
plt.figure(figsize=(10, 7))
plt.scatter(df['A'], df['LK'])
plt.xlabel('Area of Kernel')
plt.ylabel('Length of Kernel')
plt.title('Kernel Area versus Length')
plt.show()
```

The output is as follows:

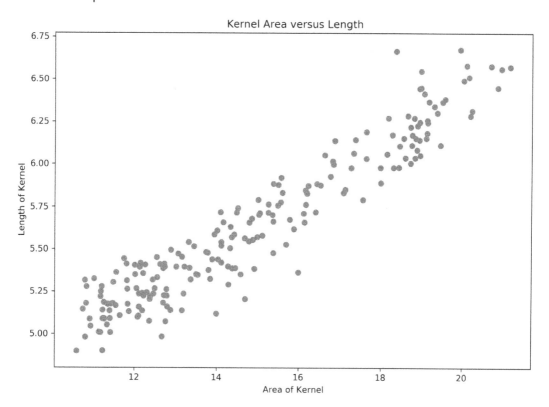

Figure 4.10: Plot of the data

5. Compute the mean value using the **pandas** method:

   ```
   df.mean()
   ```

   The output is as follows:

   ```
   A    14.847524
   LK    5.628533
   dtype: float64
   ```

6. Compute the mean value using the **numpy** method:

   ```
   np.mean(df.values, axis=0)
   ```

   The output is as follows:

   ```
   array([14.84752381,  5.62853333])
   ```

7. Compute the standard deviation value using the **pandas** method:

   ```
   df.std()
   ```

   The output is as follows:

   ```
   A    2.909699
   LK   0.443063
   dtype: float64
   ```

8. Compute the standard deviation value using the **numpy** method:

   ```
   np.std(df.values, axis=0)
   ```

   The output is as follows:

   ```
   array([2.90276331, 0.44200731])
   ```

9. Compute the variance values using the **pandas** method:

   ```
   df.var()
   ```

   The output is as follows:

   ```
   A    8.466351
   LK   0.196305
   dtype: float64
   ```

10. Compute the variance values using the **numpy** method:

    ```
    np.var(df.values, axis=0)
    ```

    The output is as follows:

    ```
    array([8.42603482, 0.19537046])
    ```

11. Compute the covariance matrix using the **pandas** method:

    ```
    df.cov()
    ```

    The output is as follows:

	A	LK
A	8.466351	1.224704
LK	1.224704	0.196305

    Figure 4.11: Covariance matrix using the pandas method

12. Compute the covariance matrix using the **numpy** method:

    ```
    np.cov(df.values.T)
    ```

    The output is as follows:

    ```
    array([[8.46635078, 1.22470367],
           [1.22470367, 0.19630525]])
    ```

Now that we know how to compute the foundational statistic values, we will turn our attention to the remaining components of PCA.

> **NOTE**
>
> To access the source code for this specific section, please refer to https://packt.live/2BHiLFz.
>
> You can also run this example online at https://packt.live/2O80UtW.

EIGENVALUES AND EIGENVECTORS

The mathematical concept of eigenvalues and eigenvectors is a very important one in the fields of physics and engineering, and they also form the final steps in computing the principal components of a dataset. The exact mathematical definition of eigenvalues and eigenvectors is outside the scope of this book, as it is quite involved and requires a reasonable understanding of linear algebra. Any square matrix A of dimensions *n x n* has a vector, *x*, of shape *n x 1* in such a way that it satisfies the following relation:

$$Ax = \lambda x$$

Figure 4.12: Equation representing PCA

Here, the term λ is a numerical value and denotes the eigenvalue, whereas *x* denotes the corresponding eigenvector. *N* denotes the order of the matrix, *A*. There will be exactly *n* eigenvalue and eigenvectors for matrix *A*. Without diving into the mathematical details of PCA, let's take a look at another way of representing the preceding equation as follows:

$$A = U\Sigma U^{-1}$$

Figure 4.13: Alternative equation representing PCA

Putting this simply into the context of PCA, we can derive the following:

- **Covariance Matrix** (*A*): As discussed in the preceding section, matrix *A* should be a square matrix before it can undergo eigenvalue decomposition. Since, in the case of our dataset, it has rows greater than the number of columns (let the shape of dataset be *m x n* where *m* is the number of rows and *n* is the number of columns). Therefore, we cannot perform eigenvalue decomposition directly. To perform eigenvalue decomposition on a rectangular matrix, it is first converted to a square matrix by computing its covariance matrix. A covariance matrix has a shape of *n x n*, that is, it is a square matrix of order '*n*'.

- **Eigenvectors** (*U*) are the components contributing information to the dataset as described in the first paragraph of this section on principal components called eigenvectors. Each eigenvector describes some amount of variability within the dataset. This variability is indicated by the corresponding eigenvalue. The larger the eigenvalue, the greater its contribution. An eigenvectors matrix has a shape of *n x n*.

- **Eigenvalues** (Σ) are the individual values that describe how much contribution each eigenvector provides to the dataset. As described previously, the single eigenvector that describes the largest contribution is referred to as the principal component, and, as such, will have the largest eigenvalue. Accordingly, the eigenvector with the smallest eigenvalue contributes the least amount of variance or information to the data. Eigenvalues are a diagonal matrix, which has the diagonal elements representing eigenvalues.

Please note that even the SVD of a covariance matrix of data produces eigenvalue decomposition, which we will see in *Exercise 4.04, scikit-learn PCA*. However, SVD uses a different process for the decomposition of the matrix. Remember that eigenvalue decomposition can be done for a square matrix only, whereas SVD can be done for a rectangular matrix as well.

> **NOTE**
>
> **Square matrix**: A square matrix has the same number of rows and columns. The number of rows in a square matrix is called the order of the matrix. A matrix that has an unequal number of rows and columns is known as a rectangular matrix.
>
> **Diagonal matrix**: A diagonal matrix has all non-diagonal elements as zero.

EXERCISE 4.02: COMPUTING EIGENVALUES AND EIGENVECTORS

As discussed previously, deriving and computing the eigenvalues and eigenvectors manually is a little involved and is not within the scope of this book. Thankfully, `numpy` provides all the functionality for us to compute these values. Again, we will use the Seeds dataset for this example:

> **NOTE**
>
> This dataset is sourced from https://archive.ics.uci.edu/ml/datasets/seeds. (UCI Machine Learning Repository [http://archive.ics.uci.edu/ml]. Irvine, CA: University of California, School of Information and Computer Science.) Citation: Contributors gratefully acknowledge the support of their work by the Institute of Agrophysics of the Polish Academy of Sciences in Lublin. The dataset can also be downloaded from https://packt.live/34gOQ0B.

1. Import the **pandas** and **numpy** packages:

```
import pandas as pd
import numpy as np
```

2. Load the dataset:

```
df = pd.read_csv('../Seed_Data.csv')
df.head()
```

The output is as follows:

	A	P	C	LK	WK	A_Coef	LKG	target
0	15.26	14.84	0.8710	5.763	3.312	2.221	5.220	0
1	14.88	14.57	0.8811	5.554	3.333	1.018	4.956	0
2	14.29	14.09	0.9050	5.291	3.337	2.699	4.825	0
3	13.84	13.94	0.8955	5.324	3.379	2.259	4.805	0
4	16.14	14.99	0.9034	5.658	3.562	1.355	5.175	0

Figure 4.14: The first five rows of the dataset

3. Again, we only require the **A** and **LK** features, so remove the other columns:

```
df = df[['A', 'LK']]
df.head()
```

The output is as follows:

	A	LK
0	15.26	5.763
1	14.88	5.554
2	14.29	5.291
3	13.84	5.324
4	16.14	5.658

Figure 4.15: The area and length of the kernel features

4. From the linear algebra module of **numpy**, use the **eig** function to compute the **eigenvalues** and **eigenvectors** characteristic vectors. Note the use of the covariance matrix of data here:

```
eigenvalues, eigenvectors = np.linalg.eig(np.cov(df.T))
```

> **NOTE**
>
> The **numpy** function, **cov**, can be used to calculate the covariance matrix of data. It produces a square matrix of order equal to the number of features of data.

5. Look at the eigenvalues; we can see that the first value is the largest, so the first eigenvector contributes the most information:

```
eigenvalues
```

The output is as follows:

```
array([8.64390408, 0.01875194])
```

6. It is handy to look at eigenvalues as a percentage of the total variance within the dataset. We will use a cumulative sum function to do this:

   ```
   eigenvalues = np.cumsum(eigenvalues)
   eigenvalues
   ```

   The output is as follows:

   ```
   array([8.64390408, 8.66265602])
   ```

7. Divide by the last or maximum value to convert eigenvalues into a percentage:

   ```
   eigenvalues /= eigenvalues.max()
   eigenvalues
   ```

   The output is as follows:

   ```
   array([0.99783531, 1.])
   ```

   We can see here that the first (or principal) component comprises 99% of the variation within the data, and, therefore, most of the information.

8. Now, let's take a look at **eigenvectors**:

   ```
   eigenvectors
   ```

   A section of the output is as follows:

   ```
   array([[ 0.98965371, -0.14347657],
          [ 0.14347657,  0.98965371]])
   ```

9. Confirm that the shape of the eigenvector matrix is in (**n x n**) format; that is, **2 x 2**:

   ```
   eigenvectors.shape
   ```

   The output is as follows:

   ```
   (2, 2)
   ```

10. So, from the eigenvalues, we saw that the principal component was the first eigenvector. Look at the values for the first eigenvector:

    ```
    P = eigenvectors[0]
    P
    ```

    The output is as follows:

    ```
    array([0.98965371, -0.14347657])
    ```

We have decomposed the dataset down into the principal components, and, using the eigenvectors, we can further reduce the dimensionality of the available data.

> **NOTE**
>
> To access the source code for this specific section, please refer to https://packt.live/3e5x3N3.
>
> You can also run this example online at https://packt.live/3f5Skrk.

In later examples, we will consider PCA and apply this technique to an example dataset.

THE PROCESS OF PCA

Now, we have all of the pieces ready to complete PCA in order to reduce the number of dimensions in a dataset.

The overall algorithm for completing PCA is as follows:

1. Import the required Python packages (**numpy** and **pandas**).
2. Load the entire dataset.
3. From the available data, select the features that you wish to use in dimensionality reduction.

> **NOTE**
>
> If there is a significant difference in the scale between the features of the dataset; for example, one feature ranges in values between 0 and 1, and another between 100 and 1,000, you may need to normalize one of the features, as such differences in magnitude can eliminate the effect of the smaller features. In such a situation, you may need to divide the larger feature by its maximum value.
>
> As an example, take a look at this:
>
> ```
> x1 = [0.1, 0.23, 0.54, 0.76, 0.78]
> x2 = [121, 125, 167, 104, 192]
> # Normalise x2 to be between 0 and 1
> x2 = (x2-np.min(x2)) / (np.max(x2)-np.min(x2))
> ```

4. Compute the **covariance** matrix of the selected (and possibly normalized) data.

5. Compute the eigenvalues and eigenvectors of the **covariance** matrix.

6. Sort the eigenvalues (and corresponding eigenvectors) from the highest to the lowest.

7. Compute the eigenvalues as a percentage of the total variance within the dataset.

8. Select the number of eigenvalues and corresponding eigenvectors. They will be required to comprise a predetermined value of a minimum composition variance.

> **NOTE**
>
> At this stage, the sorted eigenvalues represent a percentage of the total variance within the dataset. As such, we can use these values to select the number of eigenvectors required, either for the problem being solved or to sufficiently reduce the size of the dataset being applied to the model. For example, say that we required at least 90% of the variance to be accounted for within the output of PCA. We would then select the number of eigenvalues (and corresponding eigenvectors) that comprise at least 90% of the variance.

9. Multiply the dataset by the selected eigenvectors and you have completed a PCA, thereby reducing the number of features representing the data.

10. Plot the result.

Before moving on to the next exercise, note that **transpose** is a term from linear algebra that means to swap the rows with the columns and vice versa. Let's say we have a matrix of **X=[1, 2, 3]**, then, the transpose of X would be $X^T = \begin{bmatrix} 1 \\ 2 \\ 3 \end{bmatrix}$.

EXERCISE 4.03: MANUALLY EXECUTING PCA

For this exercise, we will be completing PCA manually, again using the Seeds dataset. For this example, we want to sufficiently reduce the number of dimensions within the dataset to comprise at least 75% of the available variance:

> **NOTE**
>
> This dataset is sourced from https://archive.ics.uci.edu/ml/datasets/seeds. (UCI Machine Learning Repository [http://archive.ics.uci.edu/ml]. Irvine, CA: University of California, School of Information and Computer Science.) Citation: Contributors gratefully acknowledge the support of their work by the Institute of Agrophysics of the Polish Academy of Sciences in Lublin. The dataset can also be downloaded from https://packt.live/2Xe7cxO.

1. Import the **pandas** and **numpy** packages:

   ```
   import pandas as pd
   import numpy as np
   import matplotlib.pyplot as plt
   ```

2. Load the dataset:

   ```
   df = pd.read_csv('../Seed_Data.csv')
   df.head()
   ```

   The output is as follows:

	A	P	C	LK	WK	A_Coef	LKG	target
0	15.26	14.84	0.8710	5.763	3.312	2.221	5.220	0
1	14.88	14.57	0.8811	5.554	3.333	1.018	4.956	0
2	14.29	14.09	0.9050	5.291	3.337	2.699	4.825	0
3	13.84	13.94	0.8955	5.324	3.379	2.259	4.805	0
4	16.14	14.99	0.9034	5.658	3.562	1.355	5.175	0

   Figure 4.16: The first five rows of the dataset

124 | Dimensionality Reduction Techniques and PCA

3. Again, we only require the **A** and **LK** features, so remove the other columns. In this example, we are not normalizing the selected dataset:

```
df = df[['A', 'LK']]
df.head()
```

The output is as follows:

	A	LK
0	15.26	5.763
1	14.88	5.554
2	14.29	5.291
3	13.84	5.324
4	16.14	5.658

Figure 4.17: The area and length of the kernel features

4. Compute the **covariance** matrix for the selected data. Note that we need to take the transpose of the **covariance** matrix to ensure that it is based on the number of features (2) and not samples (150):

```
data = np.cov(df.values.T)
"""
The transpose is required to ensure the covariance matrix is
based on features, not samples data
"""
data
```

The output is as follows:

```
array([[8.46635078, 1.22470367],
       [1.22470367, 0.19630525]])
```

5. Compute the eigenvectors and eigenvalues for the covariance matrix, Again, use the **full_matrices** function argument:

```
eigenvectors, eigenvalues, _ = np.linalg\
                    .svd(data, full_matrices=False)
```

6. Eigenvalues are returned, sorted from the highest value to the lowest:

   ```
   eigenvalues
   ```

   The output is as follows:

   ```
   array([8.64390408, 0.01875194])
   ```

7. Eigenvectors are returned as a matrix:

   ```
   eigenvectors
   ```

   The output is as follows:

   ```
   array([[-0.98965371, -0.14347657],
          [-0.14347657,  0.98965371]])
   ```

8. Compute the eigenvalues as a percentage of the variance within the dataset:

   ```
   eigenvalues = np.cumsum(eigenvalues)
   eigenvalues /= eigenvalues.max()
   eigenvalues
   ```

   The output is as follows:

   ```
   array([0.99783531, 1.        ])
   ```

9. As per the introduction to the exercise, we need to describe the data with at least 75% of the available variance. As per *Step 7*, the principal component comprises 99% of the available variance. As such, we require only the principal component from the dataset. What are the principal components? Let's take a look:

   ```
   P = eigenvectors[0]
   P
   ```

   The output is as follows:

   ```
   array([-0.98965371, -0.14347657])
   ```

   Now, we can apply the dimensionality reduction process. Execute a matrix multiplication of the principal component with the transpose of the dataset.

   > **NOTE**
   >
   > The dimensionality reduction process is a matrix multiplication of the selected eigenvectors and the data to be transformed.

10. Without taking the transpose of the **df.values** matrix, multiplication could not occur:

```
x_t_p = P.dot(df.values.T)
x_t_p
```

A section of the output is as follows:

```
array([-15.92897116, -15.52291615, -14.90128612, -14.46067667,
       -16.78480139, -15.00398523, -15.33617323, -14.74165693,
       -17.32640496, -17.11412321, -15.92194081, -14.66506721,
       -14.52665917, -14.42353632, -14.3843806 , -15.20679083,
       -14.57971203, -16.32066179, -15.29470516, -13.33820381,
       -14.82528704, -14.75600459, -16.52175236, -12.68660391,
       -15.68528813, -16.85939248, -13.65934747, -13.38224443,
       -14.75901759, -14.10225923, -13.8063641 , -16.15573066,
       -14.7644764 , -14.59708943, -15.71382658, -16.77232562,
```

Figure 4.18: The result of matrix multiplication

> **NOTE**
>
> The transpose of the dataset is required to execute matrix multiplication, as the **inner dimensions of the matrix must be the same** for matrix multiplication to occur. For **A** ("A dot B") to be valid, **A** must have the shape of *m x n*, and **B** must have the shape of *n x p*. In this example, the inner dimensions of **A** and **B** are both *n*. The resulting matrix would have dimensions of *m x p*.

In the following example, the output of the PCA is a single-column, 210-sample dataset. As such, we have just reduced the size of the initial dataset by half, comprising approximately 99% of the variance within the data.

11. Plot the values of the principal component:

```
plt.figure(figsize=(10, 7))
plt.plot(x_t_p)
plt.title('Principal Component of Selected Seeds Dataset')
plt.xlabel('Sample')
plt.ylabel('Component Value')
plt.show()
```

The output is as follows, and shows the new component values of the 210-sample dataset, as seen printed in the preceding step:

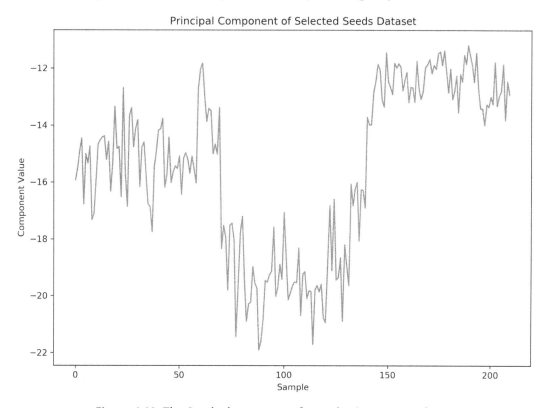

Figure 4.19: The Seeds dataset transformed using a manual PCA

In this exercise, we simply computed the covariance matrix of the dataset without applying any transformations to the dataset beforehand. If the two features have roughly the same mean and standard deviation, this is perfectly fine. However, if one feature is much larger in value (and has a somewhat different mean) than the other, then this feature may dominate the other when decomposing into components. This could have the effect of removing the information provided by the smaller feature altogether. One simple normalization technique before computing the covariance matrix would be to subtract the respective means from the features, thus centering the dataset around zero. We will demonstrate this in *Exercise 4.05, Visualizing Variance Reduction with Manual PCA*.

> **NOTE**
>
> To access the source code for this specific section, please refer to https://packt.live/3fa8X57.
>
> You can also run this example online at https://packt.live/3iOvg2P.

EXERCISE 4.04: SCIKIT-LEARN PCA

Typically, we will not complete PCA manually, especially when scikit-learn provides an optimized API with convenient methods that allow us to easily transform the data to and from the reduced-dimensional space. In this exercise, we will look at using a scikit-learn PCA on the Seeds dataset in more detail:

> **NOTE**
>
> This dataset is sourced from https://archive.ics.uci.edu/ml/datasets/seeds. (UCI Machine Learning Repository [http://archive.ics.uci.edu/ml]. Irvine, CA: University of California, School of Information and Computer Science.) Citation: Contributors gratefully acknowledge the support of their work by the Institute of Agrophysics of the Polish Academy of Sciences in Lublin. The dataset also can be downloaded from https://packt.live/2Ri6VGk.

1. Import the **pandas**, **numpy**, and **PCA** modules from the **sklearn** packages:

```
import pandas as pd
import numpy as np
import matplotlib.pyplot as plt
from sklearn.decomposition import PCA
```

2. Load the dataset:

```
df = pd.read_csv('../Seed_Data.csv')
df.head()
```

The output is as follows:

	A	P	C	LK	WK	A_Coef	LKG	target
0	15.26	14.84	0.8710	5.763	3.312	2.221	5.220	0
1	14.88	14.57	0.8811	5.554	3.333	1.018	4.956	0
2	14.29	14.09	0.9050	5.291	3.337	2.699	4.825	0
3	13.84	13.94	0.8955	5.324	3.379	2.259	4.805	0
4	16.14	14.99	0.9034	5.658	3.562	1.355	5.175	0

Figure 4.20: The first five rows of the dataset

3. Again, we only require the **A** and **LK** features, so remove the other columns. In this example, we are not normalizing the selected dataset:

```
df = df[['A', 'LK']]
df.head()
```

The output is as follows:

	A	LK
0	15.26	5.763
1	14.88	5.554
2	14.29	5.291
3	13.84	5.324
4	16.14	5.658

Figure 4.21: The area and length of the kernel features

4. Fit the data to a scikit-learn PCA model of the covariance data. Using the default values, as we have here, produces the maximum number of eigenvalues and eigenvectors possible for the dataset:

```
model = PCA()
model.fit(df.values)
```

The output is as follows:

```
PCA(copy=True, iterated_power='auto', n_components=None,
    random_state=None,
    svd_solver='auto', tol=0.0, whiten=False)
```

Here, **copy** indicates that the data fit within the model is copied before any calculations are applied. If **copy** was set to **False**, data passed to PCA is overwritten. **iterated_power** shows that the **A** and **LK** features are the number of principal components to keep. The default value is **None**, which selects the number of components as one less than the minimum of either the number of samples or the number of features. **random_state** allows the user to specify a seed for the random number generator used by the SVD solver. **svd_solver** specifies the SVD solver to be used during PCA. **tol** is the tolerance value used by the SVD solver. With **whiten**, the component vectors are multiplied by the square root of the number of samples. This will remove some information but can improve the performance of some downstream estimators.

5. The percentage of variance described by the components (eigenvalues) is contained within the **explained_variance_ratio_** property. Display the values for **explained_variance_ratio_**:

```
model.explained_variance_ratio_
```

The output is as follows:

```
array([0.99783531, 0.00216469])
```

6. Display the eigenvectors via the **components_** property:

```
model.components_
```

The output is as follows:

```
array([[0.98965371, 0.14347657]])
```

7. In this exercise, we will again only use the primary component, so we will create a new **PCA** model, this time specifying the number of components (eigenvectors/eigenvalues) to be **1**:

   ```
   model = PCA(n_components=1)
   ```

8. Use the **fit** method to fit the **covariance** matrix to the **PCA** model and generate the corresponding eigenvalues/eigenvectors:

   ```
   model.fit(df.values)
   ```

   The output is as follows:

   ```
   PCA(copy=True, iterated_power='auto', n_components=1,
       random_state=None,
       svd_solver='auto', tol=0.0, whiten=False)
   ```

   The model is fitted using a number of default parameters, as listed in the preceding output. **copy = True** is the data provided to the **fit** method, which is copied before PCA is applied. **iterated_power='auto'** is used to define the number of iterations by the internal SVD solver. **n_components=1** specifies that the PCA model is to return only the principal component. **random_state=None** specifies the random number generator to be used by the internal SVD solver if required. **svd_solver='auto'** is the type of SVD solver used. **tol=0.0** is the tolerance value for the SVD solver. **whiten=False** specifies that the eigenvectors are not to be modified. If set to **True**, whitening modifies the components further by multiplying by the square root of the number of samples and dividing by the singular values. This can help to improve the performance of later algorithm steps.

   Typically, you will not need to worry about adjusting any of these parameters, other than the number of components (**n_components**), which you can pass while declaring the PCA object as **model = PCA(n_components=1)**.

9. Display the eigenvectors using the **components_** property:

   ```
   model.components_
   ```

   The output is as follows:

   ```
   array([[0.98965371, 0.14347657]])
   ```

10. Transform the Seeds dataset into the lower space by using the **fit_transform** method of the model on the dataset. Assign the transformed values to the **data_t** variable:

    ```
    data_t = model.fit_transform(df.values)
    ```

11. Plot the transformed values to visualize the result:

    ```
    plt.figure(figsize=(10, 7))
    plt.plot(data_t)
    plt.xlabel('Sample')
    plt.ylabel('Transformed Data')
    plt.title('The dataset transformed by the principal component')
    plt.show()
    ```

The output is as follows:

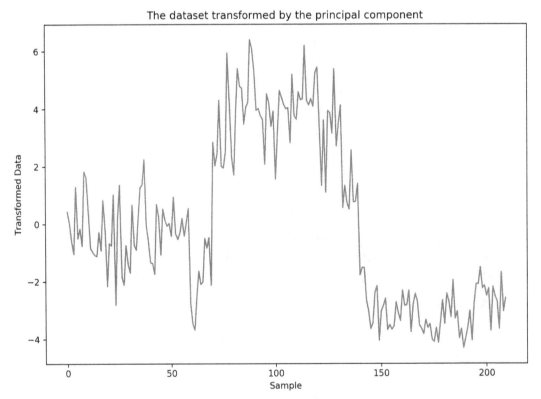

Figure 4.22: The seeds dataset transformed using the scikit-learn PCA

You have just reduced the dimensionality of the Seeds dataset using manual PCA, along with the scikit-learn API. But before we celebrate too early, compare *Figure 4.19* and *Figure 4.22*; these plots should be identical, shouldn't they? We used two separate methods to complete a PCA on the same dataset and selected the principal component for both. In the next activity, we will investigate why there are differences between the two.

> **NOTE**
>
> To access the source code for this specific section, please refer to https://packt.live/2ZQV85c.
>
> You can also run this example online at https://packt.live/2VSG99R.

ACTIVITY 4.01: MANUAL PCA VERSUS SCIKIT-LEARN

Suppose that you have been asked to port some legacy code from an older application executing PCA manually to a newer application that uses scikit-learn. During the porting process, you observe some differences between the output of the manual PCA and that of your port. Why is there a difference between the output of our manual PCA and scikit-learn? Compare the results of the two approaches on the Seeds dataset. What are the differences between them?

The aim of this activity is to truly dive into understanding how PCA works by doing it from scratch, and then comparing your implementation against the one included in scikit-learn to see whether there are any major differences:

> **NOTE**
>
> This dataset is sourced from https://archive.ics.uci.edu/ml/datasets/seeds. (UCI Machine Learning Repository [http://archive.ics.uci.edu/ml]. Irvine, CA: University of California, School of Information and Computer Science.) Citation: Contributors gratefully acknowledge the support of their work by the Institute of Agrophysics of the Polish Academy of Sciences in Lublin. The dataset can also be downloaded from https://packt.live/2JIH1qT.

1. Import the **pandas**, **numpy**, and **matplotlib** plotting libraries and the scikit-learn **PCA** model.

2. Load the dataset and select only the kernel features as per the previous exercises. Display the first five rows of the data.

3. Compute the **covariance** matrix for the data.

4. Transform the data using the scikit-learn API and only the first principal component. Store the transformed data in the **sklearn_pca** variable.

5. Transform the data using the manual PCA and only the first principal component. Store the transformed data in the **manual_pca** variable.

6. Plot the **sklearn_pca** and **manual_pca** values on the same plot to visualize the difference.

7. Notice that the two plots look almost identical, but with some key differences. What are these differences?

8. See whether you can modify the output of the manual PCA process to bring it in line with the scikit-learn version.

> **NOTE**
>
> Hint: The scikit-learn API subtracts the mean of the data prior to the transform.

Expected output: By the end of this activity, you will have transformed the dataset using both the manual and scikit-learn PCA methods. You will have produced a plot demonstrating that the two reduced datasets are, in fact, identical, and you should have an understanding of why they initially looked quite different. The final plot should look similar to the following:

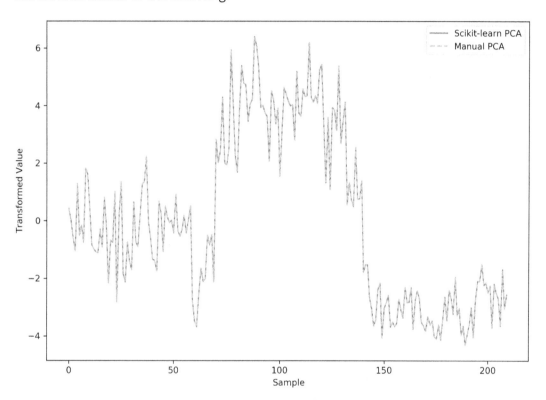

Figure 4.23: The expected final plot

This plot will demonstrate that the dimensionality reduction completed by the two methods are, in fact, the same.

> **NOTE**
>
> The solution to this activity can be found on page 437.

RESTORING THE COMPRESSED DATASET

Now that we have covered a few different examples of transforming a dataset into a lower-dimensional space, we should consider what practical effect this transformation has had on the data. Using PCA as a preprocessing step to condense the number of features in the data will result in some of the variance being discarded. The following exercise will walk us through this process so that we can see how much information has been discarded by the transformation.

EXERCISE 4.05: VISUALIZING VARIANCE REDUCTION WITH MANUAL PCA

One of the most important aspects of dimensionality reduction is understanding how much information has been removed from the dataset as a result of the dimensionality reduction process. Removing too much information will add additional challenges to later processing, while not removing enough defeats the purpose of PCA or other techniques. In this exercise, we will visualize the amount of information that has been removed from the Seeds dataset as a result of PCA:

> **NOTE**
>
> This dataset is sourced from https://archive.ics.uci.edu/ml/datasets/seeds. (UCI Machine Learning Repository [http://archive.ics.uci.edu/ml]. Irvine, CA: University of California, School of Information and Computer Science.) Citation: Contributors gratefully acknowledge the support of their work by the Institute of Agrophysics of the Polish Academy of Sciences in Lublin. The dataset also can be downloaded from https://packt.live/2RhnDFS.

1. Import the **pandas**, **numpy**, and **matplotlib** plotting libraries:

   ```
   import pandas as pd
   import numpy as np
   import matplotlib.pyplot as plt
   ```

2. Read in the **wheat kernel** features from the Seeds dataset:

   ```
   df = pd.read_csv('../Seed_Data.csv')[['A', 'LK']]
   df.head()
   ```

The output is as follows:

	A	LK
0	15.26	5.763
1	14.88	5.554
2	14.29	5.291
3	13.84	5.324
4	16.14	5.658

Figure 4.24: Kernel features

3. Center the dataset around zero by subtracting the respective means.

> **NOTE**
>
> As discussed at the end of *Exercise 4.03*, *Manually Executing PCA*, here, we are centering the data before computing the covariance matrix.

The code appears as follows:

```
means = np.mean(df.values, axis=0)
means
```

The output is as follows:

```
array([14.84752381,  5.62853333])
```

4. To calculate the data and print the results, use the following code:

```
data = df.values - means
data
```

A section of the output is as follows:

```
array([[ 4.12476190e-01,   1.34466667e-01],
       [ 3.24761905e-02,  -7.45333333e-02],
       [-5.57523810e-01,  -3.37533333e-01],
       [-1.00752381e+00,  -3.04533333e-01],
       [ 1.29247619e+00,   2.94666667e-02],
       [-4.67523810e-01,  -2.42533333e-01],
       [-1.57523810e-01,  -6.55333333e-02],
       [-7.37523810e-01,  -2.08533333e-01],
       [ 1.78247619e+00,   4.24466667e-01],
       [ 1.59247619e+00,   2.55466667e-01],
```

Figure 4.25: Section of the output

5. Use manual PCA to transform the data on the basis of the first principal component:

```
eigenvectors, eigenvalues, _ = np.linalg.svd(np.cov(data.T), \
                                 full_matrices=False)
P = eigenvectors[0]
P
```

The output is as follows:

```
array([-0.98965371, -0.14347657])
```

6. Transform the data into the lower-dimensional space by doing a dot product of the preceding **P** with a transposed version of the data matrix:

```
data_transformed = P.dot(data.T)
```

7. Reshape the principal components for later use:

```
P = P.reshape((-1, 1))
```

8. To compute the inverse transform of the reduced dataset, we need to restore the selected eigenvectors to the higher-dimensional space. To do this, we will invert the matrix. Matrix inversion is another linear algebra technique that we will only cover very briefly. A square matrix, *A*, is said to be invertible if there is another square matrix, *B*, and if *AB=BA=I*, where *I* is a special matrix known as an identity matrix, consisting of values of **1** only through the center diagonal:

```
P_transformed = np.linalg.pinv(P)
P_transformed
```

The output is as follows:

```
array([[-0.98965371, -0.14347657]])
```

9. Prepare the transformed data for use in the matrix multiplication:

```
data_transformed = data_transformed.reshape((-1, 1))
```

10. Compute the inverse transform of the reduced data and plot the result to visualize the effect of removing the variance from the data:

```
data_restored = data_transformed.dot(P_transformed)
data_restored
```

A section of the output is as follows:

```
array([[ 4.23078358e-01,  6.13364376e-02],
       [ 2.12245047e-02,  3.07705530e-03],
       [-5.93973963e-01, -8.61122915e-02],
       [-1.03002474e+00, -1.49329425e-01],
       [ 1.27005392e+00,  1.84128026e-01],
       [-4.92337409e-01, -7.13773752e-02],
       [-1.63586316e-01, -2.37161785e-02],
       [-7.51951584e-01, -1.09015341e-01],
       [ 1.80605390e+00,  2.61835450e-01],
       [ 1.59596848e+00,  2.31377993e-01],
       [ 4.16120744e-01,  6.03277467e-02],
```

Figure 4.26: The inverse transform of the reduced data

11. Add the **means** array back to the transformed data:

```
data_restored += means
```

12. Visualize the result by plotting the original and the transformed datasets:

```
plt.figure(figsize=(10, 7))
plt.plot(data_restored[:,0], data_restored[:,1], \
         linestyle=':', label='PCA restoration')
plt.scatter(df['A'], df['LK'], marker='*', label='Original')
plt.legend()
plt.xlabel('Area of Kernel')
plt.ylabel('Length of Kernel')
plt.title('Inverse transform after removing variance')
plt.show()
```

The output is as follows:

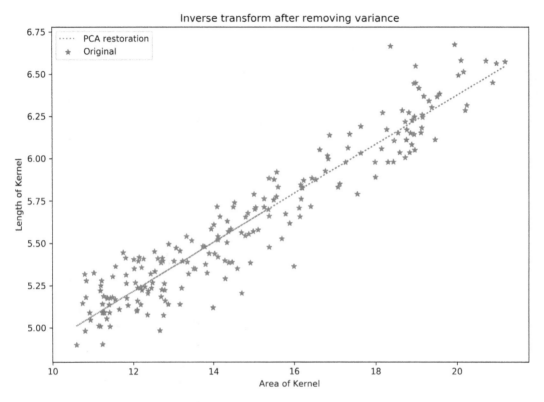

Figure 4.27: The inverse transform after removing variance

13. There are only two components of variation in this dataset. If we do not remove any of the components, what will be the result of the inverse transform? Again, transform the data into the lower-dimensional space, but this time, use all of the eigenvectors:

```
P = eigenvectors
data_transformed = P.dot(data.T)
```

14. Transpose **data_transformed** to put it in the correct shape for matrix multiplication:

```
data_transformed = data_transformed.T
```

15. Now, restore the data back to the higher-dimensional space:

```
data_restored = data_transformed.dot(P)
data_restored
```

A section of the output is as follows:

```
array([[ 4.12476190e-01,  1.34466667e-01],
       [ 3.24761905e-02, -7.45333333e-02],
       [-5.57523810e-01, -3.37533333e-01],
       [-1.00752381e+00, -3.04533333e-01],
       [ 1.29247619e+00,  2.94666667e-02],
       [-4.67523810e-01, -2.42533333e-01],
       [-1.57523810e-01, -6.55333333e-02],
       [-7.37523810e-01, -2.08533333e-01],
       [ 1.78247619e+00,  4.24466667e-01],
       [ 1.59247619e+00,  2.55466667e-01],
```

Figure 4.28: The restored data

16. Add the means back to the restored data:

```
data_restored += means
```

17. Visualize the restored data in the context of the original dataset:

```
plt.figure(figsize=(10, 7))
plt.scatter(data_restored[:,0], data_restored[:,1], \
            marker='d', label='PCA restoration', c='k')
plt.scatter(df['A'], df['LK'], marker='o', \
            label='Original', c='#1f77b4')
plt.legend()
plt.xlabel('Area of Kernel')
plt.ylabel('Length of Kernel')
plt.title('Inverse transform after removing variance')
plt.show()
```

The output is as follows:

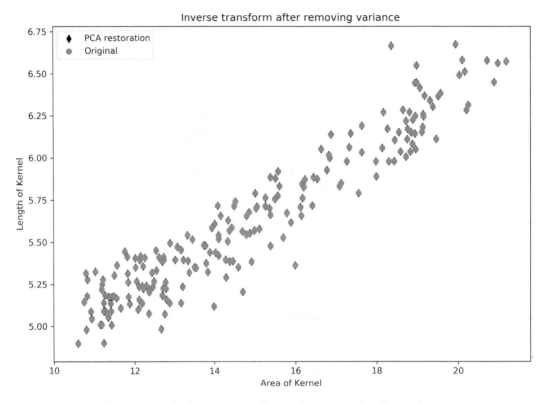

Figure 4.29: The inverse transform after removing the variance

If we compare the two plots produced in this exercise, we can see that the PCA went down, and the restored dataset is essentially a negative linear trend line between the two feature sets. We can compare this to the dataset restored from all of the available components, where we have recreated the original dataset as a whole.

> **NOTE**
>
> To access the source code for this specific section, please refer to https://packt.live/38EztBu.
>
> You can also run this example online at https://packt.live/3f8LDVC.

EXERCISE 4.06: VISUALIZING VARIANCE REDUCTION WITH SCIKIT-LEARN

In this exercise, we will again visualize the effect of reducing the dimensionality of the dataset; however, this time, we will be using the scikit-learn API. This is this method that you will commonly use in practical applications due to the power and simplicity of the scikit-learn model:

> **NOTE**
>
> This dataset is sourced from https://archive.ics.uci.edu/ml/datasets/seeds. (UCI Machine Learning Repository [http://archive.ics.uci.edu/ml]. Irvine, CA: University of California, School of Information and Computer Science.) Citation: Contributors gratefully acknowledge the support of their work by the Institute of Agrophysics of the Polish Academy of Sciences in Lublin. The dataset can also be downloaded from https://packt.live/3bVlJm4.

1. Import the **pandas**, **numpy**, and **matplotlib** plotting libraries and the **PCA** model from scikit-learn:

    ```
    import pandas as pd
    import numpy as np
    import matplotlib.pyplot as plt
    from sklearn.decomposition import PCA
    ```

2. Read in the **Wheat Kernel** features from the Seeds dataset:

    ```
    df = pd.read_csv('../Seed_Data.csv')[['A', 'LK']]
    df.head()
    ```

The output is as follows:

	A	LK
0	15.26	5.763
1	14.88	5.554
2	14.29	5.291
3	13.84	5.324
4	16.14	5.658

Figure 4.30: The Wheat Kernel features from the Seeds dataset

3. Use the scikit-learn API to transform the data on the basis of the first principal component:

```
model = PCA(n_components=1)
data_p = model.fit_transform(df.values)
```

4. Compute the inverse transform of the reduced data and plot the result to visualize the effect of removing the variance from the data:

```
data = model.inverse_transform(data_p)
plt.figure(figsize=(10, 7))
plt.plot(data[:,0], data[:,1], linestyle=':', \
         label='PCA restoration')
plt.scatter(df['A'], df['LK'], marker='*', label='Original')
plt.legend()
plt.xlabel('Area of Kernel')
plt.ylabel('Length of Kernel')
plt.title('Inverse transform after removing variance')
plt.show()
```

The output is as follows:

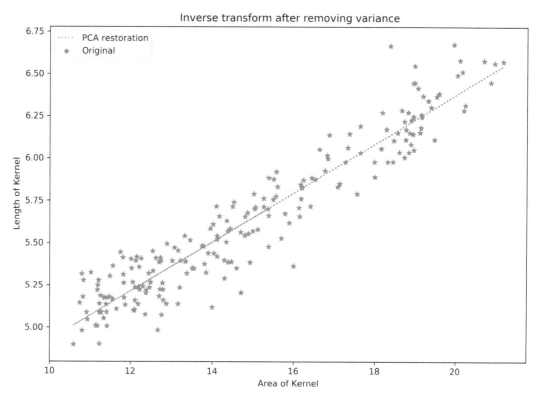

Figure 4.31: The inverse transform after removing the variance

5. There are only two components of variation in this dataset. If we do not remove any of the components, what will the result of the inverse transform be? Let's find out by computing the inverse transform and seeing how the results change without removing any components:

```
model = PCA()
data_p = model.fit_transform(df.values)
data = model.inverse_transform(data_p)
plt.figure(figsize=(10, 7))
plt.scatter(data[:,0], data[:,1], marker='d', \
            label='PCA restoration', c='k')
plt.scatter(df['A'], df['LK'], marker='o', \
            label='Original', c='#1f77b4')
plt.legend()
```

```
plt.xlabel('Area of Kernel')
plt.ylabel('Length of Kernel')
plt.title('Inverse transform after removing variance')
plt.show()
```

The output is as follows:

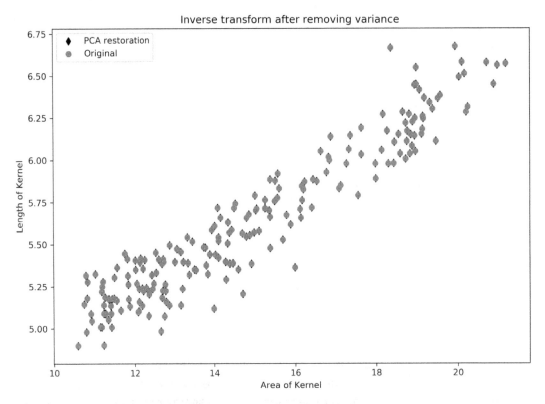

Figure 4.32: The inverse transform after removing the variance

As we can see here, if we don't remove any of the components in the PCA, it will recreate the original data when performing inverse transform. We have demonstrated the effect of removing information from the dataset and the ability to recreate the original data using all of the available eigenvectors.

> **NOTE**
>
> To access the source code for this specific section, please refer to https://packt.live/2O362zv.
>
> You can also run this example online at https://packt.live/3fdWYDU.

The previous exercises specified the reduction in dimensionality using PCA to two dimensions, partly to allow the results to be visualized easily. We can, however, use PCA to reduce the dimensions to any value less than that of the original set. The following example demonstrates how PCA can be used to reduce a dataset to three dimensions, thereby allowing visualizations.

EXERCISE 4.07: PLOTTING 3D PLOTS IN MATPLOTLIB

Creating 3D scatter plots in **matplotlib** is unfortunately not as simple as providing a series of (*x*, *y*, *z*) coordinates to a scatter plot. In this exercise, we will work through a simple 3D plotting example using the Seeds dataset:

> **NOTE**
>
> This dataset is sourced from https://archive.ics.uci.edu/ml/datasets/seeds. (UCI Machine Learning Repository [http://archive.ics.uci.edu/ml]. Irvine, CA: University of California, School of Information and Computer Science.)
>
> Citation: Contributors gratefully acknowledge the support of their work by the Institute of Agrophysics of the Polish Academy of Sciences in Lublin. The dataset can also be downloaded from https://packt.live/3c2tAhT.

1. Import **pandas** and **matplotlib**. To enable 3D plotting, you will also need to import **Axes3D**:

```
from mpl_toolkits.mplot3d import Axes3D
import pandas as pd
import matplotlib.pyplot as plt
```

148 | Dimensionality Reduction Techniques and PCA

2. Read in the dataset and select the **A**, **LK**, and **C** columns:

   ```
   df = pd.read_csv('../Seed_Data.csv')[['A', 'LK', 'C']]
   df.head()
   ```

   The output is as follows:

	A	LK	C
0	15.26	5.763	0.8710
1	14.88	5.554	0.8811
2	14.29	5.291	0.9050
3	13.84	5.324	0.8955
4	16.14	5.658	0.9034

   Figure 4.33: The first five rows of the data

3. Plot the data in three dimensions and use the **projection='3d'** argument with the **add_subplot** method to create the 3D plot:

   ```
   fig = plt.figure(figsize=(10, 7))
   # Where Axes3D is required
   ax = fig.add_subplot(111, projection='3d')
   ax.scatter(df['A'], df['LK'], df['C'])
   ax.set_xlabel('Area of Kernel')
   ax.set_ylabel('Length of Kernel')
   ax.set_zlabel('Compactness of Kernel')
   ax.set_title('Expanded Seeds Dataset')
   plt.show()
   ```

Principal Component Analysis | 149

The plot will appear as follows:

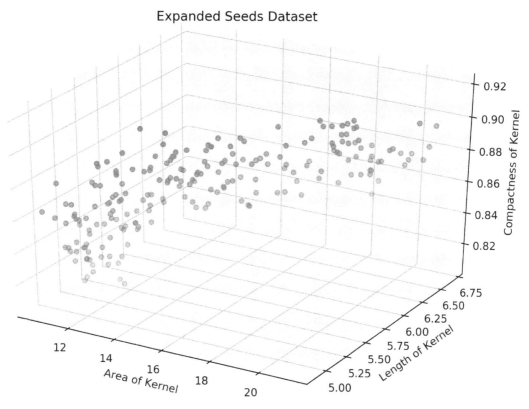

Figure 4.34: The expanded Seeds dataset

> **NOTE**
>
> While the **Axes3D** library was imported but not directly used, it is required for configuring the plot window in three dimensions. If the import of **Axes3D** was omitted, the **projection='3d'** argument would return an **AttributeError** exception.
>
> To access the source code for this specific section, please refer to https://packt.live/3gPM1J9.
>
> You can also run this example online at https://packt.live/2AFVXFr.

ACTIVITY 4.02: PCA USING THE EXPANDED SEEDS DATASET

In this activity, we are going to use the complete Seeds dataset to look at the effect of selecting a differing number of components in the PCA decomposition. This activity aims to simulate the process that is typically completed in a real-world problem as we try to determine the optimum number of components to select, attempting to balance the extent of dimensionality reduction and information loss. Therefore, we will be using the scikit-learn PCA model:

> **NOTE**
>
> This dataset is sourced from https://archive.ics.uci.edu/ml/datasets/seeds. (UCI Machine Learning Repository [http://archive.ics.uci.edu/ml]. Irvine, CA: University of California, School of Information and Computer Science.) Citation: Contributors gratefully acknowledge the support of their work by the Institute of Agrophysics of the Polish Academy of Sciences in Lublin. The dataset also can be downloaded from https://packt.live/3aPY0nj.

The following steps will help you to complete the activity:

1. Import **pandas** and **matplotlib**. To enable 3D plotting, you will also need to import **Axes3D**.

2. Read in the dataset and select the **Area of Kernel**, **Length of Kernel**, and **Compactness of Kernel** columns.

3. Plot the data in three dimensions.

4. Create a **PCA** model without specifying the number of components.

5. Fit the model to the dataset.

Principal Component Analysis | 151

6. Display the eigenvalues or **explained_variance_ratio_**.

7. We want to reduce the dimensionality of the dataset but still keep at least 90% of the variance. What is the minimum number of components required to keep 90% of the variance?

8. Create a new **PCA** model, this time specifying the number of components required to keep at least 90% of the variance.

9. Transform the data using the new model.

10. Plot the transformed data.

11. Restore the transformed data to the original dataspace.

12. Plot the restored data in three dimensions in one subplot and the original data in a second subplot to visualize the effect of removing some of the variance:

```
fig = plt.figure(figsize=(10, 14))
# Original Data
ax = fig.add_subplot(211, projection='3d')
# Transformed Data
ax = fig.add_subplot(212, projection='3d')
```

152 | Dimensionality Reduction Techniques and PCA

Expected output: The final plot will appear as follows:

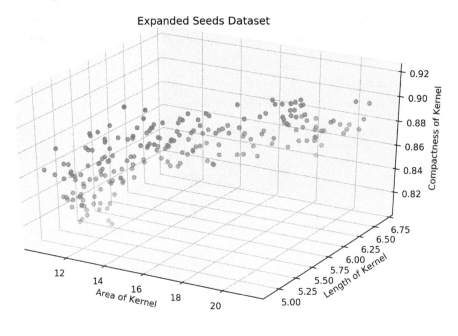

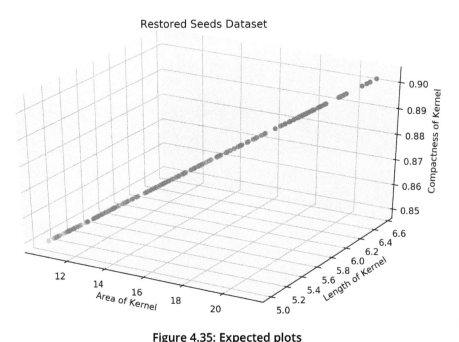

Figure 4.35: Expected plots

> **NOTE**
>
> The solution to this activity can be found on page 443.

SUMMARY

In this chapter, we covered the process of dimensionality reduction and PCA. We completed a number of exercises and developed the skills to reduce the size of a dataset by extracting only the most important components of variance within the data, using both a manual PCA process and the model provided by scikit-learn. During this chapter, we also returned the reduced datasets back to the original dataspace and observed the effect of removing the variance on the original data. Finally, we discussed a number of potential applications for PCA and other dimensionality reduction processes. In our next chapter, we will introduce neural network-based autoencoders and use the Keras package to implement them.

5
AUTOENCODERS

OVERVIEW

In this chapter, we will look at autoencoders and their applications. We will see how autoencoders are used in dimensionality reduction and denoising. We will implement an artificial neural network and an autoencoder using the Keras framework. By the end of this chapter, you will be able to implement an autoencoder model using convolutional neural networks.

INTRODUCTION

We'll continue our discussion of dimensionality reduction techniques as we turn our attention to autoencoders. Autoencoders are a particularly interesting area of focus as they provide a means of using supervised learning based on artificial neural networks but in an unsupervised context. Being based on artificial neural networks, autoencoders are an extremely effective means of performing dimensionality reduction, but also provide additional benefits. With recent increases in the availability of data, processing power, and network connectivity, autoencoders are experiencing a resurgence in usage and the study of them, the likes of which have not been seen since their origins in the late 1980s. This is also consistent with the study of artificial neural networks, which were first described and implemented as a concept in the 1960s. Presently, you would only need to conduct a cursory internet search to discover the popularity and power of neural networks.

Autoencoders can be used for de-noising images and generating artificial data samples in combination with other methods, such as recurrent or **Long Short-Term Memory** (**LSTM**) architectures, to predict sequences of data. The flexibility and power that arises from the use of artificial neural networks also enable autoencoders to form very efficient representations of the data, which can then be used either directly as an extremely efficient search method, or as a feature vector for later processing.

Consider the use of an autoencoder in an image de-noising application, where we are presented with the image on the left in *Figure 5.1*. We can see that the image is affected by the addition of some random noise. We can use a specially trained autoencoder to remove this noise, as represented in the image on the right in the following figure. In learning how to remove this noise, the autoencoder has also learned to encode the important information that composes the image and how to reconstruct (or decode) this information into a clearer version of the original image:

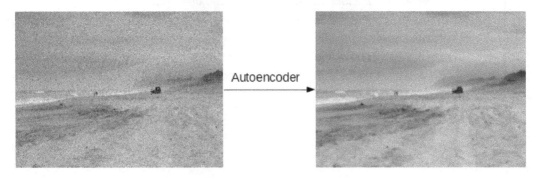

Figure 5.1: Autoencoder de-noising

> **NOTE**
>
> This image has been modified from http://www.freenzphotos.com/free-photos-of-bay-of-plenty/stormy-fishermen/ under CC0.

This example demonstrates one aspect of autoencoders that makes them useful for unsupervised learning (the encoding stage) and one that is useful in generating new images (decoding). We will delve further into these two useful stages of autoencoders and apply the output of the autoencoder to clustering the CIFAR-10 dataset (https://www.cs.toronto.edu/~kriz/cifar.html).

Here is a representation of an encoder and a decoder:

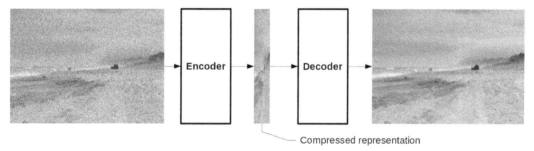

Figure 5.2: Encoder/decoder representation

FUNDAMENTALS OF ARTIFICIAL NEURAL NETWORKS

Given that autoencoders are based on artificial neural networks, an understanding of neural networks is also critical for understanding autoencoders. This section of the chapter will briefly review the fundamentals of artificial neural networks. It is important to note that there are many aspects of neural networks that are outside the scope of this book. The topic of neural networks could easily fill, and has filled, many books on its own, and this section is not to be considered an exhaustive discussion of the topic.

As described earlier, artificial neural networks are primarily used in supervised learning problems, where we have a set of input information, say a series of images, and we are training an algorithm to map the information to a desired output, such as a class or category. Consider the CIFAR-10 dataset shown in *Figure 5.3* as an example, which contains images of 10 different categories (airplane, automobile, bird, cat, deer, dog, frog, horse, ship, and truck), with 6,000 images per category.

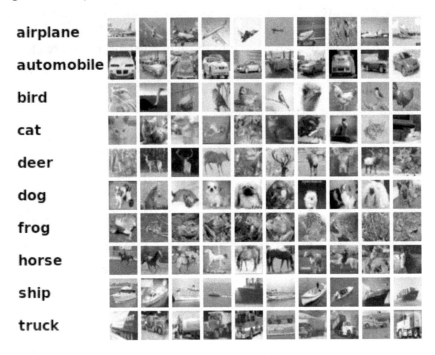

Figure 5.3: CIFAR-10 dataset

When neural networks are used in a supervised learning context, the images are fed to the network with a representation of the corresponding category labels being the desired output of the network.

The network is then trained to maximize its ability to infer or predict the correct label for a given image.

> **NOTE**
>
> This image is taken from https://www.cs.toronto.edu/~kriz/cifar.html from *Learning Multiple Layers of Features from Tiny Images* (https://www.cs.toronto.edu/~kriz/learning-features-2009-TR.pdf), Alex Krizhevsky, 2009.

THE NEURON

The artificial neural network derives its name from the biological neural networks commonly found in the brain. While the accuracy of the analogy can certainly be questioned, it is a useful metaphor to break down the concept of artificial neural networks and facilitate understanding. As with their biological counterparts, the neuron is the building block on which all neural networks are constructed, connecting a number of neurons in different configurations to form more powerful structures. Each neuron in *Figure 5.4* is composed of four individual parts: an input value, a tunable weight (theta), an activation function that operates on the product of the weight and input value, and the resulting output value:

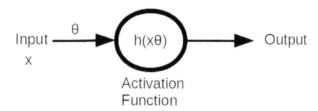

Figure 5.4: Anatomy of a neuron

The activation function is specifically chosen depending upon the objective of the neural network being designed, and there are a number of common functions, including **tanh**, **sigmoid**, **linear**, and **ReLU** (rectified linear unit). Throughout this chapter, we will use both the **sigmoid** and **ReLU** activation functions, so let's look at them in a little more detail.

THE SIGMOID FUNCTION

The sigmoid activation function is very commonly used as an output in the classification of neural networks due to its ability to shift the input values to approximate a binary output. The sigmoid function produces the following output:

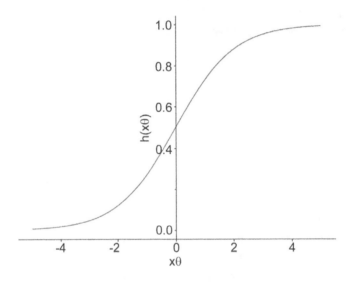

Figure 5.5: Output of the sigmoid function

We can see in the preceding figure that the output of the sigmoid function asymptotes (approaches but never reaches) 1 as x increases, and asymptotes 0 as x moves further away from 0 in the negative direction. This function is used in classification tasks as it provides close to a binary output.

We can see that sigmoid has an asymptotic nature. Due to this, as the value of the input reaches the extremities, the training process slows down (known as **vanishing gradient**). This is a bottleneck in training. Therefore, to speed up the training process, intermediary stages of neural networks use **Rectified Linear Unit (ReLU)**. However, this does have certain limitations as ReLU has the problems of dead cells and bias.

RECTIFIED LINEAR UNIT (RELU)

The rectified linear unit is a very useful activation function that's commonly used at intermediary stages of neural networks. Simply put, the value 0 is assigned to input values less than 0, and the actual value is returned for values greater than 0.

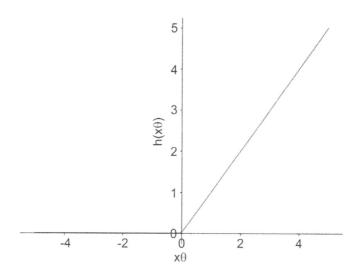

Figure 5.6: Output of ReLU

EXERCISE 5.01: MODELING THE NEURONS OF AN ARTIFICIAL NEURAL NETWORK

In this exercise, we will practically introduce a programmatic representation of the neuron in **NumPy** using the **sigmoid** function. We will keep the inputs fixed and adjust the tunable weights to investigate the effect on the neuron. To relate this framework back to a common model in supervised learning, our approach in this exercise is the same as logistic regression. Perform the following steps:

1. Import the **numpy** and **matplotlib** packages:

   ```
   import numpy as np
   import matplotlib.pyplot as plt
   ```

2. Define the **sigmoid** function as a Python function:

   ```
   def sigmoid(z):
       return np.exp(z) / (np.exp(z) + 1)
   ```

> **NOTE**
>
> Here, we're using the **sigmoid** function. You could also use the **ReLU** function. The **ReLU** activation function, while being powerful in artificial neural networks, is easy to define. It simply needs to return the input value if greater than 0; otherwise, it returns 0:
>
> ```
> def relu(x):
>     return np.max(0, x)
> ```

3. Define the inputs (**x**) and tunable weights (**theta**) for the neuron. In this example, the inputs (**x**) will be **100** numbers linearly spaced between **−5** and **5**. Set **theta = 1**:

```
theta = 1
x = np.linspace(-5, 5, 100)
x
```

The output is as follows:

```
array([-5.        , -4.8989899 , -4.7979798 , -4.6969697 , -4.5959596 ,
       -4.49494949, -4.39393939, -4.29292929, -4.19191919, -4.09090909,
       -3.98989899, -3.88888889, -3.78787879, -3.68686869, -3.58585859,
       -3.48484848, -3.38383838, -3.28282828, -3.18181818, -3.08080808,
       -2.97979798, -2.87878788, -2.77777778, -2.67676768, -2.57575758,
       -2.47474747, -2.37373737, -2.27272727, -2.17171717, -2.07070707,
       -1.96969697, -1.86868687, -1.76767677, -1.66666667, -1.56565657,
       -1.46464646, -1.36363636, -1.26262626, -1.16161616, -1.06060606,
       -0.95959596, -0.85858586, -0.75757576, -0.65656566, -0.55555556,
       -0.45454545, -0.35353535, -0.25252525, -0.15151515, -0.05050505,
        0.05050505,  0.15151515,  0.25252525,  0.35353535,  0.45454545,
        0.55555556,  0.65656566,  0.75757576,  0.85858586,  0.95959596,
        1.06060606,  1.16161616,  1.26262626,  1.36363636,  1.46464646,
        1.56565657,  1.66666667,  1.76767677,  1.86868687,  1.96969697,
        2.07070707,  2.17171717,  2.27272727,  2.37373737,  2.47474747,
        2.57575758,  2.67676768,  2.77777778,  2.87878788,  2.97979798,
        3.08080808,  3.18181818,  3.28282828,  3.38383838,  3.48484848,
        3.58585859,  3.68686869,  3.78787879,  3.88888889,  3.98989899,
        4.09090909,  4.19191919,  4.29292929,  4.39393939,  4.49494949,
        4.5959596 ,  4.6969697 ,  4.7979798 ,  4.8989899 ,  5.        ])
```

Figure 5.7: Printing the inputs

4. Compute the outputs (**y**) of the neuron:

   ```
   y = sigmoid(x * theta)
   ```

5. Plot the output of the neuron versus the input:

   ```
   fig = plt.figure(figsize=(10, 7))
   ax = fig.add_subplot(111)
   ax.plot(x, y)
   ax.set_xlabel('$x$', fontsize=22)
   ax.set_ylabel('$h(x\Theta)$', fontsize=22)
   ax.spines['left'].set_position(('data', 0))
   ax.spines['top'].set_visible(False)
   ax.spines['right'].set_visible(False)
   ax.tick_params(axis='both', which='major', labelsize=22)
   plt.show()
   ```

   In the following output, you can see the plotted **sigmoid** function – note that it passes through the origin at **0.5**:

   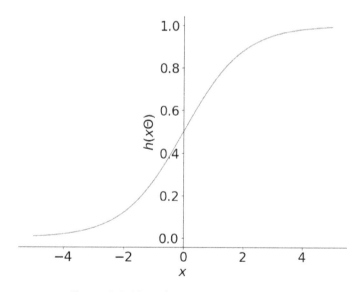

   Figure 5.8: Plot of neurons versus inputs

6. Set the tunable parameter, **theta**, to **5**, and recompute and store the output of the neuron:

   ```
   theta = 5
   y_2 = sigmoid(x * theta)
   ```

164 | Autoencoders

7. Change the tunable parameter, **theta**, to **0.2**, and recompute and store the output of the neuron:

```
theta = 0.2
y_3 = sigmoid(x * theta)
```

8. Plot the three different output curves of the neuron (**theta = 1, theta = 5, and theta = 0.2**) on one graph:

```
fig = plt.figure(figsize=(10, 7))
ax = fig.add_subplot(111)
ax.plot(x, y, label='$\Theta=1$')
ax.plot(x, y_2, label='$\Theta=5$', linestyle=':')
ax.plot(x, y_3, label='$\Theta=0.2$', linestyle='--')
ax.set_xlabel('$x\Theta$', fontsize=22)
ax.set_ylabel('$h(x\Theta)$', fontsize=22)
ax.spines['left'].set_position(('data', 0))
ax.spines['top'].set_visible(False)
ax.spines['right'].set_visible(False)
ax.tick_params(axis='both', which='major', labelsize=22)
ax.legend(fontsize=22)
plt.show()
```

The output is as follows:

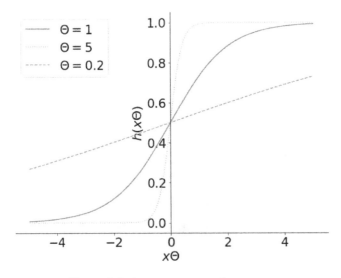

Figure 5.9: Output curves of neurons

In this exercise, we modeled the basic building block of an artificial neural network with a **sigmoid** activation function. We can see that using the **sigmoid** function increases the steepness of the gradient and means that only small values of *x* will push the output to either close to 1 or 0. Similarly, reducing **theta** reduces the sensitivity of the neuron to non-zero values and results in extreme input values being required to push the result of the output to either 0 or 1, tuning the output of the neuron.

> **NOTE**
>
> To access the source code for this specific section, please refer to https://packt.live/2AE9Kwc.
>
> You can also run this example online at https://packt.live/3e59UdK.

EXERCISE 5.02: MODELING NEURONS WITH THE RELU ACTIVATION FUNCTION

Similar to *Exercise 5.01, Modeling the Neurons of an Artificial Neural Network*, we will model a network again, this time with the ReLU activation function. In this exercise, we will develop a range of response curves for the ReLU activated neuron and describe the effect of changing the value of theta on the output of the neuron:

1. Import **numpy** and **matplotlib**:

    ```
    import numpy as np
    import matplotlib.pyplot as plt
    ```

2. Define the ReLU activation function as a Python function:

    ```
    def relu(x):
        return np.max((0, x))
    ```

3. Define the inputs (**x**) and tunable weights (**theta**) for the neuron. In this example, the inputs (**x**) will be 100 numbers linearly spaced between **-5** and **5**. Set **theta = 1**:

    ```
    theta = 1
    x = np.linspace(-5, 5, 100)
    x
    ```

The output is as follows:

```
array([-5.        , -4.8989899 , -4.7979798 , -4.6969697 , -4.5959596 ,
       -4.49494949, -4.39393939, -4.29292929, -4.19191919, -4.09090909,
       -3.98989899, -3.88888889, -3.78787879, -3.68686869, -3.58585859,
       -3.48484848, -3.38383838, -3.28282828, -3.18181818, -3.08080808,
       -2.97979798, -2.87878788, -2.77777778, -2.67676768, -2.57575758,
       -2.47474747, -2.37373737, -2.27272727, -2.17171717, -2.07070707,
       -1.96969697, -1.86868687, -1.76767677, -1.66666667, -1.56565657,
       -1.46464646, -1.36363636, -1.26262626, -1.16161616, -1.06060606,
       -0.95959596, -0.85858586, -0.75757576, -0.65656566, -0.55555556,
       -0.45454545, -0.35353535, -0.25252525, -0.15151515, -0.05050505,
        0.05050505,  0.15151515,  0.25252525,  0.35353535,  0.45454545,
        0.55555556,  0.65656566,  0.75757576,  0.85858586,  0.95959596,
        1.06060606,  1.16161616,  1.26262626,  1.36363636,  1.46464646,
        1.56565657,  1.66666667,  1.76767677,  1.86868687,  1.96969697,
        2.07070707,  2.17171717,  2.27272727,  2.37373737,  2.47474747,
        2.57575758,  2.67676768,  2.77777778,  2.87878788,  2.97979798,
        3.08080808,  3.18181818,  3.28282828,  3.38383838,  3.48484848,
        3.58585859,  3.68686869,  3.78787879,  3.88888889,  3.98989899,
        4.09090909,  4.19191919,  4.29292929,  4.39393939,  4.49494949,
        4.5959596 ,  4.6969697 ,  4.7979798 ,  4.8989899 ,  5.        ])
```

Figure 5.10: Printing the inputs

4. Compute the output (**y**):

```
y = [relu(_x * theta) for _x in x]
```

5. Plot the output of the neuron versus the input:

```
fig = plt.figure(figsize=(10, 7))
ax = fig.add_subplot(111)
ax.plot(x, y)
ax.set_xlabel('$x$', fontsize=22)
ax.set_ylabel('$h(x\Theta)$', fontsize=22)
ax.spines['left'].set_position(('data', 0))
ax.spines['top'].set_visible(False)
ax.spines['right'].set_visible(False)
ax.tick_params(axis='both', which='major', labelsize=22)
plt.show()
```

The output is as follows:

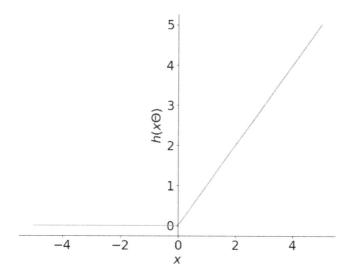

Figure 5.11: Plot of the neuron versus the input

6. Now, set **theta** = 5 and recompute and store the output of the neuron:

   ```
   theta = 5
   y_2 = [relu(_x * theta) for _x in x]
   ```

7. Now, set **theta** = 0.2 and recompute and store the output of the neuron:

   ```
   theta = 0.2
   y_3 = [relu(_x * theta) for _x in x]
   ```

8. Plot the three different output curves of the neuron (**theta** = 1, **theta** = 5, and **theta** = 0.2) on one graph:

   ```
   fig = plt.figure(figsize=(10, 7))
   ax = fig.add_subplot(111)
   ax.plot(x, y, label='$\Theta=1$')
   ax.plot(x, y_2, label='$\Theta=5$', linestyle=':')
   ax.plot(x, y_3, label='$\Theta=0.2$', linestyle='--')
   ax.set_xlabel('$x\Theta$', fontsize=22)
   ax.set_ylabel('$h(x\Theta)$', fontsize=22)
   ax.spines['left'].set_position(('data', 0))
   ```

```
ax.spines['top'].set_visible(False)
ax.spines['right'].set_visible(False)
ax.tick_params(axis='both', which='major', labelsize=22)
ax.legend(fontsize=22)
plt.show()
```

The output is as follows:

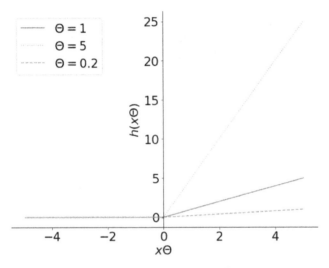

Figure 5.12: Three output curves of the neuron

In this exercise, we created a model of a ReLU-based artificial neural network neuron. We can see that the output of this neuron is very different to the sigmoid activation function. There is no saturation region for values greater than 0 because it simply returns the input value of the function. In the negative direction, there is a saturation region where only 0 will be returned if the input is less than 0. The ReLU function is an extremely powerful and commonly used activation function that has shown itself to be more powerful than the sigmoid function in some circumstances. ReLU is often a good first-choice activation function.

> **NOTE**
>
> To access the source code for this specific section, please refer to https://packt.live/2O5rnIn.
>
> You can also run this example online at https://packt.live/3iJ2Kzu.

NEURAL NETWORKS: ARCHITECTURE DEFINITION

Individual neurons aren't particularly useful in isolation; they provide an activation function and a means of tuning the output, but a single neuron would have a limited learning ability. Neurons become much more powerful when many of them are combined and connected together in a network structure. By using a number of different neurons and combining the outputs of individual neurons, more complex relationships can be established and more powerful learning algorithms can be built. In this section, we will briefly discuss the structure of a neural network and implement a simple neural network using the Keras machine learning framework (https://keras.io/). Keras is a high-level neural network API that is used on top of an existing library, such as TensorFlow or Theano. Keras makes it easy to switch between lower-level frameworks because the high-level interface it provides remains the same irrespective of the underlying library. In this book, we will be using TensorFlow as the underlying library.

The following is a simplified representation of a neural network with one hidden layer:

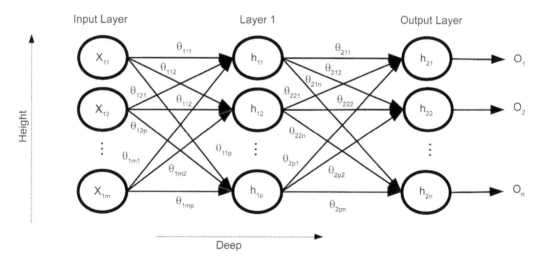

Figure 5.13: Simplified representation of a neural network

The preceding figure illustrates the structure of a two-layered, fully connected neural network. One of the first observations we can make is that there is a lot of information contained within this structure, with a high degree of connectivity, as represented by the arrows that point to and from each of the nodes. Working from the left-hand side of the image, we can see the input values to the neural network, as represented by the (x) values. In this example, we have *m* input values per sample, hence, values from x_{11} to x_{1m}. These values per sample are called attributes or the features of the data and only one sample is fed at a time into the network. These values are then multiplied by the corresponding weights of the first layer of the neural network ($\theta_{111} - \theta_{1mp}$) before being passed into the activation function of the corresponding neuron. This is known as a **feed-forward** neural network. The notation used in the preceding figure to identify the weights is θ_{ijk}, where *i* is the layer the weight belongs to, *j* is the input node number (starting with 1 at the top), and *k* is the node in the subsequent layer that the weight feeds into.

Looking at the inter-connectivity between the outputs of layer 1 (also known as the **hidden layer**) and the inputs to the output layer, we can see that there are a large number of tunable parameters (weights) that can be used to map the input to the desired output. The network of the preceding figure represents an *n* class neural network classifier, where the output for each of the *n* nodes represents the probability of the input belonging to the corresponding class.

Each layer is able to use a different activation function, as described by h_1 and h_2, thus allowing different activation functions to be mixed, in which the first layer could use ReLU, the second could use tanh, and the third could use sigmoid, for example. The final output is calculated by taking the sum of the product of the output of the previous layer with the corresponding weights with the activation function applied.

If we consider the output of the first node of layer 1, it can be calculated by multiplying the inputs by the corresponding weights, adding the result, and passing it through the activation function:

$$h_{11}(x_{11}\theta_{111} + x_{12}\theta_{121} + \ldots + x_{1m}\theta_{1m1})$$

Figure 5.14: Calculating the output of the last node

As we increase the number of layers between the input and output of the network, we increase the depth of the network. An increase in the depth also constitutes an increase in the number of trainable parameters, as well as the complexity of the relationships within the data, as described by the network. Additionally, as we add more neurons to each layer, we increase the height of the neural network. By adding more neurons, the ability of the network to describe the dataset increases as we add more trainable parameters. If too many neurons are added, the network can memorize the dataset but fails to generalize new samples. The trick in constructing neural networks is to find the balance between sufficient complexity to be able to describe the relationships within the data and not being so complicated as to memorize the training samples.

EXERCISE 5.03: DEFINING A KERAS MODEL

In this exercise, we will define a neural network architecture (similar to *Figure 5.13*) using the Keras machine learning framework to classify images for the CIFAR-10 dataset. As each input image is 32 x 32 pixels in size, the input vector will comprise 32*32 = 1,024 values. With 10 individual classes in CIFAR-10, the output of the neural network will be composed of 10 individual values, with each value representing the probability of the input data belonging to the corresponding class.

> **NOTE**
>
> The CIFAR-10 dataset (https://www.cs.toronto.edu/~kriz/cifar.html) is made up of 60,000 images across 10 classes. These 10 classes are airplane, automobile, bird, cat, deer, dog, frog, horse, ship, and truck, with 6,000 images per class. Learn more about this dataset via the preceding link.

1. For this exercise, we will require the Keras machine learning framework. If you have yet to install Keras and TensorFlow, do so using **conda** from within your Jupyter notebook:

   ```
   !conda install tensorflow keras
   ```

   Alternatively, you can install it using **pip**:

   ```
   !pip install tensorflow keras
   ```

2. We will require the **Sequential** and **Dense** classes from **keras.models** and **keras.layers**, respectively. Import these classes:

```
from keras.models import Sequential
from keras.layers import Dense
```

As described earlier, the input layer will receive 1,024 values. The second layer (Layer 1) will have 500 units and, because the network is to classify one of 10 different classes, the output layer will have 10 units. In Keras, a model is defined by passing an ordered list of layers to the **Sequential** model class.

3. This example uses the **Dense** layer class, which is a fully connected neural network layer. The first layer will use a ReLU activation function, while the output will use the **softmax** function to determine the probability of each class. Define the model:

```
model = Sequential\
        ([Dense(500, input_shape=(1024,), activation='relu'),\
          Dense(10, activation='softmax')])
```

4. With the model defined, we can use the **summary** method to confirm the structure and the number of trainable parameters (or weights) within the model:

```
model.summary()
```

The output is as follows:

```
Layer (type)                 Output Shape              Param #
=================================================================
dense_1 (Dense)              (None, 500)               512500
_____
dense_2 (Dense)              (None, 10)                5010
=================================================================
Total params: 517,510
Trainable params: 517,510
Non-trainable params: 0
```

Figure 5.15: Structure and count of trainable parameters in the model

This table summarizes the structure of the neural network. We can see that there are the two layers that we specified, with 500 units in the first layer and 10 output units in the second layer. The `Param #` column tells us how many trainable weights are available in that specific layer. The table also tells us that there are 517,510 trainable weights in total within the network.

> **NOTE**
>
> To access the source code for this specific section, please refer to https://packt.live/31WaTdR.
>
> You can also run this example online at https://packt.live/3gGEtbA.

In this exercise, we created a neural network model in Keras that contains a network of over 500,000 weights that can be used to classify the images of CIFAR-10. In the next section, we will train the model.

NEURAL NETWORKS: TRAINING

With the neural network model defined, we can begin the training process; at this stage, we will be training the model in a supervised fashion to develop some familiarity with the Keras framework before moving on to training autoencoders. Supervised learning models are trained by providing the model with both the input information as well as the known output; the goal of training is to construct a network that takes the input information and returns the known output using only the parameters of the model.

In a supervised classification example such as CIFAR-10, the input information is an image and the known output is the class that the image belongs to. During training, for each sample prediction, the errors in the feedforward network predictions are calculated using a specified error function. Each of the weights within the model is then tuned in an attempt to reduce the error. This tuning process is known as **backpropagation** because the error is propagated backward through the network from the output to the start of the network.

During backpropagation, each trainable weight is adjusted in proportion with its contribution to the overall error multiplied by a value known as the **learning rate**, which controls the rate of change in the trainable weights. Looking at the following figure, we can see that increasing the value of the learning rate can increase the speed at which the error is reduced, but risks not converging on a minimum error as we step over the values. A learning rate that's too small may lead to us running out of patience or simply not having sufficient time to find the global minimum. During neural network training, our goal is to find the global minimum of errors – basically, the point in the training at which the weights are tuned in such a way that it is impossible to minimize the number of errors any further. Thus, finding the correct learning rate is a trial-and-error process, though starting with a larger learning rate and reducing it can often be a productive method. The following figure represents the effect of the selection of the learning rate on the optimization of the cost function.

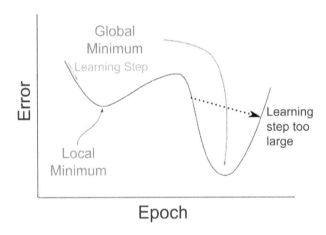

Figure 5.16: Selecting the correct learning rate

In the preceding figure, you see the learning errors over an epoch, which in this case is over time. One epoch corresponds to one complete cycle through the training dataset. Training is repeated until the error in the predictions stop reducing or the developer runs out of patience waiting for a result. In order to complete the training process, we first need to make some design decisions, the first being the most appropriate error function. There are a range of error functions available for use, from a simple mean squared difference to more complex options. Categorical cross-entropy (which is used in the following exercise) is a very useful error function for classifying more than one class.

With the error function defined, we need to choose the method of updating the trainable parameters using the error function. One of the most memory-efficient and effective update methods is **Stochastic Gradient Descent** (**SGD**). There are a number of variants of SGD, all of which involve adjusting each of the weights in accordance with their individual contribution to the calculated error. The final training design decision to be made is the performance metric by which the model is evaluated and the best architecture selected; in a classification problem, this may be the classification accuracy of the model or perhaps the model that produces the lowest error score in a regression problem. These comparisons are generally made using a cross-validation method.

EXERCISE 5.04: TRAINING A KERAS NEURAL NETWORK MODEL

Thankfully, we don't need to worry about manually programming the components of the neural network, such as backpropagation, because the Keras framework manages this for us. In this exercise, we will use Keras to train a neural network to classify a small subset of the CIFAR-10 dataset using the model architecture defined in the preceding exercise. As with all machine learning problems, the first and the most important step is to understand as much as possible about the dataset, and this will be the initial focus of the exercise:

> **NOTE**
>
> You can download the `data_batch_1` and `batches.meta` files from https://packt.live/3eexo1s.

1. Import `pickle`, `numpy`, `matplotlib`, and the `Sequential` class from `keras.models`, and import `Dense` from `keras.layers`. We'll use `pickle` for this exercise to serialize objects in Python for transfer or storage:

    ```
    import pickle
    import numpy as np
    import matplotlib.pyplot as plt
    from keras.models import Sequential
    from keras.layers import Dense

    import tensorflow.python.util.deprecation as deprecation
    deprecation._PRINT_DEPRECATION_WARNINGS = False
    ```

2. Load the sample of the CIFAR-10 dataset that is provided with the accompanying source code in the **data_batch_1** file:

   ```
   with open('data_batch_1', 'rb') as f:
       batch_1 = pickle.load(f, encoding='bytes')
   ```

3. The data is loaded as a dictionary. Display the keys of the dictionary:

   ```
   batch_1.keys()
   ```

   The output is as follows:

   ```
   dict_keys([b'batch_label', b'labels', b'data', b'filenames'])
   ```

4. Note that the keys are stored as binary strings as denoted by **b'**. We are interested in the contents of data and labels. Let's look at labels first:

   ```
   labels = batch_1[b'labels']
   labels
   ```

   A section of output is as follows, with each class number corresponding to one of the text labels (airplane, car, and so on):

$$[6, 9, 9, 4, 1, 1, 2, 7, 8, 3, 4, 7, 7,$$

Figure 5.17: Displaying the labels

5. We can see that the labels are a list of values 0-9, indicating which class each sample belongs to. Now, look at the contents of the **data** key:

    ```
    batch_1[b'data']
    ```

    The output is as follows:

    ```
    array([[ 59,  43,  50, ..., 140,  84,  72],
           [154, 126, 105, ..., 139, 142, 144],
           [255, 253, 253, ...,  83,  83,  84],
           ...,
           [ 71,  60,  74, ...,  68,  69,  68],
           [250, 254, 211, ..., 215, 255, 254],
           [ 62,  61,  60, ..., 130, 130, 131]], dtype=uint8)
    ```

    Figure 5.18: Content of the data key

6. The data key provides a NumPy array with all the image data stored within the array. What is the shape of the image data?

    ```
    batch_1[b'data'].shape
    ```

    The output is as follows:

    ```
    (10000, 3072)
    ```

7. We can see that we have 1,000 samples, but each sample is a single dimension of 3,072 samples. Aren't the images supposed to be 32 x 32 pixels? Yes, they are, but because the images are color or RGB images, they contain three channels (red, green, and blue), which means the images are 32 x 32 x 3. They are also flattened, providing 3,072 length vectors. So, we can reshape the array and then visualize a sample of images. According to the CIFAR-10 documentation, the first 1,024 samples are red, the second 1,024 are green, and the third 1,024 are blue:

    ```
    images = np.zeros((10000, 32, 32, 3), dtype='uint8')
    """
    Breaking the 3,072 samples of each single image into thirds,
    which correspond to Red, Green, Blue channels
    """
    for idx, img in enumerate(dat[b'data']):
        images[idx, :, :, 0] = img[:1024].reshape((32, 32))  # Red
        images[idx, :, :, 1] = img[1024:2048]\
                                .reshape((32, 32))  # Green
        images[idx, :, :, 2] = img[2048:].reshape((32, 32))  # Blue
    ```

178 | Autoencoders

8. Display the first 12 images, along with their labels:

```
plt.figure(figsize=(10, 7))
for i in range(12):
    plt.subplot(3, 4, i + 1)
    plt.imshow(images[i])
    plt.title(labels[i])
    plt.axis('off')
```

The following output shows a sample of low-resolution images from our dataset – which is a result of the 32 x 32 resolution we originally received:

Figure 5.19: The first 12 images

What is the actual meaning of the labels? We'll find that out in the next step.

9. Load the **batches.meta** file using the following code:

```
with open('batches.meta', 'rb') as f:
    label_strings = pickle.load(f, encoding='bytes')
label_strings
```

The output is as follows:

```
{b'num_cases_per_batch': 10000,
 b'label_names': [b'airplane',
  b'automobile',
  b'bird',
  b'cat',
  b'deer',
  b'dog',
  b'frog',
  b'horse',
  b'ship',
  b'truck'],
 b'num_vis': 3072}
```

Figure 5.20: Meaning of the labels

10. Decode the binary strings to get the actual labels:

```
actual_labels = [label.decode() for label in \
                label_strings[b'label_names']]
actual_labels
```

The output is as follows:

```
['airplane',
 'automobile',
 'bird',
 'cat',
 'deer',
 'dog',
 'frog',
 'horse',
 'ship',
 'truck']
```

Figure 5.21: Printing the actual labels

11. Print the labels for the first 12 images:

    ```
    for lab in labels[:12]:
        print(actual_labels[lab], end=', ')
    ```

    The output is as follows:

    ```
    frog, truck, truck, deer, automobile, automobile,
    bird, horse, ship, cat, deer, horse,
    ```

12. Now we need to prepare the data for training the model. The first step is to prepare the output. Currently, the output is a list of numbers 0-9, but we need each sample to be represented as a vector of 10 units as per the previous model.

    > **NOTE**
    >
    > This is known as one-hot encoding, where, for each sample, there are as many columns as there are possible classes, and the identified class is indicated by a 1 in the appropriate column. As an example, say we had the labels [3, 2, 1, 3, 1] with 4 possible classes; the corresponding one-hot encoded value would be as follows:
    >
    > ```
    > array([[0., 0., 0., 1.],
    >        [0., 0., 1., 0.],
    >        [0., 1., 0., 0.],
    >        [0., 0., 0., 1 ],
    >        [0., 1., 0., 0.]])
    > ```

    The encoded output will be a NumPy array with a shape of 10,000 x 10:

    ```
    one_hot_labels = np.zeros((images.shape[0], 10))
    for idx, lab in enumerate(labels):
        one_hot_labels[idx, lab] = 1
    ```

13. Display the one-hot encoding values for the first 12 samples:

    ```
    one_hot_labels[:12]
    ```

The output is as follows:

```
array([[0., 0., 0., 0., 0., 0., 1., 0., 0., 0.],
       [0., 0., 0., 0., 0., 0., 0., 0., 0., 1.],
       [0., 0., 0., 0., 0., 0., 0., 0., 0., 1.],
       [0., 0., 0., 0., 1., 0., 0., 0., 0., 0.],
       [0., 1., 0., 0., 0., 0., 0., 0., 0., 0.],
       [0., 1., 0., 0., 0., 0., 0., 0., 0., 0.],
       [0., 0., 1., 0., 0., 0., 0., 0., 0., 0.],
       [0., 0., 0., 0., 0., 0., 0., 1., 0., 0.],
       [0., 0., 0., 0., 0., 0., 0., 0., 1., 0.],
       [0., 0., 0., 1., 0., 0., 0., 0., 0., 0.],
       [0., 0., 0., 0., 1., 0., 0., 0., 0., 0.],
       [0., 0., 0., 0., 0., 0., 0., 1., 0., 0.]])
```

Figure 5.22: One-hot encoding values for the first 12 samples

14. The model has 1,024 inputs because it expects a 32 x 32 grayscale image. Take the average of the three channels for each image to convert it to RGB:

```
images = images.mean(axis=-1)
```

15. Display the first 12 images again:

```
plt.figure(figsize=(10, 7))
for i in range(12):
    plt.subplot(3, 4, i + 1)
    plt.imshow(images[i], cmap='gray')
    plt.title(labels[i])
    plt.axis('off')
```

The output is as follows:

Figure 5.23: Displaying the first 12 images again.

16. Finally, scale the images to be between 0 and 1, which is required for all inputs to a neural network. As the maximum value in an image is 255, we will simply divide by 255:

```
images /= 255.
```

17. We also need the images to be in the shape 10,000 x 1,024. We will select the first 7,000 samples for training and the last 3,000 samples to evaluate the model:

```
images = images.reshape((-1, 32 ** 2))
x_train = images[:7000]
y_train = one_hot_labels[:7000]
x_test = images[7000:]
y_test = one_hot_labels[7000:]
```

18. Redefine the model with the same architecture as *Exercise 5.03, Defining a Keras Model*:

    ```
    model = Sequential\
            ([Dense(500, input_shape=(1024,), activation='relu'),\
              Dense(10, activation='softmax')])
    ```

19. Now we can train the model in Keras. We first need to compile the method to specify the training parameters. We will be using categorical cross-entropy, with Adam and a performance metric of classification accuracy:

    ```
    model.compile(loss='categorical_crossentropy',\
                  optimizer='adam',\
                  metrics=['accuracy'])
    ```

20. Train the model using backpropagation for 100 epochs and the **fit** method of the model:

    ```
    model.fit(x_train, y_train, epochs=100, \
              validation_data=(x_test, y_test), \
              shuffle = False)
    ```

    The output is as follows. Please note, given the random nature of neural network training, that your results may differ slightly:

    ```
    Epoch 95/100
    7000/7000 [==============================] - 5s 738us/step - loss: 0.7421 - acc: 0.7547 - val_loss: 2.9634 - val_acc: 0.3253
    Epoch 96/100
    7000/7000 [==============================] - 5s 750us/step - loss: 0.7380 - acc: 0.7601 - val_loss: 2.9720 - val_acc: 0.3263
    Epoch 97/100
    7000/7000 [==============================] - 5s 752us/step - loss: 0.7392 - acc: 0.7559 - val_loss: 3.0003 - val_acc: 0.3230
    Epoch 98/100
    7000/7000 [==============================] - 5s 737us/step - loss: 0.7367 - acc: 0.7531 - val_loss: 3.0153 - val_acc: 0.3263
    Epoch 99/100
    7000/7000 [==============================] - 5s 726us/step - loss: 0.7280 - acc: 0.7596 - val_loss: 3.0718 - val_acc: 0.3217
    Epoch 100/100
    7000/7000 [==============================] - 5s 743us/step - loss: 0.7270 - acc: 0.7567 - val_loss: 3.0701 - val_acc: 0.3247
    <keras.callbacks.History at 0x2df1c91bf08>
    ```

    Figure 5.24: Training the model

184 | Autoencoders

> **NOTE**
>
> Here, we are using Keras to train our neural network model. The initialization of weights in a Keras layer is done randomly and cannot be controlled by any random seed. Hence, the results may vary slightly each time the code is executed.

21. We achieved approximately 75.67% classification accuracy over the training data and 32.47% classification accuracy over the validation data (seen in *Figure 5.24* as `acc: 0.7567` and `val_acc: 0.3247`) for the 10,000 samples using this network. Examine the predictions made for the first 12 samples again:

```
predictions = model.predict(images[:12])
predictions
```

The output is as follows:

```
array([[8.2426687e-04, 1.9769844e-02, 1.3154600e-04, 1.7852411e-01,
        2.0686911e-01, 5.6794225e-03, 5.0732917e-01, 9.0836524e-04,
        7.9949051e-02, 1.5147775e-05],
       [1.1431259e-06, 7.5865492e-02, 7.0075362e-07, 3.0784035e-04,
        1.7017168e-08, 3.2263887e-03, 2.8468296e-06, 1.2712204e-03,
        2.3498660e-02, 8.9582562e-01],
       [5.1413889e-07, 1.3860208e-07, 1.7348151e-08, 9.3050639e-06,
        1.1070193e-07, 8.7996384e-08, 2.4442379e-08, 8.3750974e-06,
        8.5207148e-06, 9.9997294e-01],
       [1.8286852e-02, 6.4492985e-03, 1.7885046e-01, 1.9191353e-02,
        7.5285876e-01, 2.7415346e-04, 2.3039132e-03, 1.9792159e-04,
        2.1585245e-02, 2.0951588e-06],
       [3.0686788e-03, 8.6890823e-01, 5.7748322e-09, 2.6816693e-07,
        8.4346098e-08, 4.9803628e-09, 8.9698729e-14, 1.4195305e-07,
        1.2738889e-01, 6.3358812e-04],
       [3.3242475e-02, 5.4924262e-01, 1.2958751e-02, 1.5084515e-02,
        1.1913211e-02, 7.2970563e-03, 4.6956107e-02, 4.3636712e-04,
        2.9065263e-01, 3.2216311e-02],
```

Figure 5.25: Printing the predictions

22. We can use the `argmax` method to determine the most likely class for each sample:

```
np.argmax(predictions, axis=1)
```

The output is as follows:

```
array([6, 9, 9, 4, 1, 1, 2, 7, 8, 3, 4, 7], dtype=int64)
```

23. Compare with the labels:

```
labels[:12]
```

The output is as follows:

```
[6, 9, 9, 4, 1, 1, 2, 7, 8, 3, 4, 7]
```

> **NOTE**
>
> To access the source code for this specific section, please refer to https://packt.live/2CgH25b.
>
> You can also run this example online at https://packt.live/38CKwuD.

We have now trained a neural network model in Keras. Complete the next activity to further reinforce your skills in training neural networks.

ACTIVITY 5.01: THE MNIST NEURAL NETWORK

In this activity, you will train a neural network to identify images in the MNIST dataset and reinforce your skills in training neural networks. This activity forms the basis of many neural network architectures in different classification problems, particularly in computer vision. From object detection and identification to classification, this general structure is used in a variety of applications.

These steps will help you to complete the activity:

1. Import **pickle**, **numpy**, **matplotlib**, and the **Sequential** and **Dense** classes from Keras.

2. Load the **mnist.pkl** file, which contains the first 10,000 images and the corresponding labels from the MNIST dataset that are available in the accompanying source code. The MNIST dataset is a series of 28 x 28 grayscale images of handwritten digits, 0 through 9. Extract the images and labels.

> **NOTE**
>
> You can find the **mnist.pkl** file at https://packt.live/2JOLAQB.

3. Plot the first 10 samples along with the corresponding labels.
4. Encode the labels using one-hot encoding.
5. Prepare the images for input into a neural network. As a hint, there are **two** separate steps in this process.
6. Construct a neural network model in Keras that accepts the prepared images and has a hidden layer of 600 units with a ReLU activation function and an output of the same number of units as classes. The output layer uses a `softmax` activation function.
7. Compile the model using multiclass cross-entropy, stochastic gradient descent, and an accuracy performance metric.
8. Train the model. How many epochs are required to achieve at least 95% classification accuracy on the training data?

By completing this activity, you will have trained a simple neural network to identify handwritten digits 0 through 9. You will have also developed a general framework for building neural networks for classification problems. With this framework, you can extend and modify the network for a range of other tasks. A preview of the digits you will be classifying can be seen here:

Figure 5.26: Preview of digits to be classified

> **NOTE**
> The solution to this activity can be found on page 449.

AUTOENCODERS

Autoencoders are a specifically designed neural network architecture that aims to compress the input information into lower dimensional space in an efficient yet descriptive manner. Autoencoder networks can be decomposed into two individual sub-networks or stages: an **encoding** stage and a **decoding** stage.

The following is a simplified autoencoder model using the CIFAR-10 dataset:

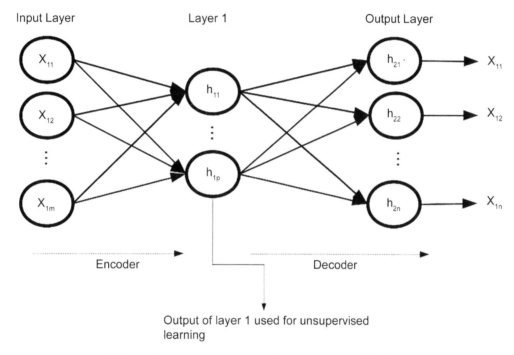

Figure 5.27: Simple autoencoder network architecture

The first, or encoding, stage takes the input information and compresses it through a subsequent layer that has fewer units than the size of the input sample. The latter stage, that is, the decoding stage, then expands the compressed form of the image and aims to return the compressed data to its original form. As such, the inputs and desired outputs of the network are the same; the network takes, say, an image in the CIFAR-10 dataset and tries to return the same image. This network architecture is shown in the preceding figure; in this image, we can see that the encoding stage of the autoencoder reduces the number of neurons to represent the information, while the decoding stage takes the compressed format and returns it to its original state. The use of the decoding stage helps to ensure that the encoder has correctly represented the information because the compressed representation is all that is provided to restore the image in its original state.

EXERCISE 5.05: SIMPLE AUTOENCODER

In this exercise, we will construct a simple autoencoder for the sample of the CIFAR-10 dataset, compressing the information stored within the images for later use.

> **NOTE**
>
> In this exercise, we will be using the **data_batch_1** file, which is a sample of the CIFAR-10 dataset. This file can be downloaded from https://packt.live/3bYi5l8.

1. Import **pickle**, **numpy**, and **matplotlib**, as well as the **Model** class, from **keras.models**, and import **Input** and **Dense** from **keras.layers**:

   ```
   import pickle
   import numpy as np
   import matplotlib.pyplot as plt
   from keras.models import Model
   from keras.layers import Input, Dense

   import tensorflow.python.util.deprecation as deprecation
   deprecation._PRINT_DEPRECATION_WARNINGS = False
   ```

2. Load the data:

   ```
   with open('data_batch_1', 'rb') as f:
       batch_1 = pickle.load(f, encoding='bytes')
   ```

3. As this is an unsupervised learning method, we are only interested in the image data. Load the image data:

   ```
   images = np.zeros((10000, 32, 32, 3), dtype='uint8')
   for idx, img in enumerate(batch_1[b'data']):
       images[idx, :, :, 0] = img[:1024].reshape((32, 32)) # Red
       images[idx, :, :, 1] = img[1024:2048]\
                              .reshape((32, 32)) # Green
       images[idx, :, :, 2] = img[2048:].reshape((32, 32)) # Blue
   ```

4. Convert the image to grayscale, scale between 0 and 1, and flatten each to a single 1,024 length vector:

```
images = images.mean(axis=-1)
images = images / 255.0
images = images.reshape((-1, 32 ** 2))
images
```

The output is as follows:

```
array([[0.24052288, 0.1751634 , 0.18431373, ..., 0.70588235, 0.46143791,
        0.3751634 ],
       [0.67712418, 0.52156863, 0.39738562, ..., 0.54248366, 0.54771242,
        0.54901961],
       [1.        , 0.99215686, 0.99215686, ..., 0.32156863, 0.32287582,
        0.32679739],
       ...,
       [0.25098039, 0.21437908, 0.27843137, ..., 0.28888889, 0.29673203,
        0.29934641],
       [0.99346405, 0.99477124, 0.85620915, ..., 0.8379085 , 1.        ,
        0.99738562],
       [0.1620915 , 0.16078431, 0.15816993, ..., 0.64705882, 0.64705882,
        0.64836601]])
```

Figure 5.28: Scaled image

5. Define the autoencoder model. As we need access to the output of the encoder stage, we will need to define the model using a slightly different method to that used previously. Define an input layer of **1024** units:

```
input_layer = Input(shape=(1024,))
```

6. Define a subsequent **Dense** layer of **256** units (a compression ratio of 1024/256 = 4) and a ReLU activation function as the encoding stage. Note that we have assigned the layer to a variable and passed the previous layer to a **call** method for the class:

```
encoding_stage = Dense(256, activation='relu')(input_layer)
```

7. Define a subsequent decoder layer using the sigmoid function as an activation function and the same shape as the input layer. The sigmoid function has been selected because the input values to the network are only between 0 and 1:

```
decoding_stage = Dense(1024, activation='sigmoid')\
                    (encoding_stage)
```

8. Construct the model by passing the first and last layers of the network to the **Model** class:

   ```
   autoencoder = Model(input_layer, decoding_stage)
   ```

9. Compile the autoencoder using a binary cross-entropy loss function and **adadelta** gradient descent:

   ```
   autoencoder.compile(loss='binary_crossentropy',\
                       optimizer='adadelta')
   ```

   > **NOTE**
   >
   > **adadelta** is a more sophisticated version of stochastic gradient descent where the learning rate is adjusted on the basis of a window of recent gradient updates. Compared to the other methods of modifying the learning rate, this prevents the gradient of very old epochs from influencing the learning rate.

10. Now, let's fit the model; again, we pass the images as the training data and as the desired output. Train for 100 epochs:

    ```
    autoencoder.fit(images, images, epochs=100)
    ```

    The output is as follows:

    ```
    Epoch 95/100
    10000/10000 [==============================] - 10s 1ms/step - loss: 0.5778: 1
    Epoch 96/100
    10000/10000 [==============================] - 9s 894us/step - loss: 0.5777
    Epoch 97/100
    10000/10000 [==============================] - 9s 933us/step - loss: 0.5778
    Epoch 98/100
    10000/10000 [==============================] - 9s 904us/step - loss: 0.5776 0s
    Epoch 99/100
    10000/10000 [==============================] - 9s 860us/step - loss: 0.5776
    Epoch 100/100
    10000/10000 [==============================] - 10s 952us/step - loss: 0.5774
    <keras.callbacks.History at 0x2c8993af888>
    ```

    Figure 5.29: Training the model

11. Calculate and store the output of the encoding stage for the first five samples:

```
encoder_output = Model(input_layer, encoding_stage)\
                 .predict(images[:5])
```

12. Reshape the encoder output to 16 x 16 (16 x 16 = 256) pixels and multiply by 255:

```
encoder_output = encoder_output.reshape((-1, 16, 16)) * 255
```

13. Calculate and store the output of the decoding stage for the first five samples:

```
decoder_output = autoencoder.predict(images[:5])
```

14. Reshape the output of the decoder to 32 x 32 and multiply by 255:

```
decoder_output = decoder_output.reshape((-1, 32,32)) * 255
```

15. Reshape the original images:

```
images = images.reshape((-1, 32, 32))
plt.figure(figsize=(10, 7))
for i in range(5):
    # Plot the original images
    plt.subplot(3, 5, i + 1)
    plt.imshow(images[i], cmap='gray')
    plt.axis('off')
    # Plot the encoder output
    plt.subplot(3, 5, i + 6)
    plt.imshow(encoder_output[i], cmap='gray')
    plt.axis('off')
    # Plot the decoder output
    plt.subplot(3, 5, i + 11)
    plt.imshow(decoder_output[i], cmap='gray')
    plt.axis('off')
```

The output is as follows:

Figure 5.30: Output of the simple autoencoder

In the preceding figure, we can see three rows of images. The first row is the original grayscale image, the second row is the corresponding autoencoder output for the original image, and finally, the third row is the reconstruction of the original image from the encoded input. We can see that the decoded images in the third row contain information about the basic shape of the image; we can see the main body of the frog and the deer, as well as the outline of the trucks and cars in the sample. Given that we only trained the model for 100 samples, this exercise would also benefit from an increase in the number of training epochs to further improve the performance of both the encoder and decoder. Now that we have the output of the autoencoder stage trained, we can use it as the feature vector for other unsupervised algorithms, such as K-means or K nearest neighbors.

> **NOTE**
>
> To access the source code for this specific section, please refer to https://packt.live/2BQH03R.
>
> You can also run this example online at https://packt.live/2Z9CMgl.

ACTIVITY 5.02: SIMPLE MNIST AUTOENCODER

In this activity, you will create an autoencoder network for the MNIST dataset contained within the accompanying source code. An autoencoder network such as the one built in this activity can be extremely useful in the preprocessing stage of unsupervised learning. The encoded information produced by the network can be used in clustering or segmentation analysis, such as image-based web searches:

1. Import **pickle**, **numpy**, and **matplotlib**, as well as the **Model**, **Input**, and **Dense** classes, from Keras.

2. Load the images from the supplied sample of the MNIST dataset that is provided with the accompanying source code (**mnist.pkl**).

   > **NOTE**
   >
   > You can download the **mnist.pkl** file from https://packt.live/2wmpyl5.

3. Prepare the images for input into a neural network. As a hint, there are **two** separate steps in this process.

4. Construct a simple autoencoder network that reduces the image size to 10 x 10 after the encoding stage.

5. Compile the autoencoder using a binary cross-entropy loss function and **adadelta** gradient descent.

6. Fit the encoder model.

7. Calculate and store the output of the encoding stage for the first five samples.

8. Reshape the encoder output to 10 x 10 (10 x 10 = 100) pixels and multiply by 255.

9. Calculate and store the output of the decoding stage for the first five samples.

10. Reshape the output of the decoder to 28 x 28 and multiply by 255.

11. Plot the original image, the encoder output, and the decoder.

194 | Autoencoders

By completing this activity, you will have successfully trained an autoencoder network that extracts the critical information from the dataset, preparing it for later processing. The output will be similar to the following:

Figure 5.31: Expected plot of the original image, the encoder output, and the decoder

> **NOTE**
> The solution to this activity can be found on page 452.

EXERCISE 5.06: MULTI-LAYER AUTOENCODER

In this exercise, we will construct a multi-layer autoencoder for the sample of the CIFAR-10 dataset, compressing the information stored within the images for later use:

> **NOTE**
> You can download the **data_batch_1** file from https://packt.live/2VcY0a9.

1. Import **pickle**, **numpy**, and **matplotlib**, as well as the **Model** class, from **keras.models**, and import **Input** and **Dense** from **keras.layers**:

    ```
    import pickle
    import numpy as np
    import matplotlib.pyplot as plt
    from keras.models import Model
    ```

```
from keras.layers import Input, Dense

import tensorflow.python.util.deprecation as deprecation
deprecation._PRINT_DEPRECATION_WARNINGS = False
```

2. Load the data:

```
with open('data_batch_1', 'rb') as f:
    dat = pickle.load(f, encoding='bytes')
```

3. As this is an unsupervised learning method, we are only interested in the image data. Load the image data as per the preceding exercise:

```
images = np.zeros((10000, 32, 32, 3), dtype='uint8')
for idx, img in enumerate(dat[b'data']):
    images[idx, :, :, 0] = img[:1024].reshape((32, 32)) # Red
    images[idx, :, :, 1] = img[1024:2048]\
                            .reshape((32, 32)) # Green
    images[idx, :, :, 2] = img[2048:].reshape((32, 32)) # Blue
```

4. Convert the image to grayscale, scale between 0 and 1, and flatten each to a single 1,024 length vector:

```
images = images.mean(axis=-1)
images = images / 255.0
images = images.reshape((-1, 32 ** 2))
images
```

The output is as follows:

```
array([[0.24052288, 0.1751634 , 0.18431373, ..., 0.70588235, 0.46143791,
        0.3751634 ],
       [0.67712418, 0.52156863, 0.39738562, ..., 0.54248366, 0.54771242,
        0.54901961],
       [1.        , 0.99215686, 0.99215686, ..., 0.32156863, 0.32287582,
        0.32679739],
       ...,
       [0.25098039, 0.21437908, 0.27843137, ..., 0.28888889, 0.29673203,
        0.29934641],
       [0.99346405, 0.99477124, 0.85620915, ..., 0.8379085 , 1.        ,
        0.99738562],
       [0.1620915 , 0.16078431, 0.15816993, ..., 0.64705882, 0.64705882,
        0.64836601]])
```

Figure 5.32: Scaled image

5. Define the multi-layer autoencoder model. We will use the same shape input as the simple autoencoder model:

   ```
   input_layer = Input(shape=(1024,))
   ```

6. We will add another layer before the 256 autoencoder stage – this time with 512 neurons:

   ```
   hidden_encoding = Dense(512, activation='relu')(input_layer)
   ```

7. We're using the same size autoencoder as in *Exercise 5.05, Simple Autoencoder*, but the input to the layer is the **hidden_encoding** layer this time:

   ```
   encoding_stage = Dense(256, activation='relu')(hidden_encoding)
   ```

8. Add a decoding hidden layer:

   ```
   hidden_decoding = Dense(512, activation='relu')(encoding_stage)
   ```

9. Use the same output stage as in the previous exercise, this time connected to the hidden decoding stage:

   ```
   decoding_stage = Dense(1024, activation='sigmoid')\
                    (hidden_decoding)
   ```

10. Construct the model by passing the first and last layers of the network to the **Model** class:

    ```
    autoencoder = Model(input_layer, decoding_stage)
    ```

11. Compile the autoencoder using a binary cross-entropy loss function and **adadelta** gradient descent:

    ```
    autoencoder.compile(loss='binary_crossentropy',\
                        optimizer='adadelta')
    ```

12. Now, let's fit the model; again, we pass the images as the training data and as the desired output. Train for 100 epochs:

    ```
    autoencoder.fit(images, images, epochs=100)
    ```

The output is as follows:

```
Epoch 95/100
10000/10000 [==============================] - 25s 2ms/step - loss: 0.5806
Epoch 96/100
10000/10000 [==============================] - 25s 3ms/step - loss: 0.5803
Epoch 97/100
10000/10000 [==============================] - 25s 3ms/step - loss: 0.5803
Epoch 98/100
10000/10000 [==============================] - 19s 2ms/step - loss: 0.5801
Epoch 99/100
10000/10000 [==============================] - 18s 2ms/step - loss: 0.5802
Epoch 100/100
10000/10000 [==============================] - 18s 2ms/step - loss: 0.5800
<keras.callbacks.History at 0x25906b92088>
```

Figure 5.33: Training the model

13. Calculate and store the output of the encoding stage for the first five samples:

    ```
    encoder_output = Model(input_stage, encoding_stage)\
                    .predict(images[:5])
    ```

14. Reshape the encoder output to 16 x 16 (16 x 16 = 256) pixels and multiply by 255:

    ```
    encoder_output = encoder_output.reshape((-1, 16, 16)) * 255
    ```

15. Calculate and store the output of the decoding stage for the first five samples:

    ```
    decoder_output = autoencoder.predict(images[:5])
    ```

16. Reshape the output of the decoder to 32 x 32 and multiply by 255:

    ```
    decoder_output = decoder_output.reshape((-1, 32, 32)) * 255
    ```

17. Plot the original image, the encoder output, and the decoder:

    ```
    images = images.reshape((-1, 32, 32))
    plt.figure(figsize=(10, 7))
    for i in range(5):
        # Plot original images
        plt.subplot(3, 5, i + 1)
        plt.imshow(images[i], cmap='gray')
        plt.axis('off')
        # Plot encoder output
        plt.subplot(3, 5, i + 6)
        plt.imshow(encoder_output[i], cmap='gray')
    ```

```
plt.axis('off')
# Plot decoder output
plt.subplot(3, 5, i + 11)
plt.imshow(decoder_output[i], cmap='gray')
plt.axis('off')
```

The output is as follows:

Figure 5.34: Output of the multi-layer autoencoder

By looking at the error score produced by both the simple and multilayer autoencoders and by comparing *Figure 5.30* and *Figure 5.34*, we can see that there is little difference between the output of the two encoder structures. The middle row of both figures show that the features learned by the two models are, in fact, different. There are a number of options we can use to improve both of these models, such as training for more epochs, using a different number of units or neurons in the layers, or using varying numbers of layers. This exercise was constructed to demonstrate how to build and use an autoencoder, but optimization is often a process of systematic trial and error. We encourage you to adjust some of the parameters of the model and investigate the different results for yourself.

> **NOTE**
>
> To access the source code for this specific section, please refer to https://packt.live/2ZbaT81.
>
> You can also run this example online at https://packt.live/2ZHvOyo.

CONVOLUTIONAL NEURAL NETWORKS

In constructing all of our previous neural network models, you would have noticed that we removed all the color information when converting the image to grayscale, and then flattened each image into a single vector of length 1,024. In doing so, we essentially threw out a lot of information that may be of use to us. The colors in the images may be specific to the class or objects in the image; additionally, we lost a lot of our spatial information pertaining to the image; for example, the position of the trailer in the truck image relative to the cab or the legs of the deer relative to the head. Convolutional neural networks do not suffer from this information loss. This is because, rather than using a flat structure of trainable parameters, they store the weights in a grid or matrix, which means that each group of parameters can have many layers in their structure. By organizing the weights in a grid, we prevent the loss of spatial information because the weights are applied in a sliding fashion across the image. Also, by having many layers, we can retain the color channels associated with the image.

In developing convolutional neural network-based autoencoders, the MaxPooling2D and Upsampling2D layers are very important. The MaxPooling 2D layer downsamples or reduces the size of an input matrix in two dimensions by selecting the maximum value within a window of the input. Say we had a 2 x 2 matrix, where three cells have a value of 1 and one single cell has a value of 2:

1	1
1	2

Figure 5.35: Demonstration of a sample matrix

If provided to the MaxPooling2D layer, this matrix would return a single value of 2, thus reducing the size of the input in both directions by one half.

The UpSampling2D layer has the opposite effect as that of the MaxPooling2D layer, increasing the size of the input rather than reducing it. The upsampling process repeats the rows and columns of the data, thus doubling the size of the input matrix. For the preceding example, you would have the 2 x 2 matrix converted into a 4 x 4 matrix, with the bottom right 4 pixels at value 2, and the rest at value 1.

EXERCISE 5.07: CONVOLUTIONAL AUTOENCODER

In this exercise, we will develop a convolutional neural network-based autoencoder and compare performance with the previous fully connected neural network autoencoder:

> **NOTE**
>
> You can download the **data_batch_1** file from https://packt.live/2x31ww3.

1. Import **pickle**, **numpy**, and **matplotlib**, as well as the **Model** class, from **keras.models**, and import **Input**, **Conv2D**, **MaxPooling2D**, and **UpSampling2D** from **keras.layers**:

   ```
   import pickle
   import numpy as np
   import matplotlib.pyplot as plt
   from keras.models import Model
   from keras.layers import Input, Conv2D, MaxPooling2D, UpSampling2D

   import tensorflow.python.util.deprecation as deprecation
   deprecation._PRINT_DEPRECATION_WARNINGS = False
   ```

2. Load the data:

   ```
   with open('data_batch_1', 'rb') as f:
       batch_1 = pickle.load(f, encoding='bytes')
   ```

3. As this is an unsupervised learning method, we are only interested in the image data. Load the image data as per the preceding exercise:

   ```
   images = np.zeros((10000, 32, 32, 3), dtype='uint8')
   for idx, img in enumerate(batch_1[b'data']):
       images[idx, :, :, 0] = img[:1024].reshape((32, 32)) # Red
       images[idx, :, :, 1] = img[1024:2048]\
                           .reshape((32, 32)) # Green
       images[idx, :, :, 2] = img[2048:].reshape((32, 32)) # Blue
   ```

4. As we are using a convolutional network, we can use the images with only rescaling:

   ```
   images = images / 255.
   ```

5. Define the convolutional autoencoder model. We will use the same shape input as an image:

```
input_layer = Input(shape=(32, 32, 3,))
```

6. Add a convolutional stage with 32 layers or filters, a 3 x 3 weight matrix, a ReLU activation function, and using the same padding, which means the output has the same length as the input image.

> **NOTE**
>
> Conv2D convolutional layers are the two-dimensional equivalent of weights in a fully connected neural network. The weights exist in a series of 2D weight filters or layers, which are then convolved with the input of the layer.

The code will look as follows:

```
hidden_encoding = Conv2D\
                 (32, # Number of filters in the weight matrix
                  (3, 3), # Shape of the weight matrix
                  activation='relu', padding='same', \
                  # Retaining dimensions between input and output \
                  )(input_layer)
```

7. Add a max pooling layer to the encoder with a 2 x 2 kernel. **MaxPooling** looks at all the values in an image, scanning through with a 2 x 2 matrix. The maximum value in each 2 x 2 area is returned, thus reducing the size of the encoded layer by a half:

```
encoded = MaxPooling2D((2, 2))(hidden_encoding)
```

8. Add a decoding convolutional layer (this layer should be identical to the previous convolutional layer):

```
hidden_decoding = \
Conv2D(32, # Number of filters in the weight matrix \
       (3, 3), # Shape of the weight matrix \
       activation='relu', \
       # Retaining dimensions between input and output \
       padding='same', \
       )(encoded)
```

9. Now we need to return the image to its original size, for which we will upsample by the same size as **MaxPooling2D**:

```
upsample_decoding = UpSampling2D((2, 2))(hidden_decoding)
```

10. Add the final convolutional stage using three layers for the RGB channels of the images:

```
decoded = \
Conv2D(3, # Number of filters in the weight matrix \
       (3, 3), # Shape of the weight matrix \
       activation='sigmoid', \
       # Retaining dimensions between input and output \
       padding='same', \
       )(upsample_decoding)
```

11. Construct the model by passing the first and last layers of the network to the **Model** class:

```
autoencoder = Model(input_layer, decoded)
```

12. Display the structure of the model:

```
autoencoder.summary()
```

The output is as follows:

```
Layer (type)                    Output Shape              Param #
=================================================================
input_1 (InputLayer)            (None, 32, 32, 3)         0
_____
conv2d_1 (Conv2D)               (None, 32, 32, 32)        896
_____
max_pooling2d_1 (MaxPooling2    (None, 16, 16, 32)        0
_____
conv2d_2 (Conv2D)               (None, 16, 16, 32)        9248
_____
up_sampling2d_1 (UpSampling2    (None, 32, 32, 32)        0
_____
conv2d_3 (Conv2D)               (None, 32, 32, 3)         867
=================================================================
Total params: 11,011
Trainable params: 11,011
Non-trainable params: 0
```

Figure 5.36: Structure of the model

> **NOTE**
>
> We have far fewer trainable parameters compared to the previous autoencoder examples. This has been a specific design decision to ensure that the example runs on a wide variety of hardware. Convolutional networks typically require a lot more processing power and often special hardware such as Graphical Processing Units (GPUs).

13. Compile the autoencoder using a binary cross-entropy loss function and **adadelta** gradient descent:

```
autoencoder.compile(loss='binary_crossentropy',\
                    optimizer='adadelta')
```

14. Now, let's fit the model; again, we pass the images as the training data and as the desired output. Instead of training for 100 epochs like before, we will use 20 epochs, since convolutional networks take a lot longer to compute:

```
autoencoder.fit(images, images, epochs=20)
```

The output is as follows:

```
Epoch 15/20
10000/10000 [==============================] - 131s 13ms/step - loss: 0.5515
Epoch 16/20
10000/10000 [==============================] - 131s 13ms/step - loss: 0.5513
Epoch 17/20
10000/10000 [==============================] - 131s 13ms/step - loss: 0.5512
Epoch 18/20
10000/10000 [==============================] - 133s 13ms/step - loss: 0.5510
Epoch 19/20
10000/10000 [==============================] - 131s 13ms/step - loss: 0.5507
Epoch 20/20
10000/10000 [==============================] - 132s 13ms/step - loss: 0.5507
<keras.callbacks.History at 0x1325a5ea948>
```

Figure 5.37: Training the model

Note that the error was already less than in the previous autoencoder exercise after the second epoch, suggesting a better encoding/decoding model. This reduced error can be mostly attributed to the fact that the convolutional neural network did not discard a lot of data, and the encoded images are 16 x 16 x 32, which is significantly larger than the previous 16 x 16 size. Additionally, we have not compressed the images *per se* as they now contain fewer pixels (16 x 16 x 32 = 8,192), but with more depth (32 x 32 x 3 = 3,072) than before. This information has been rearranged to allow more effective encoding/decoding processes.

15. Calculate and store the output of the encoding stage for the first five samples:

    ```
    encoder_output = Model(input_layer, encoded).predict(images[:5])
    ```

16. Each encoded image has a shape of 16 x 16 x 32 due to the number of filters selected for the convolutional stage. As such, we cannot visualize them without modification. We will reshape them to be 256 x 32 in size for visualization:

    ```
    encoder_output = encoder_output.reshape((-1, 256, 32))
    ```

17. Get the output of the decoder for the first five images:

    ```
    decoder_output = autoencoder.predict(images[:5])
    ```

18. Plot the original image, the mean encoder output, and the decoder:

    ```
    plt.figure(figsize=(10, 7))
    for i in range(5):
        # Plot original images
        plt.subplot(3, 5, i + 1)
        plt.imshow(images[i], cmap='gray')
        plt.axis('off')
        # Plot encoder output
        plt.subplot(3, 5, i + 6)
        plt.imshow(encoder_output[i], cmap='gray')
        plt.axis('off')
        # Plot decoder output
        plt.subplot(3, 5, i + 11)
        plt.imshow(decoder_output[i])
        plt.axis('off')
    ```

The output is as follows:

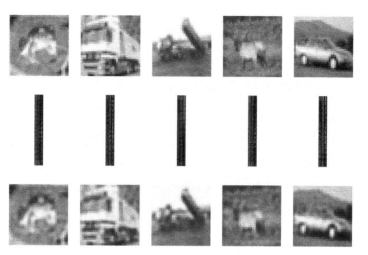

Figure 5.38: The original image, the encoder output, and the decoder output

> **NOTE**
>
> To access the source code for this specific section, please refer to https://packt.live/2VYprpq.
>
> You can also run this example online at https://packt.live/38EDgic.

ACTIVITY 5.03: MNIST CONVOLUTIONAL AUTOENCODER

In this activity, we will reinforce our knowledge of convolutional autoencoders using the MNIST dataset. Convolutional autoencoders typically achieve significantly improved performance when working with image-based datasets of a reasonable size. This is particularly useful when using autoencoders to generate artificial image samples:

1. Import **pickle**, **numpy**, and **matplotlib**, as well as the **Model** class, from **keras.models**, and import **Input**, **Conv2D**, **MaxPooling2D**, and **UpSampling2D** from **keras.layers**.

2. Load the **mnist.pkl** file, which contains the first 10,000 images and corresponding labels from the MNIST dataset, which are available in the accompanying source code.

> **NOTE**
>
> You can download the `mnist.pkl` file from https://packt.live/3e4HOR1.

3. Rescale the images to have values between 0 and 1.

4. We need to reshape the images to add a single depth channel for use with convolutional stages. Reshape the images to have a shape of 28 x 28 x 1.

5. Define an input layer. We will use the same shape input as an image.

6. Add a convolutional stage, with 16 layers or filters, a 3 x 3 weight matrix, a ReLU activation function, and using the same padding, which means the output has the same length as the input image.

7. Add a max pooling layer to the encoder with a 2 x 2 kernel.

8. Add a decoding convolutional layer.

9. Add an upsampling layer.

10. Add the final convolutional stage using one layer as per the initial image depth.

11. Construct the model by passing the first and last layers of the network to the **Model** class.

12. Display the structure of the model.

13. Compile the autoencoder using a binary cross-entropy loss function and **adadelta** gradient descent.

14. Now, let's fit the model; again, we pass the images as the training data and as the desired output. Train for 20 epochs as convolutional networks take a lot longer to compute.

15. Calculate and store the output of the encoding stage for the first five samples.

16. Reshape the encoder output for visualization, where each image is **X*Y** in size.

17. Get the output of the decoder for the first five images.

18. Reshape the decoder output to be **28 x 28** in size.

19. Reshape the original images back to be **28 x 28** in size.

20. Plot the original image, the mean encoder output, and the decoder.

At the end of this activity, you will have developed an autoencoder comprising convolutional layers within the neural network. Note the improvements made in the decoder representations. The output will be similar to the following:

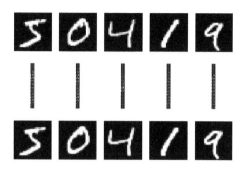

Figure 5.39: Expected original image, the encoder output, and the decoder

NOTE

The solution to this activity can be found on page 455.

SUMMARY

In this chapter, we started with an introduction to artificial neural networks, how they are structured, and the processes by which they learn to complete a particular task. Starting with a supervised learning example, we built an artificial neural network classifier to identify objects within the CIFAR-10 dataset. We then progressed to the autoencoder architecture of neural networks and learned how we can use these networks to prepare a dataset for use in an unsupervised learning problem. Finally, we completed this investigation with autoencoders, looking at convolutional neural networks and the benefits that these additional layers can provide. This chapter prepared us well for the final installment of dimensionality reduction, when we will look at using and visualizing the encoded data with t-distributed nearest neighbors (t-SNE). T-distributed nearest neighbors provides an extremely effective method for visualizing high-dimensional data even after applying reduction techniques such as PCA. T-SNE is a particularly useful method for unsupervised learning. In the next chapter, we will talk more about embeddings, which are critical tools that help us deal with high-dimensional data. As you saw with the CIFAR-10 dataset in this chapter, color image files can rapidly increase in size and slow down the performance of any neural network algorithm. By using dimensionality reduction, we can minimize the impact of high-dimensional data.

6

T-DISTRIBUTED STOCHASTIC NEIGHBOR EMBEDDING

OVERVIEW

In this chapter, we will discuss **Stochastic Neighbor Embedding** (**SNE**) and **t-Distributed Stochastic Neighbor Embedding** (**t-SNE**) as a means of visualizing high-dimensional datasets. We will implement t-SNE models in scikit-learn and explain the limitations of t-SNE. Being able to extract high-dimensional information into lower dimensions will prove helpful for visualization and exploratory analysis, as well as being helpful in conjunction with the clustering algorithms we explored in prior chapters. By the end of this chapter, we will be able to find clusters in high-dimensional data, such as user-level information or images in a low-dimensional space.

INTRODUCTION

So far, we have described a number of different methods for reducing the dimensionality of a dataset as a means of cleaning the data, reducing its size for computational efficiency, or for extracting the most important information available within the dataset. While we have demonstrated many methods for reducing high-dimensional datasets, in many cases, we are unable to reduce the number of dimensions to a size that can be visualized, that is, two or three dimensions, without excessively degrading the quality of the data. Consider the MNIST dataset that we used earlier in this book, which was a collection of digitized handwritten digits of the numbers 0 through 9. Each image is 28 x 28 pixels in size, providing 784 individual dimensions or features. If we were to reduce these 784 dimensions down to 2 or 3 for visualization purposes, we would lose almost all the available information.

In this chapter, we will discuss SNE and t-SNE as means of visualizing high-dimensional datasets. These techniques are extremely helpful in unsupervised learning and the design of machine learning systems because being able to visualize data is a powerful thing. Being able to visualize data allows relationships to be explored, groups to be identified, and results to be validated. t-SNE techniques have been used to visualize cancerous cell nuclei that have over 30 characteristics of interest, whereas data from documents can have over thousands of dimensions, sometimes even after applying techniques such as PCA.

THE MNIST DATASET

Now, we will explore SNE and t-SNE using the MNIST dataset provided with the accompanying source code as the basis of our practical examples. Before we continue, we will quickly review MNIST and the data that is within it. The complete MNIST dataset is a collection of 60,000 training and 10,000 test examples of handwritten digits of the numbers 0 to 9, represented as black and white (or grayscale) images that are 28 x 28 pixels in size (giving 784 dimensions or features) with equal numbers of each type of digit (or class) in the dataset. Due to its size and the quality of the data, MNIST has become one of the quintessential datasets in machine learning, often being used as the reference dataset for many research papers in machine learning. One of the advantages of using MNIST to explore SNE and t-SNE compared to other datasets is that while the samples contain a high number of dimensions, they can be visualized even after dimensionality reduction because they can be represented as an image. *Figure 6.1* shows a sample of the MNIST dataset:

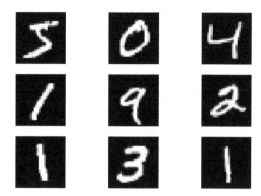

Figure 6.1: MNIST data sample

The following figure shows the same sample reduced to 30 components using PCA:

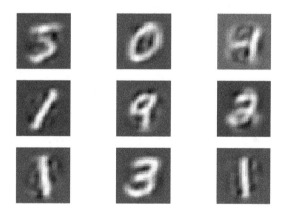

Figure 6.2: MNIST reduced using PCA to 30 components

STOCHASTIC NEIGHBOR EMBEDDING (SNE)

SNE is one of a number of different methods that fall within the category of **manifold learning**, which aims to describe high-dimensional spaces within low-dimensional manifolds or bounded areas. At first thought, this seems like an impossible task; how can we reasonably represent data in two dimensions if we have a dataset with at least 30 features? As we work through the derivation of SNE, it is hoped that you will see how this is possible. Don't worry – we will not be covering the mathematical details of this process in great depth as it is outside of the scope of this chapter. Constructing an SNE can be divided into the following steps:

1. Convert the distances between datapoints in the high-dimensional space into conditional probabilities. Say we had two points, x_i and x_j, in a high-dimensional space and we wanted to determine the probability ($p_{i|j}$) that x_j would be picked as a neighbor of x_i. To define this probability, we use a Gaussian curve. By doing this, we see that the probability is high for nearby points, while it is very low for distant points.

2. We need to determine the width of the Gaussian curve as this controls the rate of probability selection. A wide curve would suggest that many neighboring points are far away, while a narrow curve suggests that they are tightly compacted.

3. Once we project the data into the low-dimensional space, we can also determine the corresponding probability ($q_{i|j}$) between the corresponding low-dimensional data, y_i and y_j.

4. What SNE aims to do is position the data in the lower dimensions to minimize the differences between $p_{i|j}$ and $q_{i|j}$ over all the datapoints using a cost function (C). This is known as the **Kullback-Leibler** (**KL**) divergence:

$$C = \sum_i \sum_j p_{i|j} \log \frac{p_{i|j}}{q_{i|j}}$$

Figure 6.3: KL divergence

> **NOTE**
>
> For Python code to construct a Gaussian distribution, please refer to the `GaussianDist.ipynb` Jupyter notebook at https://packt.live/2UMVubU.

When Gaussian distribution is used in SNE, it reduces the dimensions of data by preserving localized patterns. To do this, SNE uses the process of gradient descent to minimize C using the standard parameters of the learning rate and epochs, as we covered in the preceding chapter when we looked at neural networks and autoencoders. SNE implements an additional term in the training process—**perplexity**. Perplexity is a selection of the effective number of neighbors used in the comparison and is relatively stable for the values of perplexity between 5 and 50. In practice, going through a process of trial and error using perplexity values within this range is recommended.

> **NOTE**
> Perplexity is covered in detail later in this chapter.

SNE provides an effective way of visualizing high-dimensional data in a low-dimensional space, though it still suffers from an issue known as **the crowding problem**. The crowding problem can occur if we have some points positioned approximately equidistantly within a region around a point, i. When these points are visualized in the lower-dimensional space, they crowd around each other, making visualization difficult. This problem is exacerbated if we try to put some more space between these crowded points, because any other points that are further away will be placed very far away within the low-dimensional space. Essentially, we are trying to balance being able to visualize close points while not losing information provided by points that are further away.

T-DISTRIBUTED SNE

t-SNE aims to address the crowding problem using a modified version of the KL divergence cost function and by substituting the Gaussian distribution with the Student's t-distribution in the low-dimensional space. The Student's t-distribution is a probability distribution much like Gaussian and is used when we have a small sample size and unknown population standard deviation. It is often used in the Student's t-test.

The modified KL cost function considers the pairwise distances in the low-dimensional space equally, while the Student's distribution employs a heavy tail in the low-dimensional space to avoid the crowding problem. In the higher-dimensional probability calculation, the Gaussian distribution is still used to ensure that a moderate distance in the higher dimensions is still represented as such in the lower dimensions. This combination of different distributions in the respective spaces allows for the faithful representation of datapoints separated by small and moderate distances.

> **NOTE**
>
> For some example code regarding how to reproduce the Student's t-distribution in Python, please refer to the Jupyter notebook at https://packt.live/2UMVubU.

Thankfully, we don't need to worry about implementing t-SNE manually because scikit-learn provides a very effective implementation in its straightforward API. What we need to remember is that both SNE and t-SNE determine the probability of two points being neighbors in both high- and low-dimensionality spaces and aim to minimize the difference in the probability between the two spaces.

EXERCISE 6.01: T-SNE MNIST

In this exercise, we will use the MNIST dataset (provided in the accompanying source code) to explore the scikit-learn implementation of t-SNE. As we described earlier, using MNIST allows us to visualize the high-dimensional space in a way that is not possible in other datasets, such as the Boston Housing Price or Iris dataset. Perform the following steps:

1. For this exercise, import **pickle**, **numpy**, **PCA**, and **TSNE** from scikit-learn, as well as **matplotlib**:

    ```
    import pickle
    import numpy as np
    import matplotlib.pyplot as plt
    from sklearn.decomposition import PCA
    from sklearn.manifold import TSNE
    np.random.seed(2)
    ```

2. Load and visualize the MNIST dataset that is provided with the accompanying source code:

> **NOTE**
>
> You can find the **mnist.pkl** file at https://packt.live/3aRuNIH.

The code is as follows:

```
with open('mnist.pkl', 'rb') as f:
    mnist = pickle.load(f)
plt.figure(figsize=(10, 7))
for i in range(9):
    plt.subplot(3, 3, i + 1)
    plt.imshow(mnist['images'][i], cmap='gray')
    plt.title(mnist['labels'][i])
    plt.axis('off')
plt.show()
```

The output is as follows:

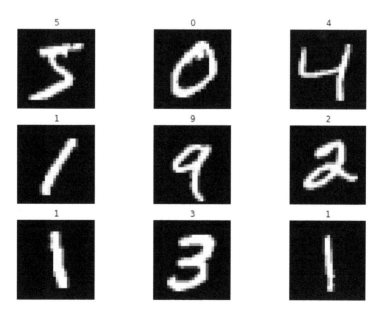

Figure 6.4: Output after loading the dataset

This demonstrates that MNIST has been successfully loaded.

3. In this exercise, we will use PCA on the dataset to extract the first 30 components.

> **NOTE**
>
> The scikit-learn PCA API requires that the data that's passed to the **fit** method is in the form required (number of samples, number of features, and so on). As such, we need to reshape the MNIST images as they are in the form (number of samples, feature 1, feature 2). Hence, we will make use of the **reshape** method in the following source code.

The code will look as follows:

```
model_pca = PCA(n_components=30)
mnist_pca = model_pca.fit(mnist['images'].reshape((-1, 28 ** 2)))
```

4. Visualize the effect of reducing the dataset to 30 components. To do this, we must transform the dataset into the lower-dimensional space and then use the **inverse_transform** method to return the data to its original size for plotting. We will, of course, need to reshape the data before and after the transform process:

```
mnist_30comp = model_pca.transform\
               (mnist['images'].reshape((-1, 28 ** 2)))
mnist_30comp_vis = model_pca.inverse_transform(mnist_30comp)
mnist_30comp_vis = mnist_30comp_vis.reshape((-1, 28, 28))
plt.figure(figsize=(10, 7))
for i in range(9):
    plt.subplot(3, 3, i + 1)
    plt.imshow(mnist_30comp_vis[i], cmap='gray')
    plt.title(mnist['labels'][i])
    plt.axis('off')
plt.show()
```

The output is as follows:

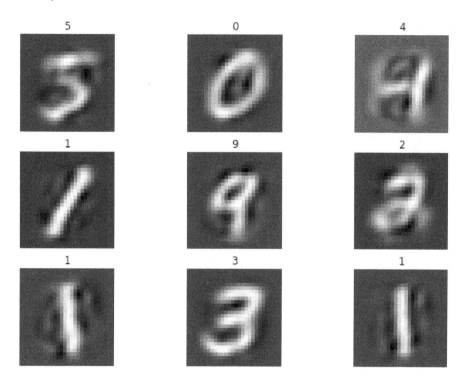

Figure 6.5: Visualizing the effect of reducing the dataset

Note that while we have lost some clarity in the images, for the most part, the numbers are still clearly visible due to the dimension reduction process. It is interesting to note, however, that the number four (4) seems to have been the most visually affected by this process. Perhaps much of the discarded information from the PCA process contained information specific to the samples of four (4).

5. Now, we will apply t-SNE to the PCA-transformed data to visualize the 30 components in a two-dimensional space. We can construct a t-SNE model in scikit-learn using the standard model API interface. We will start off by using the default values that specify that we are embedding the 30 dimensions into two for visualization using a perplexity of 30, a learning rate of 200, and 1,000 iterations. We will specify a **random_state** value of 0 and set **verbose** to 1:

```
model_tsne = TSNE(random_state=0, verbose=1)
model_tsne
```

The output is as follows:

```
TSNE(angle=0.5, early_exaggeration=12.0, init='random', learning_rate=200.0,
    method='barnes_hut', metric='euclidean', min_grad_norm=1e-07,
    n_components=2, n_iter=1000, n_iter_without_progress=300, n_jobs=None,
    perplexity=30.0, random_state=0, verbose=1)
```

Figure 6.6: Applying t-SNE to PCA-transformed data

In the preceding screenshot, we can see a number of configuration options that are available for the t-distributed SNE model, with some more important than the others. We will focus on the values of **learning_rate**, **n_components**, **n_iter**, **perplexity**, **random_state**, and **verbose**. For **learning_rate**, as we discussed previously, t-SNE uses stochastic gradient descent to project the high-dimensional data into a low-dimensional space. The learning rate controls the speed at which the process is executed. If the learning rate is too high, the model may fail to converge on a solution, and if it's too slow, it may take a very long time to reach it (if at all). A good rule of thumb is to start with the default; if you find the model producing NaNs (not-a-number values), you may need to reduce the learning rate. Once you are happy with the model, it is also wise to reduce the learning rate and let it run for longer (increase **n_iter**) as you may get a slightly better result. **n_components** is the number of dimensions in the embedding (or visualization space). More often than not, you would like a two-dimensional plot of the data, so you just need the default value of **2**. Now, **n_iter** is the maximum number of iterations of gradient descent. **perplexity** is the number of neighbors to use when visualizing the data.

Typically, a value between 5 and 50 will be appropriate, knowing that larger datasets typically require more perplexity than smaller ones. **random_state** is an important variable for any model or algorithm that initializes its values randomly at the start of training. The random number generators provided within computer hardware and software tools are not, in fact, truly random; they are actually pseudo-random number generators. They give a good approximation of randomness but are not truly random. Random numbers within computers start with a value known as a seed and are then produced in a complicated manner after that. By providing the same seed at the start of the process, the same "random numbers" are produced each time the process is run. While this sounds counter-intuitive, it is great for reproducing machine learning experiments as you won't see any difference in performance solely due to the initialization of the parameters at the start of training. This can provide more confidence that a change in performance is due to the considered change to the model or training; for example, the architecture of the neural network.

> **NOTE**
>
> Producing true random sequences is actually one of the hardest tasks to achieve with a computer. Computer software and hardware is designed so that the instructions that are provided are executed in exactly the same way each time they are run so that you get the same result. Random differences in execution, while being ideal for producing sequences of random numbers, would be a nightmare in terms of automating tasks and debugging problems.

verbose is the verbosity level of the model and describes the amount of information that's printed to the screen during the model fitting process. A value of 0 indicates no output, while 1 or greater indicates increasing levels of detail in the output.

6. Use t-SNE to transform the decomposed dataset of MNIST:

   ```
   mnist_tsne = model_tsne.fit_transform(mnist_30comp)
   ```

   The output is as follows:

```
[t-SNE] Computing 91 nearest neighbors...
[t-SNE] Indexed 10000 samples in 0.158s...
[t-SNE] Computed neighbors for 10000 samples in 13.757s...
[t-SNE] Computed conditional probabilities for sample 1000 / 10000
[t-SNE] Computed conditional probabilities for sample 2000 / 10000
[t-SNE] Computed conditional probabilities for sample 3000 / 10000
[t-SNE] Computed conditional probabilities for sample 4000 / 10000
[t-SNE] Computed conditional probabilities for sample 5000 / 10000
[t-SNE] Computed conditional probabilities for sample 6000 / 10000
[t-SNE] Computed conditional probabilities for sample 7000 / 10000
[t-SNE] Computed conditional probabilities for sample 8000 / 10000
[t-SNE] Computed conditional probabilities for sample 9000 / 10000
[t-SNE] Computed conditional probabilities for sample 10000 / 10000
[t-SNE] Mean sigma: 279.559349
[t-SNE] KL divergence after 250 iterations with early exaggeration: 85.301758
[t-SNE] KL divergence after 1000 iterations: 1.699996
```

Figure 6.7: Transforming the decomposed dataset

The output provided during the fitting process provides an insight into the calculations being completed by scikit-learn. We can see that it is indexing and computing neighbors for all the samples and is then determining the conditional probabilities of being neighbors for the data in batches of 10. At the end of the process, it provides a mean standard deviation value of **304.9988** with KL divergence after 250 and 1,000 iterations of gradient descent.

7. Now, visualize the number of dimensions in the returned dataset:

```
mnist_tsne.shape
```

The output is as follows:

```
10000,2
```

We have successfully reduced the 784 dimensions down to 2 for visualization, so what does it look like?

8. Create a scatter plot of the two-dimensional data produced by the model:

```
plt.figure(figsize=(10, 7))
plt.scatter(mnist_tsne[:,0], mnist_tsne[:,1], s=5)
plt.title('Low Dimensional Representation of MNIST');
```

The output is as follows:

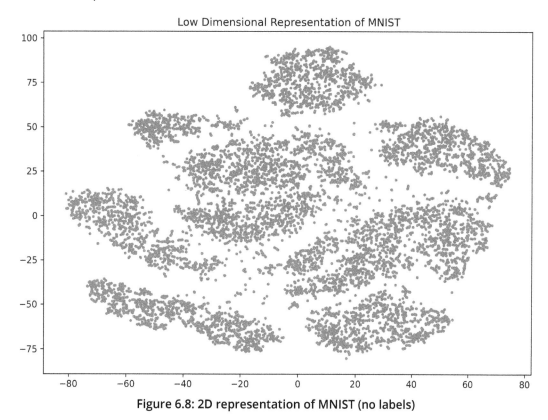

Figure 6.8: 2D representation of MNIST (no labels)

In the preceding plot, we can see that we have represented the MNIST data in two dimensions, but we can also see that it seems to be grouped together. There are a number of different clusters or clumps of data congregated together and separated from other clusters by some white space. There also seem to be about nine different groups of data. All these observations suggest that there is a relationship within and between the individual clusters.

9. Plot the two-dimensional data that's been grouped by the corresponding image labels and use markers to separate the individual labels.

> **NOTE**
>
> The marker parameter corresponds to the shapes that can be seen for individual points on the plot. They don't show up in detail on the plots in this chapter since there are many samples and hence the resolution is lost with the crops.

Along with the data, add the image labels to the plot to investigate the structure of the embedded data:

```python
MARKER = ['o', 'v', '1', 'p' ,'*', '+', 'x', 'd', '4', '.']
plt.figure(figsize=(10, 7))
plt.title('Low Dimensional Representation of MNIST');
for i in range(10):
    selections = mnist_tsne[mnist['labels'] == i]
    plt.scatter(selections[:,0], selections[:,1], alpha=0.2, \
                marker=MARKER[i], s=5);
    x, y = selections.mean(axis=0)
    plt.text(x, y, str(i), fontdict={'weight': 'bold', \
                                     'size': 30})
plt.show()
```

The output is as follows:

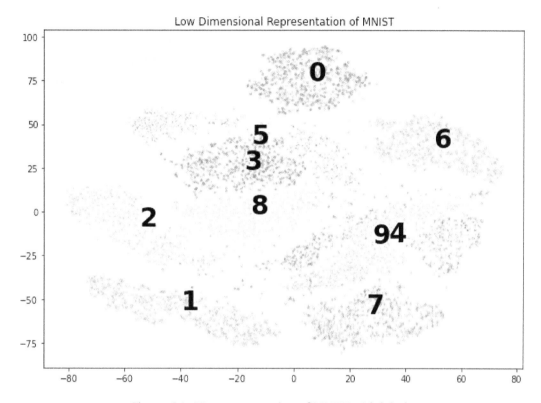

Figure 6.9: 2D representation of MNIST with labels

The preceding plot is very interesting. Here, we can see that the clusters correspond to each of the different image classes (zero through nine) within the dataset. In an unsupervised fashion, that is, without providing the labels in advance, a combination of PCA and t-SNE has been able to separate and group the individual classes within the MNIST dataset. What is particularly interesting is that there seems to be some confusion within the data regarding the number four images and the number nine images, as well as for the five and three images; the two clusters somewhat overlap. This makes sense if we look at the number nine and number four PCA images we extracted from *Step 4* of *Exercise 6.01, t-SNE MNIST*:

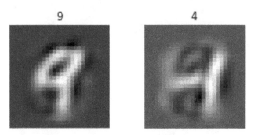

Figure 6.10: PCA images of nine

They do, in fact, look quite similar; perhaps this is due to the uncertainty in the shape of the number four. Looking at the image that follows, we can see from the four on the left-hand side that the two vertical lines almost join, while the four on the right-hand side has the two lines parallel:

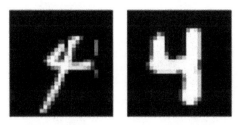

Figure 6.11: Shape of number four

The other interesting feature to note in *Figure 6.9* is the edge cases, which are shown in color in the Jupyter notebooks. Around the edges of each cluster, we can see that some samples would be misclassified in the traditional supervised learning sense but represent samples that may have more in common with other clusters than their own. Let's take a look at an example; there are a number of samples of the number three that are quite far from the correct cluster.

10. Get the index of all the number threes in the dataset:

    ```
    threes = np.where(mnist['labels'] == 3)[0]
    threes
    ```

    The output is as follows:

    ```
    array([   7,   10,   12, ..., 9974, 9977, 9991], dtype=int64)
    ```

11. Find the threes that were plotted with an **x** value of less than 0:

    ```
    tsne_threes = mnist_tsne[threes]
    far_threes = np.where(tsne_threes[:,0]< -30)[0]
    far_threes
    ```

    The output is as follows:

    ```
    array([   0,   11,   14,   17,   18,   19,   21,   22,   25,   29,   30,
             31,   32,   34,   35,   37,   39,   41,   42,   43,   51,   54,
             55,   56,   58,   60,   63,   66,   67,   68,   74,   76,   78,
             79,   80,   94,   96,   98,   99,  101,  102,  105,  107,  110,
            114,  116,  120,  122,  123,  126,  128,  133,  137,  142,  143,
            144,  145,  151,  152,  158,  169,  170,  171,  183,  184,  188,
            207,  227,  229,  230,  232,  235,  237,  238,  239,  240,  243,
            244,  245,  247,  263,  292,  294,  295,  303,  307,  308,  313,
            315,  319,  327,  335,  337,  346,  368,  370,  373,  379,  385,
            387,  388,  389,  398,  406,  410,  421,  422,  424,  427,  431,
            432,  433,  437,  438,  439,  457,  467,  493,  495,  497,  498,
            503,  505,  506,  507,  512,  517,  523,  554,  555,  557,  560,
            561,  563,  564,  567,  569,  571,  574,  579,  581,  583,  585,
            588,  593,  594,  595,  596,  598,  600,  625,  628,  648,  653,
            659,  662,  663,  666,  670,  677,  679,  689,  701,  708,  714,
            717,  721,  724,  726,  729,  734,  751,  776,  777,  781,  785,
            786,  787,  789,  791,  792,  797,  812,  816,  823,  825,  826,
            828,  829,  830,  832,  834,  836,  838,  842,  844,  847,  848,
            852,  853,  854,  857,  858,  864,  879,  881,  892,  893,  897,
            911,  936,  938,  955,  967,  973,  974, 1008, 1013, 1017, 1021,
           1023, 1024, 1028, 1030], dtype=int64)
    ```

    Figure 6.12: The threes with an x value less than zero

12. Display the coordinates to find one that is reasonably far from the three cluster:

    ```
    tsne_threes[far_threes]
    ```

    The output is as follows:

    ```
    array([[-32.119     ,  16.190784 ],
           [-33.198112  ,  26.013874 ],
           [-33.488453  ,  27.551619 ],
           [-34.523907  ,  26.634068 ],
           [-30.826197  ,  14.385183 ],
           [-37.32901   ,  15.137681 ],
           [-37.04024   ,  15.271422 ],
           [-38.383896  ,  15.578539 ],
           [-31.579344  ,  27.94891  ],
           [-33.46005   ,  27.690805 ],
    ```

    Figure 6.13: Coordinates away from the three cluster

13. Choose a sample with a reasonably high negative value as an **x** coordinate. In this example, we will select the second sample, which is sample **11**.

    > **NOTE**
    >
    > This index number has been arbitrarily chosen and is just used for example purposes. It does not have significance in and of itself.

    Display the image for the second sample as follows:

    ```
    plt.imshow(mnist['images'][11], cmap='gray')
    plt.axis('off');
    plt.show()
    ```

The output is as follows:

Figure 6.14: Image of sample 11

Looking at this sample image and the corresponding t-SNE coordinates, that is, approximately (-33, 26), it is not surprising that this sample lies near the cluster of eights and fives as there are quite a few features that are common to both of those numbers in this image. In this example, we applied a simplified SNE, demonstrating some of its efficiencies as well as possible sources of confusion and the output of unsupervised learning.

> **NOTE**
>
> To access the source code for this specific section, please refer to https://packt.live/3iDsCNf
>
> You can also run this example online at https://packt.live/3gBdrSK

ACTIVITY 6.01: WINE T-SNE

In this activity, we will reinforce our knowledge of t-SNE using the Wine dataset. By completing this activity, you will be able to build-SNE models for your own custom applications. The Wine dataset (https://archive.ics.uci.edu/ml/datasets/Wine) is a collection of attributes regarding the chemical analysis of wine from Italy from three different producers, but the same type of wine for each producer. This information could be used as an example to verify the validity of a bottle of wine made from the grapes from a specific region in Italy. The 13 attributes are Alcohol, Malic acid, Ash, Alkalinity of ash, Magnesium, Total phenols, Flavanoids, Nonflavanoid phenols, Proanthocyanins, Color intensity, Hue, OD280/OD315 of diluted wines, and Proline.

Each sample contains a class identifier (1 – 3).

> **NOTE**
>
> This dataset is sourced from https://archive.ics.uci.edu/ml/machine-learning-databases/wine/ (UCI Machine Learning Repository [http://archive.ics.uci.edu/ml]. Irvine, CA: University of California, School of Information and Computer Science). It can also be downloaded from https://packt.live/3e1JOcY.

These steps will help you complete this activity:

1. Import **pandas**, **numpy**, and **matplotlib**, as well as the **t-SNE** and **PCA** models from scikit-learn.

2. Load the Wine dataset using the **wine.data** file included in the accompanying source code and display the first five rows of data.

> **NOTE**
>
> You can delete columns within pandas DataFrames by using the **del** keyword. Simply pass **del** the DataFrame and the selected column within the square root.

3. The first column contains the labels; extract this column and remove it from the dataset.

4. Execute PCA to reduce the dataset to the first six components.

5. Determine the amount of variance within the data described by these six components.

6. Create a t-SNE model using a specified random state and a **verbose** value of 1.

7. Fit the PCA data to the t-SNE model.

8. Confirm that the shape of the t-SNE fitted data is two-dimensional.

9. Create a scatter plot of the two-dimensional data.

10. Create a secondary scatter plot of the two-dimensional data with the class labels applied to visualize any clustering that may be present.

By the end of this activity, you will have constructed a t-SNE visualization of the Wine dataset using its six components and identified some relationships in the location of the data within the plot. The final plot will look similar to the following:

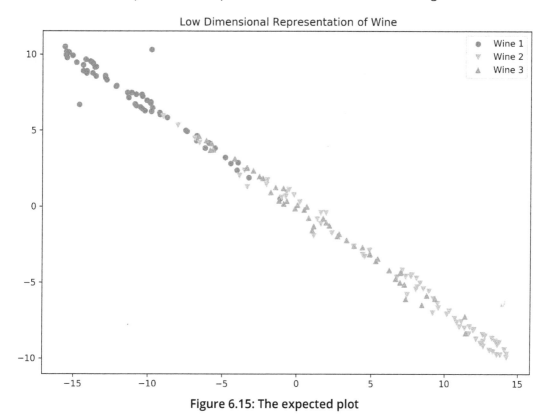

Figure 6.15: The expected plot

> **NOTE**
> The solution to this activity can be found on page 460.

INTERPRETING T-SNE PLOTS

Now that we are able to use t-distributed SNE to visualize high-dimensional data, it is important to understand the limitations of such plots and what aspects are important in interpreting and generating them. In this section, we will highlight some of the important features of t-SNE and demonstrate how care should be taken when using this visualization technique.

PERPLEXITY

As we described in the introduction to t-SNE, the perplexity values specify the number of nearest neighbors to be used when computing the conditional probability. The selection of this value can make a significant difference to the end result; with a low value of perplexity, local variations in the data dominate because a small number of samples are used in the calculation. Conversely, a large value of perplexity considers more global variations as many more samples are used in the calculation. Typically, it is worth trying a range of different values to investigate the effect of perplexity. Again, values between 5 and 50 tend to work quite well.

EXERCISE 6.02: T-SNE MNIST AND PERPLEXITY

In this exercise, we will try a range of different values for perplexity and look at the effect in the visualization plot:

1. Import **pickle**, **numpy**, and **matplotlib**, as well as **PCA** and **t-SNE** from scikit-learn:

```
import pickle
import numpy as np
import matplotlib.pyplot as plt
from sklearn.decomposition import PCA
from sklearn.manifold import TSNE
np.random.seed(2)
```

2. Load the MNIST dataset.

> **NOTE**
>
> You can find the **mnist.pkl** file at https://packt.live/2wpnWHs.

The code is as follows:

```
with open('mnist.pkl', 'rb') as f:
    mnist = pickle.load(f)
```

3. Using PCA, select only the first 30 components of variance from the image data:

```
model_pca = PCA(n_components=30)
mnist_pca = model_pca.fit_transform\
            (mnist['images'].reshape((-1, 28 ** 2)))
```

4. In this exercise, we are investigating the effect of perplexity on the t-SNE manifold. Iterate through a model/plot loop with a perplexity of 3, 30, and 300:

    ```
    MARKER = ['o', 'v', '1', 'p' ,'*', '+', 'x', 'd', '4', '.']
    for perp in [3, 30, 300]:
        model_tsne = TSNE(random_state=0, verbose=1, perplexity=perp)
        mnist_tsne = model_tsne.fit_transform(mnist_pca)
        plt.figure(figsize=(10, 7))
        plt.title(f'Low Dimensional Representation of MNIST \
    (perplexity = {perp})')
        for i in range(10):
            selections = mnist_tsne[mnist['labels'] == i]
            plt.scatter(selections[:,0], selections[:,1],\
                    alpha=0.2, marker=MARKER[i], s=5)
            x, y = selections.mean(axis=0)
            plt.text(x, y, str(i), \
                    fontdict={'weight': 'bold', 'size': 30})
    plt.show()
    ```

    The output is as follows:

```
[t-SNE] Computing 10 nearest neighbors...
[t-SNE] Indexed 10000 samples in 0.270s...
[t-SNE] Computed neighbors for 10000 samples in 10.503s...
[t-SNE] Computed conditional probabilities for sample 1000 / 10000
[t-SNE] Computed conditional probabilities for sample 2000 / 10000
[t-SNE] Computed conditional probabilities for sample 3000 / 10000
[t-SNE] Computed conditional probabilities for sample 4000 / 10000
[t-SNE] Computed conditional probabilities for sample 5000 / 10000
[t-SNE] Computed conditional probabilities for sample 6000 / 10000
[t-SNE] Computed conditional probabilities for sample 7000 / 10000
[t-SNE] Computed conditional probabilities for sample 8000 / 10000
[t-SNE] Computed conditional probabilities for sample 9000 / 10000
[t-SNE] Computed conditional probabilities for sample 10000 / 10000
[t-SNE] Mean sigma: 164.678925
[t-SNE] KL divergence after 250 iterations with early exaggeration: 96.849312
[t-SNE] KL divergence after 1000 iterations: 1.855248
```

Figure 6.16: Iterating through a model

NOTE

The preceding output has been truncated for presentation purposes. Standard outputs like this would typically be much longer. However, it has been included as it is important to keep an eye on such outputs while the model is training.

Note the KL divergence in each of the three different perplexity values, along with the increase in the average standard deviation (variance). By looking at the following t-SNE plots with class labels, we can see that with a low perplexity value, the clusters are nicely contained with relatively few overlaps. However, there is almost no space between the clusters. As we increase the perplexity, the space between the clusters improves with reasonably clear distinctions at a perplexity of 30. As the perplexity increases to 300, we can see that the clusters of eight and five, along with nine, four, and seven, are starting to converge.

Let's start with a low perplexity value:

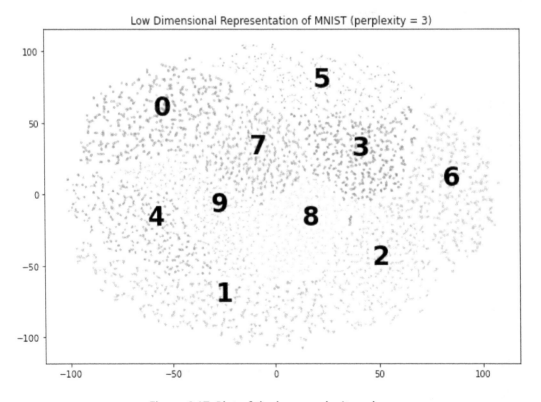

Figure 6.17: Plot of the low perplexity value

Interpreting t-SNE Plots | 233

> **NOTE**
>
> The plotting function in *Step 4* would result in this plot. The subsequent outputs are the plots for varying values of perplexity.

Increasing the perplexity by a factor of 10 shows much clearer clusters:

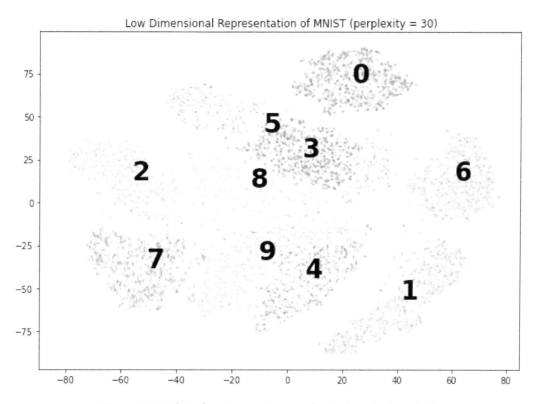

Figure 6.18: Plot after increasing perplexity by a factor of 10

234 | t-Distributed Stochastic Neighbor Embedding

By increasing the perplexity to 300, we start to merge more of the labels together:

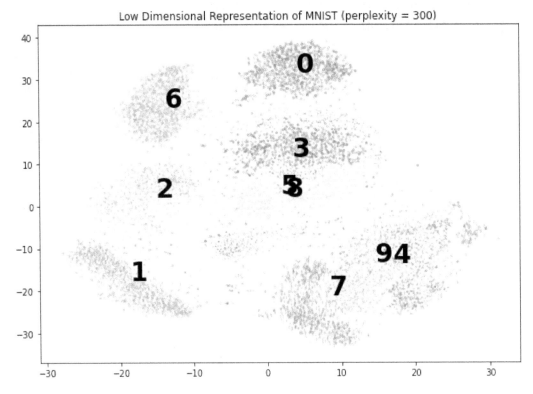

Figure 6.19: Increasing the perplexity value to 300

In this exercise, we developed our understanding of the effect of perplexity and the sensitivity of this value to the overall result. A small perplexity value can lead to a more homogenous mix of locations with very little space between them. Increasing the perplexity separates the clusters more effectively, but an excessive value leads to overlapping clusters.

> **NOTE**
>
> To access the source code for this specific section, please refer to https://packt.live/3gl0zdp.
>
> You can also run this example online at https://packt.live/3gDcjxR

ACTIVITY 6.02: T-SNE WINE AND PERPLEXITY

In this activity, we will use the Wine dataset to further reinforce the influence of perplexity on the t-SNE visualization process. In this activity, we will try to determine whether we can identify the source of the wine based on its chemical composition. The t-SNE process provides an effective means of representing and possibly identifying the sources.

> **NOTE**
>
> This dataset is sourced from https://archive.ics.uci.edu/ml/machine-learning-databases/wine/ (UCI Machine Learning Repository [http://archive.ics.uci.edu/ml]. Irvine, CA: University of California, School of Information and Computer Science). It can be downloaded from https://packt.live/3aPOmRJ.

1. Import **pandas**, **numpy**, and **matplotlib**, as well as the **t-SNE** and **PCA** models from scikit-learn.

2. Load the Wine dataset and inspect the first five rows.

3. The first column provides the labels; extract these from the DataFrame and store them in a separate variable. Ensure that the column is removed from the DataFrame.

4. Execute PCA on the dataset and extract the first six components.

5. Construct a loop that iterates through the perplexity values (1, 5, 20, 30, 80, 160, 320). For each loop, generate a t-SNE model with the corresponding perplexity and print a scatter plot of the labeled wine classes. Note the effect of different perplexity values.

236 | t-Distributed Stochastic Neighbor Embedding

By the end of this activity, you will have generated a two-dimensional representation of the Wine dataset and inspected the resulting plot for clusters or groupings of data. The plot for perplexity value 320 looks as follows:

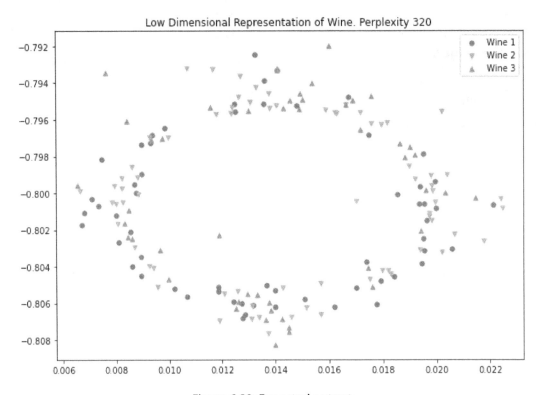

Figure 6.20: Expected output

> **NOTE**
>
> The solution to this activity can be found on page 464.

ITERATIONS

The final parameter we will experiment with is the number of iterations, which, as per our investigation of autoencoders, is simply the number of training epochs to apply to gradient descent. Thankfully, the number of iterations is a reasonably simple parameter to adjust and often requires only a certain amount of patience as the position of the points in the low-dimensional space stabilize in their final locations.

EXERCISE 6.03: T-SNE MNIST AND ITERATIONS

In this exercise, we will look at the influence of a range of different iteration parameters that have been applied to the t-SNE model and highlight some indicators that perhaps more training is required. Again, the value of these parameters is highly dependent on the dataset and the volume of data that's available for training. We will use MNIST in this example:

1. Import **pickle**, **numpy**, and **matplotlib**, as well as **PCA** and **t-SNE** from scikit-learn:

```
import pickle
import numpy as np
import matplotlib.pyplot as plt
from sklearn.decomposition import PCA
from sklearn.manifold import TSNE
np.random.seed(2)
```

2. Load the MNIST dataset:

> **NOTE**
>
> You can find the **mnist.pkl** file at https://packt.live/2xXRJao.

The code is as follows:

```
with open('mnist.pkl', 'rb') as f:
    mnist = pickle.load(f)
```

3. Using PCA, select only the first 30 components of variance from the image data:

```
model_pca = PCA(n_components=30)
mnist_pca = model_pca.fit_transform(mnist['images']\
                                    .reshape((-1, 28 ** 2)))
```

4. In this exercise, we are investigating the effect of iterations on the t-SNE manifold. Iterate through a model/plot loop with iteration and iterate with the progress values **250**, **500**, and **1000**:

```
MARKER = ['o', 'v', '1', 'p' ,'*', '+', 'x', 'd', '4', '.']
for iterations in [250, 500, 1000]:
    model_tsne = TSNE(random_state=0, verbose=1, \
                    n_iter=iterations, \
                    n_iter_without_progress=iterations)
    mnist_tsne = model_tsne.fit_transform(mnist_pca)
```

The output is as follows:

```
[t-SNE] Computing 91 nearest neighbors...
[t-SNE] Indexed 10000 samples in 0.156s...
[t-SNE] Computed neighbors for 10000 samples in 16.714s...
[t-SNE] Computed conditional probabilities for sample 1000 / 10000
[t-SNE] Computed conditional probabilities for sample 2000 / 10000
[t-SNE] Computed conditional probabilities for sample 3000 / 10000
[t-SNE] Computed conditional probabilities for sample 4000 / 10000
[t-SNE] Computed conditional probabilities for sample 5000 / 10000
[t-SNE] Computed conditional probabilities for sample 6000 / 10000
[t-SNE] Computed conditional probabilities for sample 7000 / 10000
[t-SNE] Computed conditional probabilities for sample 8000 / 10000
[t-SNE] Computed conditional probabilities for sample 9000 / 10000
[t-SNE] Computed conditional probabilities for sample 10000 / 10000
[t-SNE] Mean sigma: 276.467915
[t-SNE] KL divergence after 250 iterations with early exaggeration: 85.662193
```

Figure 6.21: Iterating through a model

5. Plot the results:

```
        plt.figure(figsize=(10, 7))
        plt.title(f'Low Dimensional Representation of MNIST \
(iterations = {iterations})')
        for i in range(10):
            selections = mnist_tsne[mnist['labels'] == i]
            plt.scatter(selections[:,0], selections[:,1], \
                        alpha=0.2, marker=MARKER[i], s=5);
            x, y = selections.mean(axis=0)
            plt.text(x, y, str(i), fontdict={'weight': 'bold', \
                                            'size': 30})
    plt.show()
```

A reduced number of iterations limits the extent to which the algorithm can find relevant neighbors, leading to ill-defined clusters:

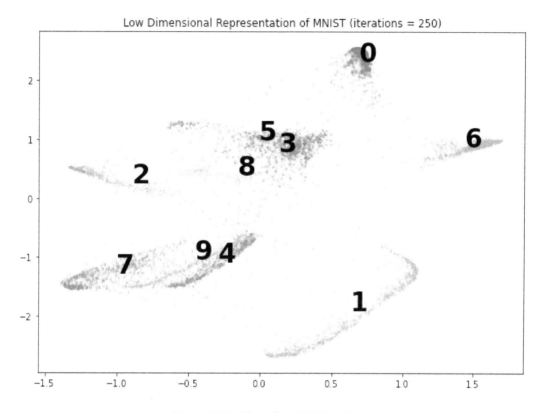

Figure 6.22: Plot after 250 iterations

240 | t-Distributed Stochastic Neighbor Embedding

Increasing the number of iterations provides the algorithm with enough time to adequately project the data:

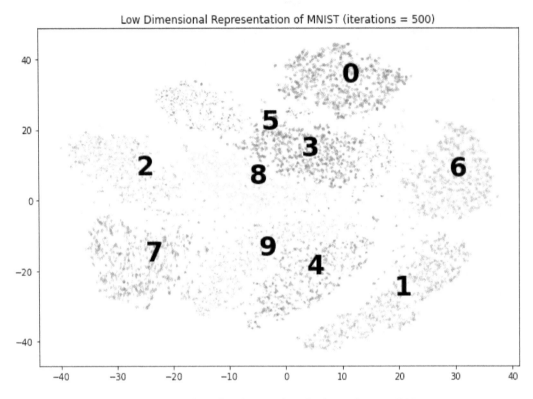

Figure 6.23: Plot after increasing the iterations to 500

Interpreting t-SNE Plots | 241

Once the clusters have settled, increased iterations have an extremely small effect and essentially lead to increased training time:

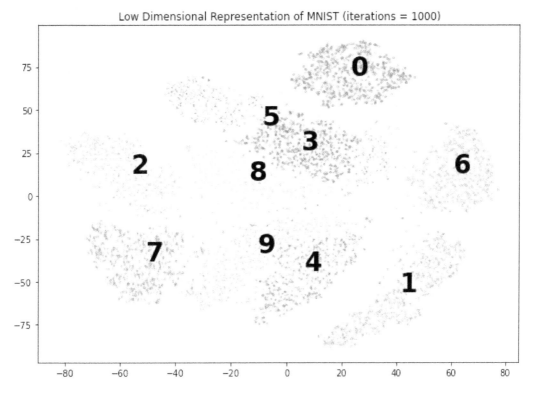

Figure 6.24: Plot after 1,000 iterations

Looking at the previous plots, we can see that the cluster positions with iteration values of 500 and 1,000 are stable and relatively unchanged between the plots. The most interesting plot is that of an iteration value of 250, where it seems as though the clusters are still in a process of motion, making their way to the final positions. As such, there is sufficient evidence to suggest that an iteration value of 500 is sufficient.

> **NOTE**
>
> To access the source code for this specific section, please refer to https://packt.live/2Zaw1uZ
>
> You can also run this example online at https://packt.live/3gCOiHf

ACTIVITY 6.03: T-SNE WINE AND ITERATIONS

In this activity, we will investigate the effect of the number of iterations on the visualization of the Wine dataset. This is a process that's commonly used during the exploration phase of data processing, cleaning, and understanding the relationships in the data. Depending on the dataset and the type of analysis, we may need to try a number of different iterations, such as the ones we will look at in this activity.

As we mentioned previously, this process is extremely helpful for projecting high-dimensional data down to a lower, more understandable number of dimensions. In this case, our dataset has 13 features; however, in the real world, you can have datasets with hundreds or even thousands of features. A common instance of this would be person-level data, which can have any number of demographic- or action-related features, which can make regular off-the-shelf analysis impossible. t-SNE is a helpful tool for working high-dimensional data into a more intuitive state.

> **NOTE**
>
> This dataset is sourced from https://archive.ics.uci.edu/ml/machine-learning-databases/wine/ (UCI Machine Learning Repository [http://archive.ics.uci.edu/ml]. Irvine, CA: University of California, School of Information and Computer Science). It can be downloaded from https://packt.live/2xXgHXo.

These steps will help you complete this activity:

1. Import **pandas**, **numpy**, and **matplotlib**, as well as the **t-SNE** and **PCA** models from scikit-learn.

2. Load the Wine dataset and inspect the first five rows.

3. The first column provides the labels; extract these from the DataFrame and store them in a separate variable. Ensure that the column is removed from the DataFrame.

4. Execute PCA on the dataset and extract the first six components.

5. Construct a loop that iterates through the iteration values (**250**, **500**, **1000**). For each loop, generate a t-SNE model with the corresponding number of iterations and an identical number of iterations without progress values.

6. Construct a scatter plot of the labeled wine classes. Note the effect of different iteration values.

By completing this activity, we will have seen the effect of modifying the iteration parameter of the model. This is an important parameter in ensuring that the data has settled into a somewhat final position in the low-dimensional space.

> **NOTE**
> The solution to this activity can be found on page 473.

FINAL THOUGHTS ON VISUALIZATIONS

As we conclude this chapter, there are a couple of important aspects to note regarding visualizations. The first is that the size of the clusters or the relative space between clusters may not actually provide any real indication of proximity. As we discussed earlier in this chapter, a combination of Gaussian and Student's t-distributions is used with SNE to represent high-dimensional data in a low-dimensional space. As such, there is no guarantee of a linear relationship in distance since t-SNE balances the positions of localized and global data structures. The actual distance between the points in local structures may be visually very close within the representation, but still might be some distance away in the high-dimensional space.

This property also has additional consequences in that, sometimes, random data can appear as if it has some structure, and that it is often required to generate multiple visualizations using different values of perplexity, learning rate, number of iterations, and random seed values.

SUMMARY

In this chapter, we were introduced to t-distributed SNEs as a means of visualizing high-dimensional information that may have been produced from prior processes, such as PCA or autoencoders. We discussed the means by which t-SNEs produce this representation and generated a number of them using the MNIST and Wine datasets and scikit-learn. In this chapter, we were able to look at some of the power of unsupervised learning because PCA and t-SNE were able to cluster the classes of each image without knowing the ground truth result. In the next chapter, we will build on this practical experience by looking into applications of unsupervised learning, including basket analysis and topic modeling.

7
TOPIC MODELING

OVERVIEW

In this chapter, we will perform basic cleaning techniques for textual data and then model the cleaned data to derive relevant topics. You will evaluate **Latent Dirichlet Allocation** (**LDA**) models and execute **non-negative matrix factorization** (**NMF**) models. Finally, you will interpret the results of topic models and identify the best topic model for the given scenario. We will see how topic modeling provides insights into the underlying structure of documents. By the end of this chapter, you will be able to build fully functioning topic models to derive value and insights for your business.

246 | Topic Modeling

INTRODUCTION

In the last chapter, the discussion focused on preparing data for modeling using dimensionality reduction and autoencoding. Large feature sets can be problematic when it comes to modeling because of multicollinearity and extensive computation and can thereby hinder real-time prediction. Dimensionality reduction using principal component analysis is one antidote to that problem. Similarly, autoencoders seek to find optimal feature encodings. You can think of autoencoders as a means of identifying quality interaction terms for the dataset. Let's now move past dimensionality reduction and look at some real-world modeling techniques.

Topic modeling is one facet of **Natural Language Processing** (**NLP**), the field of computer science exploring the syntactic and semantic analysis of natural language, which has been increasing in popularity with the increased availability of textual datasets. NLP can deal with language in almost any form, including text, speech, and images. Besides topic modeling, sentiment analysis, entity recognition, and object character recognition are noteworthy NLP applications.

Nowadays, the data being collected and analyzed comes less frequently in standard tabular forms and more often in less structured forms, such as documents, images, and audio files. As such, successful data science practitioners need to be fluent in the methodologies used to handle these diverse datasets.

Here is a demonstration of identifying words in a text and assigning them to topics:

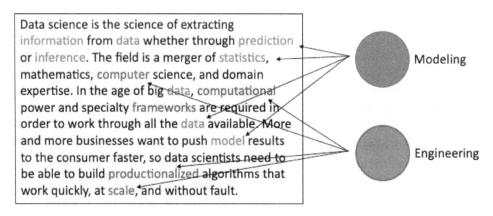

Figure 7.1: Example of identifying words in a text and assigning them to topics

Your immediate question is probably *what are topics?* Let's answer that question with an example. You could imagine, or perhaps have noticed, that on days when major events take place (such as national elections, natural disasters, or sports championships), the posts on social media websites tend to focus on those events. Posts generally reflect, in some way, the day's events, and they will do so in varying ways. Posts can, and will, have a number of divergent viewpoints that can be clustered into high-level topics. If we had tweets about the World Cup final, the topics of those tweets could cover divergent viewpoints, ranging from the quality of the refereeing to fan behavior. In the United States, the president delivers an annual speech in mid to late January called the State of the Union. With sufficient numbers of social media posts, we would be able to infer or predict high-level reactions (topics) to the speech from the social media community by grouping posts using the specific keywords contained in them. Topic models are important because they serve the same role for textual data that classic summary statistics serve for numeric data. That is, they provide a meaningful summarization of data. Let's return to the State of the Union example. The quick look here would be ascertaining the major points of the speech that either resonate with or miss the viewership.

TOPIC MODELS

Topic models fall into the unsupervised learning bucket because, almost always, the topics being identified are not known in advance. So, no target exists on which we can perform regression or classification modeling. In terms of unsupervised learning, topic models most resemble clustering algorithms, specifically k-means clustering. You'll recall that, in k-means clustering, the number of clusters is established first, and then the model assigns each data point to one of the predetermined number of clusters. The same is generally true of topic models. We select the number of topics at the start, and then the model isolates the words that form that number of topics. This is a great jumping-off point for a high-level topic modeling overview.

Before that, let's check that the correct environment and libraries are installed and ready for use. The following table lists the required libraries and their main purposes:

Library	Use
`langdetect`	Used to detect the language of any text
`matplotlib.pyplot`	Used to do basic plotting
`nltk`	Used to do a variety of NLP tasks
`numpy`	Used to work with arrays and matrices
`pandas`	Used to work with DataFrames
`pyLDAvis`	Used to visualize the results of latent Dirichlet allocation models
`pyLDAvis.sklearn`	Used to run pyLDAvis with sklearn models
`regex`	Used to write and execute regular expressions
`sklearn`	Used to build machine learning models

Figure 7.2: Table showing different libraries and their use

If any or all of these libraries are not currently installed, install the required packages via the command line using **pip**; for example, **pip install langdetect**.

Step 3 of the forthcoming exercise covers the installation of word dictionaries from the `nltk` package. Word dictionaries are simply collections of words that are curated for a specific use. The stop words word dictionary, installed below, contains the common words in the English language that do not clarify context, meaning, or intention. These common words could include *the*, *an*, *a*, and *in*. The word net word dictionary provides word mappings that help in the lemmatization process – explained below. The word mappings link words such as *run*, *running*, and *ran* together as all essentially meaning the same thing. At a high level, word dictionaries provide data scientists with a means of preparing text data for analysis without having an in-depth knowledge of linguistics or spending an enormous amount of time defining word lists or word mappings.

> **NOTE**
>
> In the exercises and activities below, the results can differ slightly from what is shown because of the optimization algorithms that support both Latent Dirichlet Allocation and Non-negative Matrix Factorization. Many of the functions do not have a seed setting capability.

EXERCISE 7.01: SETTING UP THE ENVIRONMENT

To check whether the environment is ready for topic modeling, we will perform several steps. The first of these involves loading all the libraries that will be needed in this chapter:

1. Open a new Jupyter notebook.

2. Import the requisite libraries:

   ```
   import langdetect
   import matplotlib.pyplot
   import nltk
   import numpy
   import pandas
   import pyLDAvis
   import pyLDAvis.sklearn
   import regex
   import sklearn
   ```

   Note that not all of these packages are used for cleaning the data; some of them are used in the actual modeling. But it is useful to import all of the required libraries at once, so let's take care of all library importing now.

   Libraries not yet installed will return the following error:

   ```
   ModuleNotFoundError                       Traceback (most recent call last)
   <ipython-input-3-a62286ae48f9> in <module>
         4 import numpy
         5 import pandas
   ----> 6 import pyLDAvis
         7 import pyLDAvis.sklearn
         8 import regex

   ModuleNotFoundError: No module named 'pyLDAvis'
   ```

   Figure 7.3: Library not installed error

   If this error is returned, install the relevant libraries via the command line as previously discussed. Once successfully installed, rerun the library import process using **import**.

3. Certain textual data cleaning and preprocessing processes require word dictionaries. Here, we'll install two of these dictionaries. If the **nltk** library is imported, execute the following code:

```
nltk.download('wordnet')
nltk.download('stopwords')
```

The output is as follows:

```
[nltk_data] Downloading package wordnet to
[nltk_data]     C:\Users\rutujay\AppData\Roaming\nltk_data...
[nltk_data]   Unzipping corpora\wordnet.zip.
[nltk_data] Downloading package stopwords to
[nltk_data]     C:\Users\rutujay\AppData\Roaming\nltk_data...
[nltk_data]   Unzipping corpora\stopwords.zip.
```

True

Figure 7.4: Importing libraries and downloading dictionaries

4. Run **matplotlib** and specify inline so that the plots print inside the notebook:

```
%matplotlib inline
```

The notebook and environment are now set and ready for data loading.

> **NOTE**
>
> To access the source code for this specific section, please refer to https://packt.live/34gLGKa.
>
> You can also run this example online at https://packt.live/3fbWQES.
>
> You must execute the entire Notebook in order to get the desired result.

A HIGH-LEVEL OVERVIEW OF TOPIC MODELS

When it comes to analyzing large volumes of potentially related text data, topic models are one go-to approach. By 'related', we mean that the documents describe similar topics. To run any topic model, the only data required are the documents themselves. No additional data (meta or otherwise) is required.

In the simplest terms, topic models identify the abstract topics (also known as themes) in a collection of documents (referred to as a **corpus**), using the words contained in the documents. That is, if a sentence contains the words *salary*, *employee*, and *meeting*, it would be safe to assume that that sentence is about, or that its topic is, *work*. It is of note that the documents making up the corpus need not be documents as traditionally defined – think letters or contracts. A document could be anything containing text, including tweets, news headlines, or transcribed speech.

Topic models assume that words in the same document are related and use that assumption to define abstract topics by finding groups of words that repeatedly appear in close proximity. In this way, these models are classic pattern recognition algorithms in which the detected patterns are made up of words. The general topic modeling algorithm has four main steps:

1. Determine the number of topics.
2. Scan the documents and identify co-occurring words or phrases.
3. Auto-learn groups (or clusters) of words characterizing the documents.
4. Output abstract topics characterizing the corpus as word groupings.

As *Step 1* notes, the number of topics needs to be selected before fitting the model. Selecting an appropriate number of topics can be tricky, but, as is the case with most machine learning models, this parameter can be optimized by fitting several models using different numbers of topics and selecting the best model based on a performance metric. We'll dive into this process again later.

The following is the generic topic modeling workflow:

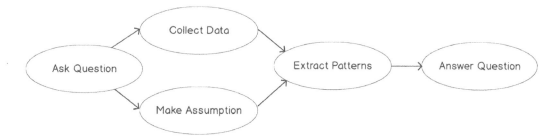

Figure 7.5: The generic topic modeling workflow

It is important to optimize the number of topics parameter, as this parameter can majorly impact topic coherence. This is because the model finds groups of words that best fit the corpus under the constraint of a predefined number of topics. If the number of topics is too high, the topics become inappropriately narrow. Overly specific topics are referred to as **over-cooked**. Likewise, if the number of topics is too low, the topics become generic and vague. These types of topics are considered **under-cooked**. Over-cooked and under-cooked topics can sometimes be fixed by decreasing or increasing the number of topics, respectively. In practice, a frequent and unavoidable result of topic models is that, frequently, at least one topic will be problematic.

A key aspect of topic models is that they do not produce specific one-word or one-phrase topics, but rather collections of words, each of which represents an abstract topic. Recall the imaginary sentence about *work* from before. The topic model built to identify the topics of some hypothetical corpus to which that sentence belongs would not return the word *work* as a topic. It would instead return a collection of words, such as *paycheck, employee,* and *boss*—words that describe the topic and from which the one-word or one-phrase topic could be inferred. This is because topic models understand word **proximity**, not context. The model has no idea what *paycheck, employee,* and *boss* mean; it only knows that these words, generally, whenever they appear, appear in close proximity to one another:

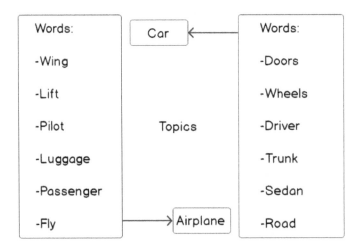

Figure 7.6: Inferring topics from word groupings

Topic models can be used to predict the topic(s) belonging to unseen documents, but if you are going to make predictions, it is important to recognize that topic models only know the words used to train them. That is, if the unseen documents have words that were not in the training data, the model will not be able to process those words even if they link to one of the topics identified in the training data. Because of this fact, topic models tend to be used more for exploratory analysis and inference than for prediction.

Each topic model outputs two matrices. The first matrix contains words against topics. It lists each word related to each topic with some quantification of the relationship. Given the number of words being considered by the model, each topic is only going to be described by a relatively small number of words.

Words can either be assigned to one topic or to multiple topics with differing quantifications. Whether words are assigned to one or multiple topics depends on the algorithm. Similarly, the second matrix contains documents against topics. It maps each document to each topic by some quantification of the relationship of each document topic combination.

When discussing topic modeling, it is important to continually reinforce the fact that the word groups representing topics are not related conceptually; they are related only by proximity. The frequent proximity of certain words in the documents is enough to define topics because of an assumption stated previously—that all words in the same document are related.

However, this assumption may either not be true or the words may be too generic to form coherent topics. Interpreting abstract topics involves balancing the innate characteristics of text data with the generated word groupings. Text data, and language in general, is highly variable, complex, and contextual, which means any generalized result needs to be consumed cautiously.

This is not to downplay or invalidate the results of the model. Given thoroughly cleaned documents and an appropriate number of topics, word groupings, as we will see, can be a good guide as to what is contained in a corpus and can effectively be incorporated into larger data systems.

We have discussed some of the limitations of topic models already, but there are some additional points to consider. The noisy nature of text data can lead topic models to assign words unrelated to one of the topics to that topic.

Again, consider the sentence about *work* from before. The word *meeting* could appear in the word grouping representing the topic *work*. It is also possible that the word *long* could be in that group, but the word *long* is not directly related to *work*. *Long* may be in the group because it frequently appears in close proximity to the word *meeting*. Therefore, *long* would probably be considered to be falsely (or spuriously) correlated to *work* and should probably be removed from the topic grouping, if possible. Spuriously correlated words in word groupings can cause significant problems when analyzing the data.

This is not necessarily a flaw in the model. It is, instead, a characteristic that, given noisy data, the model could extract quirks from the data that might negatively impact the results. Spurious correlations could be the result of how, where, or when the data was collected. If the documents were collected only in some specific geographic region, words associated with that region could be incorrectly, albeit accidentally, linked to one or many of the word groupings output from the model.

Note that, with additional words in the word group, we could be attaching more documents to that topic than should be attached. It should be straightforward that, if we shrink the number of words belonging to a topic, then that topic will be assigned to fewer documents. Keep in mind that this is not a bad thing. We want each word grouping to contain only words that make sense so that we assign the appropriate topics to the appropriate documents.

There are many topic modeling algorithms, but perhaps the two best known are **Latent Dirichlet Allocation (LDA)** and **Non-Negative Matrix Factorization (NMF)**. We will discuss both in detail later on.

BUSINESS APPLICATIONS

Despite its limitations, topic modeling can provide actionable insights that drive business value if used correctly and in the appropriate context. Let's now review some of the biggest applications of topic models.

One of the use cases is exploratory data analysis on new text data where the underlying structure of the dataset is unknown. This is the equivalent to plotting and computing summary statistics for an unseen dataset featuring numeric and categorical variables whose characteristics need to be understood before more sophisticated analyses can be reasonably performed. With the results of topic modeling, the usability of this dataset in future modeling exercises is ascertainable. For example, if the topic model returns clear and distinct topics, then that dataset would be a great candidate for further clustering-type analyses.

Determining topics creates an additional variable that can be used to sort, categorize, and/or chunk data. If our topic model returns cars, farming, and electronics as abstract topics, we could filter our large text dataset down to just the documents with farming as a topic. Once filtered, we could perform further analyses, including sentiment analysis, another round of topic modeling, or any other analysis we could think up. Beyond defining the topics present in a corpus, topic modeling returns a lot of other information indirectly that could be used to further break a large dataset down and understand its characteristics.

Among those characteristics is topic prevalence. Think about performing an analysis on an open response survey that is designed to gauge the response to a product. We could imagine the topic model returning topics in the form of sentiment. One group of words might be *good, excellent, recommend,* and *quality*, while the other might be *garbage, broken, poor,* and *disappointing*.

Given this style of survey, the topics themselves may not be that surprising, but what would be interesting is that we could count the number of documents containing each topic and glean useful insights from it. From the counts, we could say things like x-percent of the survey respondents had a positive reaction to the product, while only y-percent of the respondents had a negative reaction. Essentially, what we would have created is a rough version of a sentiment analysis.

Currently, the most frequent use of a topic model is as a component of a recommendation engine. The emphasis today is on personalization—delivering products to consumers that are specifically designed and curated for those individuals. Take websites, news or otherwise, devoted to the propagation of articles. Companies such as Yahoo and Medium need customers to keep reading in order to stay in business, and one way to keep customers reading is to feed them articles that they would be more inclined to read. This is where topic modeling comes in. Using a corpus made up of articles previously read by an individual, a topic model would essentially tell us what types of articles said subscriber likes to read. The company could then go to its inventory and find articles with similar topics and send them to the individual via their account page or email. This is custom curation to facilitate simplicity and ease of use while also maintaining engagement.

Before we get into prepping data for our model, let's quickly load and explore the data.

EXERCISE 7.02: DATA LOADING

In this exercise, we will load the data and format it. We will execute this exercise in the same notebook that we executed in *Exercise 7.01, Setting up the Environment*. It is incredibly important to understand as thoroughly as possible the dataset with which we are going to work. That process of understanding starts with knowing what the data looks like at a high level, how big the data is, what columns are present, and identifying what aspects of the dataset might be helpful in solving the problem we've been tasked with solving. We answer these basic questions below.

> **NOTE**
>
> This data is downloaded from https://archive.ics.uci.edu/ml/datasets/News+Popularity+in+Multiple+Social+Media+Platforms (UCI Machine Learning Repository [http://archive.ics.uci.edu/ml]. Irvine, CA: University of California, School of Information and Computer Science).
>
> Citation: Nuno Moniz and Luís Torgo. "Multi-Source Social Feedback of Online News Feeds".CoRR [arXiv:1801.07055 [cs.SI]] (2018).
>
> The dataset can also be downloaded from https://packt.live/2Xin2HC.

This is the only file that is required for this exercise. Once downloaded and saved locally, the data can be loaded into the notebook.

1. Define the path to the data and load it using **pandas**:

```
path = "News_Final.csv"
df = pandas.read_csv(path, header=0)
```

> **NOTE**
>
> Add the file to the same folder where you have opened your notebook.

2. Examine the data briefly by executing the following code:

```
def dataframe_quick_look(df, nrows):
    print("SHAPE:\n{shape}\n".format(shape=df.shape))
    print("COLUMN NAMES:\n{names}\n".format(names=df.columns))
    print("HEAD:\n{head}\n".format(head=df.head(nrows)))
dataframe_quick_look(df, nrows=2)
```

This user-defined function returns the shape of the data (the number of rows and columns), the column names, and the first two rows of the data:

```
SHAPE:
(93239, 11)

COLUMN NAMES:
Index(['IDLink', 'Title', 'Headline', 'Source', 'Topic', 'PublishDate',
       'SentimentTitle', 'SentimentHeadline', 'Facebook', 'GooglePlus',
       'LinkedIn'],
      dtype='object')

HEAD:
     IDLink                                              Title  \
0   99248.0        Obama Lays Wreath at Arlington National Cemetery
1   10423.0             A Look at the Health of the Chinese Economy

                                            Headline       Source     Topic  \
0  Obama Lays Wreath at Arlington National Cemete...    USA TODAY     obama
1  Tim Haywood, investment director business-unit...    Bloomberg   economy

           PublishDate  SentimentTitle  SentimentHeadline  Facebook  \
0  2002-04-02 00:00:00        0.000000          -0.053300        -1
1  2008-09-20 00:00:00        0.208333          -0.156386        -1

   GooglePlus  LinkedIn
0          -1        -1
1          -1        -1
```

Figure 7.7: Raw data

This is a much larger dataset in terms of features than is needed to run the topic models.

3. Notice that one of the columns, named **Topic**, actually contains the information that any topic model would try to ascertain. Briefly look at the topic data provided, so that when you finally generate your own topics, the results can be compared directly. Run the following line to print the unique topic values and their number of occurrences:

```
print("TOPICS:\n{topics}\n".format(topics=df["Topic"]\
    .value_counts()))
```

The output is as follows:

```
TOPICS:
economy      33928
obama        28610
microsoft    21858
palestine     8843
Name: Topic, dtype: int64
```

4. Now, extract the headline data and transform the extracted data into a list object. Print the first five elements of the list and the list length to confirm that the extraction was successful:

```
raw = df["Headline"].tolist()
print("HEADLINES:\n{lines}\n".format(lines=raw[:5]))
print("LENGTH:\n{length}\n".format(length=len(raw)))
```

The output is as follows:

```
HEADLINES:
['Obama Lays Wreath at Arlington National Cemetery. President Barack Obama has laid a wreath at the Tomb of the Unknowns to honor', 'Tim Haywood, investment director business-unit head for fixed income at Gam, discusses the China beige book and the state of the economy.', "Nouriel Roubini, NYU professor and chairman at Roubini Global Economics, explains why the global economy isn't facing the same conditions", "Finland's economy expanded marginally in the three months ended December, after contracting in the previous quarter, preliminary figures from Statistics Finland showed Monday. ", 'Tourism and public spending continued to boost the economy in January, in light of contraction in private consumption and exports, according to the Bank of Thailand data. ']

LENGTH:
93239
```

Figure 7.8: A list of headlines

With the data now loaded and correctly formatted, let's talk about textual data cleaning and then jump into some actual cleaning and preprocessing. For instructional purposes, the cleaning process will initially be built and executed on only one headline. Once we have established the process and tested it on the example headline, we will go back and run the process on every headline.

> **NOTE**
>
> To access the source code for this specific section, please refer to https://packt.live/34gLGKa.
>
> You can also run this example online at https://packt.live/3fbWQES.
>
> You must execute the entire Notebook in order to get the desired result.

CLEANING TEXT DATA

A key component of all successful modeling exercises is a clean dataset that has been appropriately and sufficiently preprocessed for the specific data type and analysis being performed. Text data is no exception, as it is virtually unusable in its raw form. It does not matter what algorithm is being run: if the data isn't properly prepared, the results will be at best meaningless and at worst misleading. As the saying goes, *garbage in, garbage out.* For topic modeling, the goal of data cleaning is to isolate the words in each document that could be relevant by removing everything that could be obstructive.

Data cleaning and preprocessing is almost always specific to the dataset, meaning that each dataset will require a unique set of cleaning and preprocessing steps selected to specifically handle the issues in it. With text data, cleaning and preprocessing steps can include language filtering, removing URLs and screen names, lemmatizing, and stop word removal, among others. We will explore these in detail in the upcoming sections and implement these ideas in the forthcoming exercises, where a dataset featuring news headlines will be cleaned for topic modeling.

DATA CLEANING TECHNIQUES

To reiterate a previous point, the goal of cleaning text for topic modeling is to isolate the words in each document that could be relevant to finding the abstract topics of the corpus. This means removing common words, short words (generally more common), numbers, and punctuation. No hard and fast process exists for cleaning data, so it is important to understand the typical problem points in the type of data being cleaned and do extensive exploratory work.

Let's now discuss some of the text data cleaning techniques that we will employ. One of the first things that needs to be done when doing any modeling task involving text is to determine the language(s) of the text. In this dataset, most of the headlines are English, so we will remove the non-English headlines for simplicity. Building models on non-English text data requires additional skill sets, the least of which is fluency in the language being modeled.

The next crucial step in data cleaning is to remove all elements of the documents that are either not relevant to word-based models or are potential sources of noise that could obscure the results. Elements needing removal could include website addresses, punctuation, numbers, and stop words. **Stop words** are basically simple, commonly used words (including *we*, *are*, and *the*). It is important to note that there is no definitive dictionary of stop words; instead, every dictionary varies slightly. Despite the differences, each dictionary contains a number of common words that are assumed to be topic agnostic. Topic models try to identify words that are both frequent and infrequent enough to be descriptive of an abstract topic.

The removal of website addresses has a similar motivation. Specific website addresses will appear very rarely, but even if one specific website address appears enough to be linked to a topic, website addresses are not interpretable in the same way as words. Removing irrelevant information from the documents reduces the amount of noise that could either prevent model convergence or obscure results.

Lemmatization, like language detection, is an important component of all modeling activities involving text. It is the process of reducing words to their base form as a way to group words that should all be the same but are not because of various changes in the tense or the part of speech. Consider the words *running*, *runs*, and *ran*. All three of these words have the base form of *run*. A great aspect of lemmatizing is that it looks at all the words in a sentence (in other words, it considers the context), before determining how to alter each word. Lemmatization, like most of the preceding cleaning techniques, simply reduces the amount of noise in the data, so that we can identify clean and interpretable topics.

Now, with a basic knowledge of textual cleaning techniques, let's apply these techniques to real-world data.

EXERCISE 7.03: CLEANING DATA STEP BY STEP

In this exercise, we will learn how to implement some key techniques for cleaning text data. Each technique will be explained as we work through the exercise. After every cleaning step, the example headline is output using **print**, so we can watch the evolution from raw data to model-ready data:

1. Select the sixth headline as the example on which we will build and test the cleaning process. The sixth headline is not a random choice; it was selected because it contains specific problems that will be addressed during the cleaning process:

```
example = raw[5]
print(example)
```

The output is as follows:

```
Over 100 attendees expected to see latest version of Microsoft Dynamics SL and Dynamics GP (PRWeb February 29, 2016) Read the full story at http://www.prweb.com/releases/2016/03/prweb13238571.htm
```

Figure 7.9: The sixth headline

2. Use the **langdetect** library to detect the language of each headline. If the language is anything other than English (**en**), remove that headline from the dataset. The **detect** function simply detects the language of the text that is passed into it. When the function fails to detect a language, which it periodically does, simply set the language to **none** for removal later on:

```
def do_language_identifying(txt):
    try: the_language = langdetect.detect(txt)
    except: the_language = 'none'
    return the_language
print("DETECTED LANGUAGE:\n{lang}\n"\
      .format(lang=do_language_identifying(example)))
```

The output is as follows:

```
DETECTED LANGUAGE:
en
```

3. Split the string containing the headline into pieces, called **tokens**, using the white spaces. The returned object is a list of words and numbers that make up the headline. Breaking the headline string into tokens makes the cleaning and preprocessing process simpler. There are multiple types of tokenizers available. Note that NLTK itself provides various types of tokenizers. Each of the tokenizers considers different ways to split the sentence into tokens. The simplest one is splitting the text based on white spaces.

```
example = example.split(" ")
print(example)
```

The output is as follows:

```
['Over', '100', 'attendees', 'expected', 'to', 'see', 'latest', 'version', 'of', 'Microsoft', 'Dynamics', 'SL', 'and', 'Dynamics', 'GP', '(PRWeb', 'February', '29,', '2016)', 'Read', 'the', 'full', 'story', 'at', 'http://www.prweb.com/releases/2016/03/prweb13238571.htm', '']
```

Figure 7.10: String split using white spaces

4. Identify all URLs using a regular expression search for tokens containing **http://** or **https://**. Replace the URLs with the **'URL'** string:

```
example = ['URL' if bool(regex.search("http[s]?://", i)) \
           else i for i in example]
print(example)
```

The output is as follows:

```
['Over', '100', 'attendees', 'expected', 'to', 'see', 'latest', 'version', 'of', 'Microsoft', 'Dynamics', 'SL', 'and', 'Dynamics', 'GP', '(PRWeb', 'February', '29,', '2016)', 'Read', 'the', 'full', 'story', 'at', 'URL', '']
```

Figure 7.11: URLs replaced with the URL string

5. Replace all punctuation and newline symbols (**\n**) with empty strings using regular expressions:

```
example = [regex.sub("[^\\w\\s]|\n", "", i) for i in example]
print(example)
```

The output is as follows:

```
['Over', '100', 'attendees', 'expected', 'to', 'see', 'latest', 'version', 'of', 'Microsoft', 'Dynamics', 'SL', 'and', 'Dynamics', 'GP', 'PRWeb', 'February', '29', '2016', 'Read', 'the', 'full', 'story', 'at', 'URL', '']
```

Figure 7.12: Punctuation replaced with empty strings using regular expressions

6. Replace all numbers with empty strings using regular expressions:

    ```
    example = [regex.sub("^[0-9]*$", "", i) for i in example]
    print(example)
    ```

    The output is as follows:

    ```
    ['Over', '', 'attendees', 'expected', 'to', 'see', 'latest', 'version', 'of', 'Microsoft', 'Dynami
    cs', 'SL', 'and', 'Dynamics', 'GP', 'PRWeb', 'February', '', '', 'Read', 'the', 'full', 'story',
    'at', 'URL', '']
    ```

    Figure 7.13: Numbers replaced with empty strings

7. Change all uppercase letters to lowercase. Converting everything to lowercase is not a mandatory step, but it does help reduce complexity. With everything lowercase, there is less to keep track of and therefore less chance of error:

    ```
    example = [i.lower() if i not in ["URL"] else i for i in example]
    print(example)
    ```

    The output is as follows:

    ```
    ['over', '', 'attendees', 'expected', 'to', 'see', 'latest', 'version', 'of', 'microsoft', 'dynami
    cs', 'sl', 'and', 'dynamics', 'gp', 'prweb', 'february', '', '', 'read', 'the', 'full', 'story',
    'at', 'URL', '']
    ```

    Figure 7.14: Uppercase letters converted to lowercase

8. Remove the **'URL'** string that was added as a placeholder in *Step 4*. The previously added **'URL'** string is not actually needed for modeling. If it seems harmless to leave it in, consider that the **'URL'** string could appear naturally in a headline and we do not want to artificially boost its number of appearances. Also, the **'URL'** string does not appear in every headline, so by leaving it in, we could be unintentionally creating a connection between the **'URL'** strings and a topic:

    ```
    example = [i for i in example if i not in ["URL",""]]
    print(example)
    ```

    The output is as follows:

    ```
    ['over', 'attendees', 'expected', 'to', 'see', 'latest', 'version', 'of', 'microsoft', 'dynamics',
    'sl', 'and', 'dynamics', 'gp', 'prweb', 'february', 'read', 'the', 'full', 'story', 'at']
    ```

    Figure 7.15: String URL removed

264 | Topic Modeling

9. Load in the **stopwords** dictionary from **nltk** and print it:

   ```
   list_stop_words = nltk.corpus.stopwords.words("english")
   list_stop_words = [regex.sub("[^\\w\\s]", "", i) \
                      for i in list_stop_words]
   print(list_stop_words)
   ```

   The output is as follows:

   ```
   ['i', 'me', 'my', 'myself', 'we', 'our', 'ours', 'ourselves', 'you', 'youre', 'youve', 'youll', 'youd', 'your', 'yours', 'yourself', 'yourselves', 'he', 'him', 'his', 'himself', 'she', 'shes', 'her', 'hers', 'herself', 'it', 'its', 'its', 'itself', 'they', 'them', 'their', 'theirs', 'themselves', 'what', 'which', 'who', 'whom', 'this', 'that', 'thatll', 'these', 'those', 'am', 'is', 'are', 'was', 'were', 'be', 'been', 'being', 'have', 'has', 'had', 'having', 'do', 'does', 'did', 'doing', 'a', 'an', 'the', 'and', 'but', 'if', 'or', 'because', 'as', 'until', 'while', 'of', 'at', 'by', 'for', 'with', 'about', 'against', 'between', 'into', 'through', 'during', 'before', 'after', 'above', 'below', 'to', 'from', 'up', 'down', 'in', 'out', 'on', 'off', 'over', 'under', 'again', 'further', 'then', 'once', 'here', 'there', 'when', 'where', 'why', 'how', 'all', 'any', 'both', 'each', 'few', 'more', 'most', 'other', 'some', 'such', 'no', 'nor', 'not', 'only', 'own', 'same', 'so', 'than', 'too', 'very', 's', 't', 'can', 'will', 'just', 'don', 'dont', 'should', 'shouldve', 'now', 'd', 'll', 'm', 'o', 're', 've', 'y', 'ain', 'aren', 'arent', 'couldn', 'couldnt', 'didn', 'didnt', 'doesn', 'doesnt', 'hadn', 'hadnt', 'hasn', 'hasnt', 'haven', 'havent', 'isn', 'isnt', 'ma', 'mightn', 'mightnt', 'mustn', 'mustnt', 'needn', 'neednt', 'shan', 'shant', 'shouldn', 'shouldnt', 'wasn', 'wasnt', 'weren', 'werent', 'won', 'wont', 'wouldn', 'wouldnt']
   ```

   Figure 7.16: List of stop words

   Before using the dictionary, it is important to reformat the words to match the formatting of our headlines. That involves confirming that everything is lowercase and without punctuation.

10. Now that we have correctly formatted the **stopwords** dictionary, use it to remove all stop words from the headline:

    ```
    example = [i for i in example if i not in list_stop_words]
    print(example)
    ```

    The output is as follows:

    ```
    ['attendees', 'expected', 'see', 'latest', 'version', 'microsoft', 'dynamics', 'sl', 'dynamics', 'gp', 'prweb', 'february', 'read', 'full', 'story']
    ```

    Figure 7.17: Stop words removed from the headline

11. Perform lemmatization by defining a function that can be applied to each headline individually. Lemmatizing requires the **wordnet** dictionary to be loaded. The **morphy** function takes each individual word in a text and returns its standard form if it recognizes it. For example, if the word input is *running* or *ran*, the **morphy** function would return *run*:

```
def do_lemmatizing(wrd):
    out = nltk.corpus.wordnet.morphy(wrd)
    return (wrd if out is None else out)
example = [do_lemmatizing(i) for i in example]
print(example)
```

The output is as follows:

```
['attendee', 'expect', 'see', 'latest', 'version', 'microsoft', 'dynamics', 'sl', 'dynamics', 'gp', 'prweb', 'february', 'read', 'full', 'story']
```

Figure 7.18: Output after performing lemmatization

12. Remove all words with a length of four or less from the list of tokens. The assumption around this step is that short words are, in general, more common and therefore will not drive the types of insights we are looking to extract from the topic models. Note that removing words of certain lengths is not a technique that should be used all the time; it is for specific cases only. For example, short words can sometimes be very indicative of topics such as in the case of identifying animals (for example, dog, cat, bird).

```
example = [i for i in example if len(i) >= 5]
print(example)
```

The output is as follows:

```
['attendee', 'expect', 'latest', 'version', 'microsoft', 'dynamics', 'dynamics', 'prweb', 'february', 'story']
```

Figure 7.19: Headline number six post-cleaning

Now that we have worked through the cleaning and preprocessing steps individually on one headline, we need to apply those steps to every one of the nearly 100,000 headlines. The most efficient way to do that is to write a function that contains all the steps outlined above and apply that function to every document in the corpus in some iterative fashion. That process is undertaken in the next exercise.

> **NOTE**
>
> To access the source code for this specific section, please refer to https://packt.live/34gLGKa.
>
> You can also run this example online at https://packt.live/3fbWQES.
>
> You must execute the entire Notebook in order to get the desired result.

EXERCISE 7.04: COMPLETE DATA CLEANING

In this exercise, we will consolidate *Steps 2* to *12* from *Exercise 7.03, Cleaning Data Step by Step*, into one function that we can apply to every headline. The function will take one headline in string format as an input and the output will be a cleaned headline as a list of tokens. The topic models require that documents be formatted as strings instead of as lists of tokens, so in *Step 4*, the lists of tokens are converted back into strings:

1. Define a function that contains all the individual steps of the cleaning process from *Exercise 7.03, Cleaning Data Step by step*:

`Exercise7.01-Exercise7.12.ipynb`

```
def do_headline_cleaning(txt):
    # identify language of tweet
    # return null if language not English
    lg = do_language_identifying(txt)
    if lg != 'en':
        return None
    # split the string on whitespace
    out = txt.split(" ")
    # identify urls
    # replace with URL
    out = ['URL' if bool(regex.search("http[s]?://", i)) \
        else i for i in out]
    # remove all punctuation
    out = [regex.sub("[^\\w\\s]|\n", "", i) for i in out]
    # remove all numerics
    out = [regex.sub("^[0-9]*$", "", i) for i in out]
```

The complete code for this step can be found at https://packt.live/34gLGKa.

2. Execute the function on each headline. The **map** function in Python is a nice way to apply a user-defined function to each element of a list. Convert the **map** object to a list and assign it to the **clean** variable. The **clean** variable is a list of lists:

```
tick = time()
clean = list(map(do_headline_cleaning, raw))
print(time()-tick)
```

3. In **do_headline_cleaning**, **None** is returned if the language of the headline is detected as being any language other than English. The elements of the final cleaned list should only be lists, not **None**, so remove all **None** types. Use **print** to display the first five cleaned headlines and the length of the **clean** variable:

```
clean = list(filter(None.__ne__, clean))
print("HEADLINES:\n{lines}\n".format(lines=clean[:5]))
print("LENGTH:\n{length}\n".format(length=len(clean)))
```

The output is as follows:

```
HEADLINES:
[['obama', 'wreath', 'arlington', 'national', 'cemetery', 'president', 'barack', 'obama', 'wreath',
'unknown', 'honor'], ['haywood', 'investment', 'director', 'businessunit', 'income', 'discus', 'chin
a', 'beige', 'state', 'economy'], ['nouriel', 'roubini', 'professor', 'chairman', 'roubini', 'globa
l', 'economics', 'explain', 'global', 'economy', 'facing', 'conditions'], ['finland', 'economy', 'ex
pand', 'marginally', 'three', 'month', 'december', 'contracting', 'previous', 'quarter', 'preliminar
y', 'figure', 'statistics', 'finland', 'monday'], ['tourism', 'public', 'spending', 'continue', 'boo
st', 'economy', 'january', 'light', 'contraction', 'private', 'consumption', 'export', 'accord', 'th
ailand']]

LENGTH:
92946
```

Figure 7.20: Example headlines and the length of the headline list

4. For every individual headline, concatenate the tokens using a white space separator. The headlines should now be an unstructured collection of words, nonsensical to the human reader, but ideal for topic modeling:

```
clean_sentences = [" ".join(i) for i in clean]
print(clean_sentences[0:10])
```

The cleaned headlines should resemble the following:

```
['obama wreath arlington national cemetery president barack obama wreath unknown honor', 'haywood
investment director businessunit income discus china beige state economy', 'nouriel roubini profes
sor chairman roubini global economics explain global economy facing conditions', 'finland economy
expand marginally three month december contracting previous quarter preliminary figure statistics
finland monday', 'tourism public spending continue boost economy january light contraction private
consumption export accord thailand', 'attendee expect latest version microsoft dynamics dynamics p
rweb february story', 'ramallah february palestine liberation organization sectretarygeneral ereka
t thursday express concern kenyan president uhuru kenyattas visit jerusalem jordan valley', 'first
michelle obama speak state white house washington wednesday interactive student workshop musical l
egacy charles student school community organization across country participate quotin performance
white housequot series', 'hancock county early monday morning family years', 'delhi feb29 technolo
gy giant microsoft target rival apple series focusing windows gross windows machine']
```

Figure 7.21: Headlines cleaned for modeling

> **NOTE**
>
> To access the source code for this specific section, please refer to https://packt.live/34gLGKa.
>
> You can also run this example online at https://packt.live/3fbWQES.
>
> You must execute the entire Notebook in order to get the desired result.

To recap, what the cleaning and preprocessing work effectively does is strip out the noise from the data so that the model can hone in on elements of the data that could actually drive insights. For example, words that are agnostic to any topic should not be informing topics, but by accident alone, if left in, could be.

In an effort to avoid what we could call *fake signal*, we remove those words. Likewise, since topic models cannot discern context, punctuation is irrelevant and is therefore removed. Even if the model could find the topics without removing the noise from the data, the uncleaned data could have thousands to millions of extra words and random characters to parse (depending on the number of documents in the corpus), which could significantly increase the computational demands. So, data cleaning is an integral part of topic modeling. You will practice this in the following activity.

ACTIVITY 7.01: LOADING AND CLEANING TWITTER DATA

In this activity, we will load and clean Twitter data for modeling to be done in subsequent activities. Our usage of the headline data is ongoing, so let's complete this activity in a separate Jupyter notebook, but with all the same requirements and imported libraries.

The goal is to take the raw tweet data, clean it, and produce the same output that we did in *Step 4* of the previous exercise. The output should be a list whose length is similar to the number of rows in the raw data file, but potentially not equal to it. This is because tweets can get dropped in the cleaning process for many reasons, such as the tweet being written in a language other than English. Each element of the list should represent one tweet and should contain just the words in the tweet that might be relevant to topic formation.

Here are the steps to complete the activity:

1. Import the necessary libraries.

2. Load the LA Times health Twitter data (`latimeshealth.txt`) from https://packt.live/2Xje5xF.

   > **NOTE**
   >
   > This dataset is sourced from https://archive.ics.uci.edu/ml/datasets/Health+News+in+Twitter (UCI Machine Learning Repository [http://archive.ics.uci.edu/ml]. Irvine, CA: University of California, School of Information and Computer Science).
   >
   > Citation: Karami, A., Gangopadhyay, A., Zhou, B., & Kharrazi, H. (2017). Fuzzy approach topic discovery in health and medical corpora. International Journal of Fuzzy Systems, 1-12.
   >
   > It is also available on GitHub at https://packt.live/2Xje5xF.

3. Run a quick exploratory analysis to ascertain data size and structure.

4. Extract the tweet text and convert it to a list object.

5. Write a function to perform language detection and tokenization on white spaces, and then replace the screen names and URLs with **SCREENNAME** and **URL**, respectively. The function should also remove punctuation, numbers, and the **SCREENNAME** and **URL** replacements. Convert everything to lowercase, except **SCREENNAME** and **URL**. It should remove all stop words, perform lemmatization, and keep words with five or more letters only.

6. Apply the function defined in *Step 5* to every tweet.

7. Remove elements of the output list equal to **None**.

8. Turn the elements of each tweet back into a string. Concatenate using white space.

9. Keep the notebook open for future activities.

   > **NOTE**
   >
   > All the activities in this chapter need to be performed in the same notebook.

The output will be as follows:

```
['running shoes extra', 'class crunch intense workout pulley system', 'thousand natural product', 'n
atural product explore beauty supplement', 'fitness weekend south beach spark activity', 'kayla harr
ison sacrifice', 'sonic treatment alzheimers disease', 'ultrasound brain restore memory alzheimers n
eedle onlyso farin mouse', 'apple researchkit really medical research', 'warning chantix drink takin
g might remember']
```

Figure 7.22: Tweets cleaned for modeling

> **NOTE**
>
> The solution to this activity can be found on page 478.

LATENT DIRICHLET ALLOCATION

In 2003, David Blei, Andrew Ng, and Michael Jordan published their article on the topic modeling algorithm known as **Latent Dirichlet Allocation (LDA)**. LDA is a generative probabilistic model. This means that the modeling process starts with the text and works backward through the process that is assumed to have generated it in order to identify the parameters of interest. In this case, it is the topics that generated the data that are of interest. The process discussed here is the most basic form of LDA, but for learning, it is also the most comprehensible.

There are M documents available for topic modeling within the corpus. Each document can be considered as the sequence of N words, i.e., a sequence ($w_1, w_2, ... w_N$).

For each document in the corpus, the assumed generative process is:

1. Select $N \sim Poisson(\lambda)$, where N is the number of words and λ is the parameter controlling the Poisson distribution.

2. Select $\theta \sim Dirichlet(\alpha)$, where θ is the distribution of topics.

3. For each N words, W_n, select topic $z_n \sim Multinomial(\theta)$, and select word W_n from $P(w_n | z_n, \beta)$.

Let's go through the generative process in a bit more detail. The preceding three steps repeat for every document in the corpus. The initial step is to choose the number of words in the document by sampling from, in most cases, the *Poisson* distribution. It is important to note that, because N is independent of the other variables, the randomness associated with its generation is mostly ignored in the derivation of the algorithm.

Coming after the selection of N is the generation of the topic mixture or distribution of topics, unique to each document. Think of this as a per-document list of topics with probabilities representing the amount of the document represented by each topic. Consider three topics: A, B, and C. An example document could be 100% topic A, 75% topic B and 25% topic C, or an infinite number of other combinations.

Lastly, the specific words in the document are selected via a probability statement conditioned on the selected topic and the distribution of words for that topic. Note that documents are not really generated in this way, but it is a reasonable proxy.

This process can be thought of as a distribution over distributions. A document is selected from the collection (distribution) of documents, and one topic is selected (via the multinomial distribution) from the probability distribution of topics for that document, generated by the Dirichlet distribution.

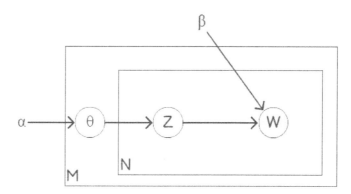

Figure 7.23: Graphical representation of LDA

The most straightforward way to build the formula representing the LDA solution is through a graphical representation. This particular representation is referred to as a plate notation graphical model, as it uses plates to represent the two iterative steps in the process.

You will recall that the generative process was executed for every document in the corpus, so the outermost plate (labeled M) represents iterating over each document. Similarly, the iteration over words in *Step 3* is represented by the innermost plate of the diagram, labeled N.

The circles represent the parameters, distributions, and results. The shaded circle, labeled w, is the selected word, which is the only known piece of data and, as such, is used to reverse-engineer the generative process. Besides w, the other four variables in the diagram are defined as follows:

- α: Hyperparameter for the topic-document Dirichlet distribution.
- β: Distribution of words for each topic.
- Z: This is the latent variable for the topic.
- θ: This is the latent variable for the distribution of topics for each document.

α and β control the frequency of topics in documents and the frequency of words in topics. If α increases, the documents become increasingly similar as the number of topics in each document increases. On the other hand, if α decreases, the documents become increasingly dissimilar as the number of topics in each document decreases. The β parameter behaves similarly. If β increases, more words from the document are used to model a topic while a lower β causes a smaller number of words to be used for a topic. Given the complexity of the distributions in LDA, there is no direct solution, so some sort of approximation algorithm is required to generate the results. The standard approximation algorithm for LDA is discussed in the next section.

VARIATIONAL INFERENCE

The big issue with LDA is that the evaluation of the conditional probabilities (the distributions) is unmanageable, so instead of computing them directly, the probabilities are approximated. Variational inference is one of the simpler approximation algorithms, but it has an extensive derivation that requires significant knowledge of probability. In order to spend more time on the application of LDA, this section will give some high-level details on how variational inference is applied in this context but will not fully explore the algorithm.

Let's take a moment to work through the variational inference algorithm intuitively. Start by randomly assigning each word in each document in the corpus to one of the topics. Then, for each document and each word in each document separately, calculate two proportions. Those proportions would be the proportion of words in the document that are currently assigned to the topic, *P(Topic|Document)* and the proportion of assignments across all documents of a specific word to the topic, *P(Word|Topic)*. Multiply the two proportions and use the resulting proportion to assign the word to a new topic. Repeat this process until a steady state is reached where topic assignments are not changing significantly. These assignments are then used to estimate the within-document topic mixture and the within-topic word mixture.

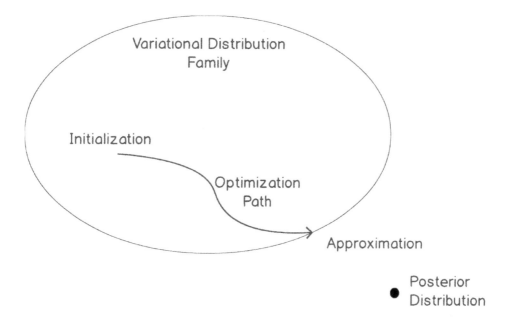

Figure 7.24: The variational inference process

The thought process behind variational inference is that, if the actual distribution is intractable, then a simpler distribution, let's call it the variational distribution, very close to true distribution, which is tractable, should be found so that inference becomes possible. In other words, since inferring the actual distribution is impossible due to the complexity of the actual distribution, we try instead to find a simpler distribution that is an excellent approximation of the actual distribution.

Let's take a momentary break from the theory for an example. Variational inference is like trying to view animals at a crowded zoo. The animals at the zoo are in an enclosed habitat, which, in this example, is the posterior distribution. Visitors cannot actually get into the habitat, so the visitors have to settle for viewing the habitat from the closest possible position, which is the posterior approximation (i.e. the best approximation of the habitat). If there are a lot of people at the zoo, it can be difficult to get to that optimal vantage point. People generally start at the back of the crowd and strategically move their way toward that optimal vantage point. The path the visitors follow to move from the back of the crowd to the optimal vantage point is the optimization path. Variational inference is simply the process of getting as close to the desired point as possible knowing that the desired point cannot actually be reached.

To start, select a family of distributions (i.e. binomial, gaussian, exponential, and so on), q, conditioned on new variational parameters. The parameters are optimized so that the original distribution, which is actually the posterior distribution for those people familiar with Bayesian statistics, and the variational distribution are as close as possible. The variational distribution will be close enough to the original posterior distribution to be used as a proxy, making any inference done on it applicable to the original posterior distribution. The generic formula for the family of distributions, q, is as follows:

$$q(\theta, z|\gamma) = q(\theta|\gamma) \prod_{n=1}^{N} q(z_n|\phi n)$$

Figure 7.25: Formula for the family of distributions, q

There is a large collection of potential variational distributions that can be used as an approximation for the posterior distribution. An initial variational distribution is selected from the collection, which acts as the starting point for an optimization process that iteratively moves closer and closer to the optimal distribution. The optimal parameters are the parameters of the distribution that best approximate the posterior. The similarity of the two distributions is measured using **Kullback-Leibler (KL)** divergence. KL divergence represents the expected amount of error generated if we approximate one distribution with another. The distribution with optimal parameters will have the smallest KL divergence when measured against the true distribution.

Once the optimal distribution has been identified, which means the optimal parameters have been identified, it can be leveraged to produce the output matrices and execute any required inference.

BAG OF WORDS

Text cannot be passed directly into any machine learning algorithm; it first needs to be encoded numerically. A straightforward way of working with text in machine learning is via a bag-of-words model, which removes all information regarding the order of the words and focuses strictly on the degree of presence (meaning the count or frequency) of each word.

The Python `sklearn` library can be leveraged to transform the cleaned vector created in the previous exercise into the structure that the LDA model requires. Since LDA is a probabilistic model, we do not want to do any scaling or weighting of the word occurrences; instead, we opt to input just the raw counts.

The input to the bag-of-words model will be the list of cleaned strings that were returned from *Exercise 7.04, Complete Data Cleaning*. The output will be the document number, the word as its numeric encoding, and a count of the number of times that word appears in that document. These three items will be presented as a tuple and an integer.

The tuple will be something like (0, 325), where 0 is the document number and 325 is the numerically encoded word. Note that 325 will be the encoding of that word across all documents. The integer would then be the count. The bag-of-words models we will be running in this chapter are from `sklearn` and are called `CountVectorizer` and `TfIdfVectorizer`. The first model returns the raw counts and the second returns a scaled value, which we will discuss a bit later.

A critical note is that the results of both topic models being covered in this chapter can vary from run to run, even when the data is the same, because of randomness. Neither the probabilities in LDA nor the optimization algorithms are deterministic, so do not be surprised if your results differ slightly from the results shown from here on out. In the next exercise, we will run the count vectorizer to numerically encode our documents, so that we can continue on to topic modeling using LDA.

EXERCISE 7.05: CREATING A BAG-OF-WORDS MODEL USING THE COUNT VECTORIZER

In this exercise, we will run the **CountVectorizer** in **sklearn** to convert our previously created cleaned vector of headlines into a bag-of-words data structure. In addition, we will define some variables that will be used throughout the modeling process:

1. Define **number_words**, **number_docs**, and **number_features**. The first two variables control the visualization of the LDA results. The **number_features** variable controls the number of words that will be kept in the feature space:

   ```
   number_words = 10
   number_docs = 10
   number_features = 1000
   ```

2. Run the count vectorizer and print the output. There are three crucial inputs, which are **max_df**, **min_df**, and **max_features**. These parameters further filter the number of words in the corpus down to those that will most likely influence the model.

   Words that only appear in a small number of documents are too rare to be attributable to any topic, so **min_df** is used to throw away words that appear in fewer than the specified number of documents. Words that appear in too many documents are not specific enough to be linked to specific topics, so **max_df** is used to throw away words that appear in more than the specified percentage of documents.

   Lastly, we do not want to overfit the model, so the number of words used to fit the model is limited to the most frequently occurring specified number (**max_features**) of words:

   ```
   vectorizer1 = sklearn.feature_extraction.text\
                 .CountVectorizer(analyzer="word",\
                                  max_df=0.5,\
                                  min_df=20,\
                                  max_features=number_features)
   clean_vec1 = vectorizer1.fit_transform(clean_sentences)
   print(clean_vec1[0])
   ```

The output is as follows:

```
(0, 408)    1
(0, 88)     1
(0, 644)    1
(0, 558)    1
(0, 573)    2
```

Figure 7.26: The bag-of-words data structure

3. Extract the feature names and the words from the vectorizer. The model is only fed the numerical encodings of the words, so having the feature names vector merge with the results will make interpretation easier:

```
feature_names_vec1 = vectorizer1.get_feature_names()
```

This exercise involved the enumeration of the documents for use in the LDA model. The required format is a bag of words. That is, a bag-of-words model is simply a listing of all the words that appear in each document with a count of the number of times each word appears in each specific document. Having accomplished this task using **sklearn**, it is time to explore the process of evaluating LDA models.

> **NOTE**
>
> To access the source code for this specific section, please refer to https://packt.live/34gLGKa.
>
> You can also run this example online at https://packt.live/3fbWQES.
>
> You must execute the entire Notebook in order to get the desired result.

PERPLEXITY

Models generally have metrics that can be leveraged to evaluate their performance. Topic models are no different, although performance, in this case, has a slightly different definition. In regression and classification, predicted values can be compared to actual values from which clear measures of performance can be calculated.

With topic models, prediction is less reliable, because the model only knows the words it was trained on and new documents may not contain any of those words, despite featuring the same topics. Due to that difference, topic models are evaluated using a metric specific to language models, called **perplexity**.

Perplexity, abbreviated to PP, measures the number of different equally most probable words that can follow any given word on average. Let's consider two words as an example: *the* and *announce*. The word *the* can preface an enormous number of equally most probable words, while the number of equally most probable words that can follow the word *announce* is significantly less—albeit still a large number.

The idea is that words that, on average, can be followed by a smaller number of equally most probable words are more specific and can be more tightly tied to topics. As such, lower perplexity scores imply better language models. Perplexity is very similar to entropy, but perplexity is typically used because it is easier to interpret. As we will see momentarily, it can be used to select the optimal number of topics. With m being the number of words in the sequence of words, perplexity is defined as:

$$PP = \widehat{P}(w_1, \ldots, w_m)^{-1/m}$$

Figure 7.27: Formula of perplexity

In this formula, $w_1, \ldots, w_m$ are the words making up some document in the test dataset. The joint probability of those words, $P(w_1, \ldots, w_m)$, is a measure of how well the test document fits in the existing model. Higher probabilities suggest stronger models. The probability is raised to the $-1/m$ power to normalize the score by the number of words in each document and to make lower values more optimal. Both these changes increase the interpretability of the score. The perplexity score, like root mean squared error, is not very meaningful as a standalone metric. It tends to be used as a comparison metric. That is, several models are built for which perplexity scores are calculated and compared to identify the best model with which to move forward.

As stated previously, LDA has two required inputs. The first is the documents themselves, and the second is the number of topics. Selecting an appropriate number of topics can be very tricky. One approach to finding the optimal number of topics is to search over several numbers of topics and select the number of topics that corresponds to the smallest perplexity score. In machine learning, this approach is referred to as grid search. In the next exercise, we will put grid search to work to find the optimal number of topics.

EXERCISE 7.06: SELECTING THE NUMBER OF TOPICS

In this exercise, we use the perplexity scores for LDA models fit on varying numbers of topics to determine the number of topics with which to move forward. Keep in mind that the original dataset had the headlines sorted into four topics. Let's see whether this approach returns four topics:

1. Define a function that fits an LDA model on various numbers of topics and computes the perplexity score. Return two items: a DataFrame that has the number of topics with its perplexity score and the number of topics with the minimum perplexity score as an integer:

   ```
   def perplexity_by_ntopic(data, ntopics):
       output_dict = {"Number Of Topics": [], \
                      "Perplexity Score": []}
       for t in ntopics:
           lda = sklearn.decomposition.LatentDirichletAllocation(\
               n_components=t, \
               learning_method="online", \
               random_state=0)
           lda.fit(data)
           output_dict["Number Of Topics"].append(t)
           output_dict["Perplexity Score"]\
               .append(lda.perplexity(data))
       output_df = pandas.DataFrame(output_dict)
       index_min_perplexity = output_df["Perplexity Score"]\
           .idxmin()
       output_num_topics = output_df.loc[\
           index_min_perplexity,  # index \
           "Number Of Topics"  # column
       ]
       return (output_df, output_num_topics)
   ```

2. Execute the function defined in *Step 1*. The **ntopics** input is a list of numbers of topics that can be of any length and contain any values. Print out the DataFrame:

   ```
   df_perplexity, optimal_num_topics = \
   perplexity_by_ntopic(clean_vec1, ntopics=[1, 2, 3, 4, 6, 8, 10])
   print(df_perplexity)
   ```

The output is as follows:

	Number Of Topics	Perplexity Score
0	1	510.040098
1	2	457.833534
2	3	409.122014
3	4	433.104230
4	6	507.485482
5	8	560.085934
6	10	577.715983

Figure 7.28: DataFrame containing the number of topics and perplexity score

3. Plot the perplexity scores as a function of the number of topics. This is just another way to view the results contained in the DataFrame from *Step 2*:

```
df_perplexity.plot.line("Number Of Topics", "Perplexity Score")
```

The plot appears as follows:

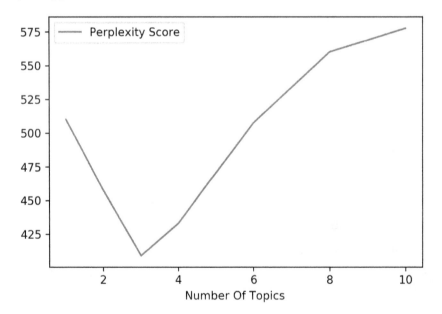

Figure 7.29: Line plot view of perplexity as a function of the number of topics

As the DataFrame and plot show, the optimal number of topics using perplexity is three. Having the number of topics set to four yielded the second-lowest perplexity. Thus, while the results did not exactly match the information contained in the original dataset, the results are close enough to engender confidence in the grid search approach to identify the optimal number of topics. There could be several reasons that the grid search returned three instead of four, which we will dig into in an upcoming exercise.

> **NOTE**
>
> To access the source code for this specific section, please refer to https://packt.live/34gLGKa.
>
> You can also run this example online at https://packt.live/3fbWQES.
>
> You must execute the entire Notebook in order to get the desired result.

Now that we've selected the optimal number of topics, we will use that number of topics to build our official LDA model. That model will then be used to create visualizations and define the list of topics present in the corpus.

EXERCISE 7.07: RUNNING LDA

In this exercise, we'll implement LDA and examine the results. LDA outputs two matrices. The first is the topic-document matrix and the second is the word-topic matrix. We will look at these matrices as returned from the model and as nicely formatted tables that are easier to digest:

1. Fit an LDA model using the optimal number of topics found in *Exercise 7.06, Selecting the Number of Topics*:

```
lda = sklearn.decomposition.LatentDirichletAllocation\
      (n_components=optimal_num_topics,\
       learning_method="online",\
       random_state=0)
lda.fit(clean_vec1)
```

The output is as follows:

```
LatentDirichletAllocation(batch_size=128, doc_topic_prior=None,
                         evaluate_every=-1, learning_decay=0.7,
                         learning_method='online', learning_offset=10.0,
                         max_doc_update_iter=100, max_iter=10,
                         mean_change_tol=0.001, n_components=3, n_jobs=None,
                         perp_tol=0.1, random_state=0, topic_word_prior=None,
                         total_samples=1000000.0, verbose=0)
```

Figure 7.30: The LDA model

2. Output the topic-document matrix and its shape to confirm that it aligns with the number of topics and the number of documents. Each row of the matrix is the per-document distribution of topics:

```
lda_transform = lda.transform(clean_vec1)
print(lda_transform.shape)
print(lda_transform)
```

The output is as follows:

```
(92946, 3)
[[0.04761958 0.90419577 0.04818465]
 [0.04258906 0.04751535 0.90989559]
 [0.16656181 0.04309434 0.79034385]
 ...
 [0.0399815  0.51492894 0.44508955]
 [0.06918206 0.86099065 0.06982729]
 [0.48210053 0.30502833 0.21287114]]
```

3. Output the word-topic matrix and its shape to confirm that it aligns with the number of features (words) specified in *Exercise 7.05, Creating a Bag-of-Words Model Using the Count Vectorizer*, and the number of topics input. Each row is basically the prevalence of assignments to that topic of each word. The prevalence score can be transformed into the per-topic distribution of words:

```
lda_components = lda.components_
print(lda_components.shape)
print(lda_components)
```

The output is as follows:

```
(3, 1000)
[[3.35570079e-01 1.98879573e+02 9.82489014e+00 ... 3.35388004e-01
  2.04173562e+02 4.03130268e-01]
 [2.74824227e+02 3.94662558e-01 3.63412044e-01 ... 3.45944379e-01
  1.77517291e+02 4.61625408e+02]
 [3.37041234e-01 7.36749100e+01 2.05707096e+02 ... 2.31714093e+02
  1.21765267e+02 7.71397922e-01]]
```

4. Define a function that formats the two output matrices into easy-to-read tables:

Exercise7.01-Exercise7.12.ipynb

```
def get_topics(mod, vec, names, docs, ndocs, nwords):
    # word to topic matrix
    W = mod.components_
    W_norm = W / W.sum(axis=1)[:, numpy.newaxis]
    # topic to document matrix
    H = mod.transform(vec)
    W_dict = {}
    H_dict = {}
```

The complete code for this step can be found at https://packt.live/34gLGKa.

The function may be tricky to navigate, so let's walk through it. Start by creating the *W* and *H* matrices, which includes converting the assignment counts of *W* into the per-topic distribution of words. Then, iterate over the topics. Inside each iteration, identify the top words and documents associated with each topic. Convert the results into two DataFrames.

5. Execute the function defined in *Step 4*:

```
W_df, H_df = get_topics(mod=lda, \
                       vec=clean_vec1, \
                       names=feature_names_vec1, \
                       docs=raw, \
                       ndocs=number_docs, \
                       nwords=number_words)
```

6. Print out the word-topic DataFrame. It shows the top 10 words (by distribution value) that are associated with each topic. From this DataFrame, we can identify the abstract topics that the word groupings represent. More on abstract topics will follow:

```
print(W_df)
```

The output is as follows:

```
                    Topic0                  Topic1                  Topic2
Word0  (0.1051, microsoft)       (0.0878, obama)       (0.0987, economy)
Word1    (0.0241, windows)   (0.0761, president)      (0.0339, economic)
Word2    (0.0235, company)      (0.0437, barack)        (0.0172, growth)
Word3  (0.019, microsofts)    (0.0159, palestine)       (0.0146, global)
Word4    (0.0161, announce)      (0.0143, state)   (0.0124, government)
Word5       (0.0144, today)     (0.0137, obamas)        (0.0113, china)
Word6      (0.0108, release) (0.0131, washington)         (0.011, world)
Word7        (0.009, update) (0.0124, palestinian)      (0.0106, percent)
Word8     (0.0088, business)      (0.0113, house)      (0.0097, country)
Word9       (0.0076, surface)     (0.0104, white)       (0.0095, market)
```

Figure 7.31: Word-topic table

7. Print out the topic-document DataFrame. This shows the 10 documents to which each topic is most closely related. The values are from the per-document distribution of topics:

```
print(H_df)
```

The output is as follows:

```
                                                       Topic0  \
Doc0  (0.9776, That appears to be the thinking behin...
Doc1  (0.9764, """I think this will raise expectatio...
Doc2  (0.9756, France's fragile economy has cooled i...
Doc3  (0.9755, (Adds Obama, Kerry remarks, U.N. aid ...
Doc4  (0.9754, Economy Secretary Keith Brown is to t...
Doc5  (0.9754, Software maker Microsoft Corp (MSFT.O...
Doc6  (0.9752, This past month, Microsoft released a...
Doc7  (0.975, The Israeli army said troops shot dead...
Doc8  (0.9749, The president of the United States we...
Doc9  (0.9749, In a partnership that would have seem...

                                                       Topic1  \
Doc0  (0.9798,  Microsoft's Lumia 950 and 950 XL hav...
Doc1  (0.9779, Microsoft is once again challenging U...
Doc2  (0.9779, After seeing Kinect fail to connect w...
Doc3  (0.9779, Microsoft on Thursday posted revenue ...
Doc4  (0.9779, Last quarter, Microsoft hit a major m...
Doc5  (0.9779, JERUSALEM: Sheikh Raed Salah, leader ...
Doc6  (0.9775, Washington (CNN) President Barack Oba...
Doc7  (0.9775, Kantor says South Africa's monetary p...
Doc8  (0.9773, Malia Obama is 17 and probably wants ...
Doc9  (0.9773, WASHINGTON - President Obama sounded ...

                                                       Topic2
Doc0  (0.9782, Microsoft CEO Satya Nadella discussed...
Doc1  (0.9782, Microsoft's latest Windows Phone, the...
Doc2  (0.978, Microsoft Nano Server is a new type of...
Doc3  (0.9779, Microsoft said Monday that it is buyi...
Doc4  (0.9779, Microsoft Corp announced 'Project Sco...
Doc5  (0.9779, The UK is facing a digital skills cri...
Doc6  (0.9778, Microsoft's CSEC will combine company...
Doc7  (0.9777, Microsoft's work in the blockchain sp...
Doc8  (0.9772, AUSTIN -- The South by Southwest conf...
Doc9  (0.9772, JOHANNESBURG """ Cabinet says the rec...
```

Figure 7.32: Topic-document table

The results of the word-topic DataFrame show that the abstract topics are Barack Obama, the economy, and Microsoft. What is interesting is that the word grouping describing the economy contains references to Palestine. All four topics specified in the original dataset are represented in the word-topic DataFrame output, but not in the fully distinct manner expected. We could be facing one of two problems.

First, the topic referencing both the economy and Palestine could be under-cooked, which means increasing the number of topics may fix the issue. The other potential problem is that LDA does not handle correlated topics well. In *Exercise 7.09, Trying Four Topics*, we will try expanding the number of topics, which will give us a better idea of why one of the word groupings is seemingly a mixture of topics.

> **NOTE**
>
> To access the source code for this specific section, please refer to https://packt.live/34gLGKa.
>
> You can also run this example online at https://packt.live/3fbWQES.
>
> You must execute the entire Notebook in order to get the desired result.

VISUALIZATION

The output of LDA models in Python using `sklearn` can be difficult to interpret in raw form. As is the case in most modeling exercises, visualizations can be a great benefit when it comes to interpreting and communicating model results. One Python library, `pyLDAvis`, integrates directly with the `sklearn` model object to produce straightforward graphics. This visualization tool returns a histogram showing the words that are the most closely related to each topic and a biplot, frequently used in PCA, where each circle corresponds to a topic. From the biplot, we know how prevalent each topic is across the entire corpus, which is reflected by the area of the circle, and the similarity of the topics, which is reflected by the closeness of the circles.

The ideal scenario is to have the circles spread throughout the plot and be of reasonable and consistent size. That is, we want the topics to be distinct and to appear uniformly across the corpus. In addition to the `pyLDAvis` graphics, we will leverage the t-SNE model, discussed in a prior chapter, to produce a two-dimensional representation of the topic-document matrix, a matrix where each row represents one document and each column represents the probability of that topic describing the document.

Latent Dirichlet Allocation | 287

Having completed the LDA model fitting, let's create some graphics to help us dig into the results.

EXERCISE 7.08: VISUALIZING LDA

Visualization is a helpful tool for exploring the results of topic models. In this exercise, we will look at three different visualizations. Those visualizations are basic histograms and specialty visualizations using t-SNE and PCA:

1. Run and display **pyLDAvis**. This plot is interactive. Clicking on each circle updates the histogram to show the top words related to that specific topic. The following is one view of this interactive plot:

   ```
   lda_plot = pyLDAvis.sklearn\
              .prepare(lda, clean_vec1, vectorizer1, R=10)
   pyLDAvis.display(lda_plot)
   ```

   The plot appears as follows:

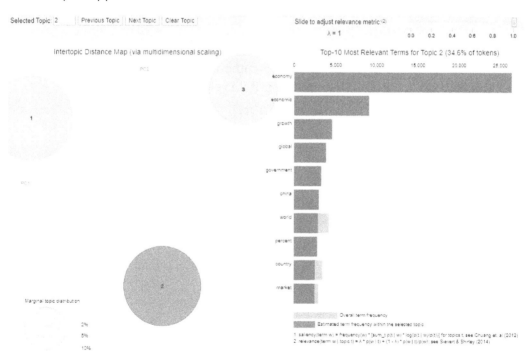

Figure 7.33: A histogram and biplot for the LDA model

288 | Topic Modeling

2. Define a function that fits a t-SNE model and then plots the results. After defining it, the pieces of the function will be described in detail, so that the steps are clear:

`Exercise7.01-Exercise7.12.ipynb`

```python
def plot_tsne(data, threshold):
    # filter data according to threshold
    index_meet_threshold = numpy.amax(data, axis=1) >= threshold
    lda_transform_filt = data[index_meet_threshold]
    # fit tsne model
    # x-d -> 2-d, x = number of topics
    tsne = sklearn.manifold.TSNE(n_components=2, \
                                 verbose=0, \
                                 random_state=0, \
                                 angle=0.5, \
                                 init='pca')
    tsne_fit = tsne.fit_transform(lda_transform_filt)
    # most probable topic for each headline
    most_prob_topic = []
```

The complete code for this step can be found at https://packt.live/34gLGKa.

Step 1: The function starts by filtering down the topic-document matrix using an input threshold value. There are tens of thousands of headlines, and any plot incorporating all the headlines is going to be difficult to read and therefore not helpful. So, this function only plots a document if one of the distribution values is greater than or equal to the input threshold value:

```python
index_meet_threshold = numpy.amax(data, axis=1) >= threshold
lda_transform_filt = data[index_meet_threshold]
```

Step 2: Once the data is filtered down, run t-SNE, where the number of components is two, so that we can plot the results in two dimensions:

```python
tsne = sklearn.manifold.TSNE(n_components=2, \
                             verbose=0, \
                             random_state=0, \
                             angle=0.5, \
                             init='pca')
tsne_fit = tsne.fit_transform(lda_transform_filt)
```

Step 3: Create a vector with an indicator of which topic is most related to each document. This vector will be used to color-code the plot by topic:

```
most_prob_topic = []
for i in range(tsne_fit.shape[0]):
    most_prob_topic.append(lda_transform_filt[i].argmax())
```

Step 4: To understand the distribution of topics across the corpus and the impact of threshold filtering, the function returns the length of the topic vector as well as the topics themselves with the number of documents to which that topic has the largest distribution value:

```
print("LENGTH:\n{}\n".format(len(most_prob_topic)))
unique, counts = numpy.unique(numpy.array(most_prob_topic), \
                              return_counts=True)
print("COUNTS:\n{}\n".format(numpy.asarray((unique, counts)).T))
```

Step 5: Create and return the plot:

```
color_list = ['b', 'g', 'r', 'c', 'm', 'y', 'k']
for i in list(set(most_prob_topic)):
    indices = [idx for idx, val in enumerate(most_prob_topic) \
               if val == i]
    matplotlib.pyplot.scatter(x=tsne_fit[indices, 0], \
                              y=tsne_fit[indices, 1], \
                              s=0.5, c=color_list[i], \
                              label='Topic' + str(i), \
                              alpha=0.25)
matplotlib.pyplot.xlabel('x-tsne')
matplotlib.pyplot.ylabel('y-tsne')
matplotlib.pyplot.legend(markerscale=10)
```

3. Execute the function:

```
plot_tsne(data=lda_transform, threshold=0.75)
```

The output is as follows:

```
LENGTH:
58059

COUNTS:
[[    0 15103]
 [    1 23904]
 [    2 19052]]
```

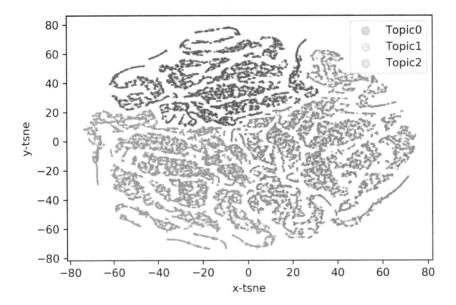

Figure 7.34: t-SNE plot with metrics around the distribution of the topics across the corpus

The visualizations show that the LDA model with three topics is producing good results overall. In the biplot, the circles are of a medium size, which suggests that the topics appear consistently across the corpus and the circles also have good spacing. The t-SNE plot shows clear clusters supporting the separation between the circles represented in the biplot. The only glaring issue, which was previously discussed, is that one of the topics has words that do not seem to belong to that topic.

> **NOTE**
>
> To access the source code for this specific section, please refer to https://packt.live/34gLGKa.
>
> You can also run this example online at https://packt.live/3fbWQES.
>
> You must execute the entire Notebook in order to get the desired result.

In the next exercise, let's rerun the LDA using four topics.

EXERCISE 7.09: TRYING FOUR TOPICS

In this exercise, LDA is run with the number of topics set to four. The motivation for doing this is to try and solve what might be an under-cooked topic from the three-topic LDA model that has words related to both Palestine and the economy. We will run through the steps first and then explore the results at the end:

1. Run an LDA model with the number of topics equal to four:

    ```
    lda4 = sklearn.decomposition.LatentDirichletAllocation(\
           n_components=4,  # number of topics data suggests \
           learning_method="online", \
           random_state=0)
    lda4.fit(clean_vec1)
    ```

Topic Modeling

The output is as follows:

```
LatentDirichletAllocation(batch_size=128, doc_topic_prior=None,
                         evaluate_every=-1, learning_decay=0.7,
                         learning_method='online', learning_offset=10.0,
                         max_doc_update_iter=100, max_iter=10,
                         mean_change_tol=0.001, n_components=4, n_jobs=None,
                         perp_tol=0.1, random_state=0, topic_word_prior=None,
                         total_samples=1000000.0, verbose=0)
```

Figure 7.35: The LDA model

2. Execute the **get_topics** function defined earlier to produce the more readable word-topic and topic-document tables:

```
W_df4, H_df4 = get_topics(mod=lda4, \
                         vec=clean_vec1, \
                         names=feature_names_vec1, \
                         docs=raw, \
                         ndocs=number_docs, \
                         nwords=number_words)
```

3. Print the word-topic table:

```
print(W_df4)
```

The output is as follows:

```
            Topic0                    Topic1                  Topic2   \
Word0  (0.0451, economy)         (0.0995, obama)        (0.1033, economy)
Word1  (0.0334, economic)        (0.0863, president)    (0.0234, economic)
Word2  (0.0233, minister)        (0.0496, barack)       (0.0233, growth)
Word3   (0.0184, could)          (0.018, palestine)     (0.0194, world)
Word4   (0.0165, would)          (0.0155, obamas)       (0.0168, government)
Word5  (0.0146, european)        (0.0148, state)        (0.0147, china)
Word6   (0.0141, global)         (0.0148, washington)   (0.0144, percent)
Word7   (0.013, clinton)         (0.0141, palestinian)  (0.0128, month)
Word8   (0.0113, britain)        (0.0128, house)        (0.0125, quarter)
Word9   (0.0112, prime)          (0.0118, white)        (0.0115, market)

            Topic3
Word0   (0.122, microsoft)
Word1   (0.0279, windows)
Word2   (0.0273, company)
Word3  (0.0221, microsofts)
Word4   (0.0217, announce)
Word5    (0.0116, today)
Word6   (0.0116, release)
Word7   (0.0105, update)
Word8   (0.0091, business)
Word9   (0.0089, surface)
```

Figure 7.36: The word-topic table using the four-topic LDA model

4. Print the document-topic table:

   print(H_df4)

The output is as follows:

```
                                                          Topic0  \
    Doc0  (0.9496, Since 1977 our economy the guiding pr...
    Doc1  (0.9494, Industrial production fell more than ...
    Doc2  (0.9362, David Cameron today denied he was 'an...
    Doc3  (0.9358, One of them is Microsoft, which has r...
    Doc4  (0.9347, Following the United Kingdom's decisi...
    Doc5  (0.9346, US President Barack Obama will on Tue...
    Doc6  (0.9316, If you've been holding out to get the...
    Doc7  (0.9307, (CNN) President Barack Obama is makin...
    Doc8  (0.9305, WASHINGTON — President Obama expresse...
    Doc9  (0.9299, BERLIN--Uncertainty following the U.K...

                                                          Topic1  \
    Doc0  (0.9749, WASHINGTON - President Obama sounded ...
    Doc1  (0.9747, WASHINGTON — President Barack Obama s...
    Doc2  (0.9732, Palestinian Prime Minister Hamdallah ...
    Doc3  (0.973, Lombardo and partner Sonny Ward have j...
    Doc4  (0.973, Human intelligence bested by Google an...
    Doc5  (0.973, Chinese Premier Li Keqiang pledged the...
    Doc6  (0.972, (Los Angeles Times) Obama administrati...
    Doc7  (0.9719, Yet another poll is out showing that ...
    Doc8  (0.9714, Microsoft logo Microsoft (NASDAQ:MSFT...
    Doc9  (0.9705, Barack Obama on Tuesday again urged C...

                                                          Topic2  \
    Doc0  (0.9757, Microsoft's CSEC will combine company...
    Doc1  (0.9749, AUSTIN -- The South by Southwest conf...
    Doc2  (0.9748, India's economy grew at 7.6% in the y...
    Doc3  (0.9748, President Obama got taken behind the ...
    Doc4  (0.9748, The US economy grew at an annual rate...
    Doc5  (0.9748, The US economy grew at an annual rate...
    Doc6  (0.9748, &quot;Colorado has such a dynamic eco...
    Doc7  (0.9748, Welcome to Palestine Today, a service...
    Doc8  (0.9748, Like the video, """Boo Hoo Palestine,...
    Doc9  (0.9739, The US economy grew at a 2% pace in t...

                                                          Topic3
    Doc0  (0.9728, This past month, Microsoft released a...
    Doc1  (0.9721, The 1010data Ecom Insights Panel Appl...
    Doc2  (0.9721, Economist with the University of Ghan...
    Doc3  (0.9721, In a partnership that would have seem...
    Doc4  (0.9718, President Barack Obama congratulated ...
    Doc5  (0.9714, Democrats are increasingly fearful th...
    Doc6  (0.9711, BENGALURU: The global tech giant Micr...
    Doc7  (0.9706, Republican Donald Trump will not be p...
    Doc8  (0.9706, &quot;We have a lot of people trying ...
    Doc9  (0.9702, Microsoft has agreed to transfer some...
```

Figure 7.37: The document-topic table using the four-topic LDA model

5. Display the results of the LDA model using `pyLDAvis`:

```
lda4_plot = pyLDAvis.sklearn\
        .prepare(lda4, clean_vec1, vectorizer1, R=10)
pyLDAvis.display(lda4_plot)
```

The plot is as follows:

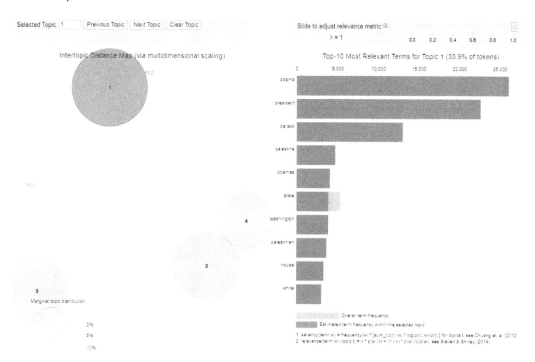

Figure 7.38: A histogram and biplot describing the four-topic LDA model

Looking at the word-topic table, we see that the four topics found by this model align with the four topics specified in the original dataset. Those topics are Barack Obama, Palestine, Microsoft, and the economy. The question now is, why did the model built using four topics have a higher perplexity score than the model with three topics? That answer comes from the visualization produced in *Step 5*.

The biplot has circles of reasonable size, but two of those circles are quite close together, which suggests that those two topics (Microsoft and the economy) are very similar. In this case, the similarity actually makes intuitive sense. Microsoft is a major global company that impacts and is impacted by the economy. The next step, if we were to make one, would be to run the t-SNE plot to check whether the clusters in the t-SNE plot overlap.

> **NOTE**
>
> To access the source code for this specific section, please refer to https://packt.live/34gLGKa.
>
> You can also run this example online at https://packt.live/3fbWQES.
>
> You must execute the entire Notebook in order to get the desired result.

Let's now apply our knowledge of LDA to another dataset.

ACTIVITY 7.02: LDA AND HEALTH TWEETS

In this activity, we'll apply LDA to the health tweets data loaded and cleaned in *Activity 7.01, Loading and Cleaning Twitter Data*. Remember to use the same notebook used in that activity. Once the steps have been executed, discuss the results of the model. Do these word groupings make sense?

For this activity, let's imagine that we are interested in acquiring a high-level understanding of the major public health topics. That is, what people are talking about in the world of health. We have collected some data that could shed light on this inquiry. The easiest way to identify the major topics in the dataset, as we have discussed, is topic modeling.

Here are the steps to complete the activity:

1. Specify the **number_words**, **number_docs**, and **number_features** variables.
2. Create a bag-of-words model and assign the feature names to another variable for use later on.
3. Identify the optimal number of topics.
4. Fit the LDA model using the optimal number of topics.

Latent Dirichlet Allocation | 297

5. Create and print the word-topic table.
6. Print the document-topic table.
7. Create a biplot visualization.
8. Keep the notebook open for future modeling.

The output will be as follows:

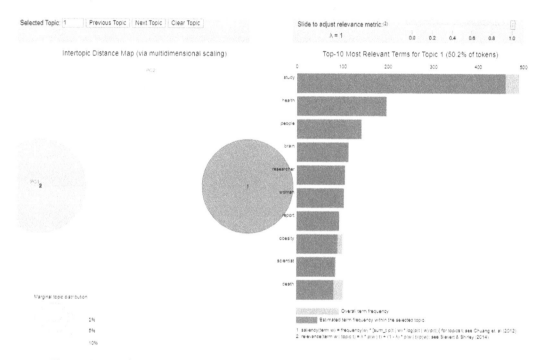

Figure 7.39: A histogram and biplot for the LDA model trained on health tweets

> **NOTE**
> The solution to this activity can be found on page 482.

EXERCISE 7.10: CREATING A BAG-OF-WORDS MODEL USING TF-IDF

In this exercise, we will create a bag-of-words model using TF-IDF:

1. Run the TF-IDF vectorizer and print out the first few rows:

```
vectorizer2 = sklearn.feature_extraction.text.TfidfVectorizer\
               (analyzer="word", \
                max_df=0.5, \
                min_df=20, \
                max_features=number_features, \
                smooth_idf=False)
clean_vec2 = vectorizer2.fit_transform(clean_sentences)
print(clean_vec2[0])
```

The output is as follows:

```
(0, 573)     0.4507821468344468
(0, 558)     0.46659328459850813
(0, 644)     0.2348350665788214
(0, 88)      0.2807186318482775
(0, 408)     0.6671849099770276
```

Figure 7.40: Output of the TF-IDF vectorizer

2. Return the feature names (the actual words in the corpus dictionary) to use when analyzing the output. You will recall that we did the same thing when we ran **CountVectorizer** in *Exercise 7.05, Creating a Bag-of-Words Model Using the Count Vectorizer*:

```
feature_names_vec2 = vectorizer2.get_feature_names()
feature_names_vec2
```

A section of the output is as follows:

```
['abbas',
 'ability',
 'accelerate',
 'accept',
 'access',
 'accord',
 'account',
 'accused',
 'achieve',
```

```
'acknowledge',
'acquire',
'acquisition',
'across',
'action',
'activist',
'activity',
'actually',
```

In this exercise, we summarized the corpus in the form of a bag-of-words model. Weights were computed for each document word combination. This bag of words output will return later on during the fitting on our next topic model. The next section will introduce NMF.

> **NOTE**
>
> To access the source code for this specific section, please refer to https://packt.live/34gLGKa.
>
> You can also run this example online at https://packt.live/3fbWQES.
>
> You must execute the entire Notebook in order to get the desired result.

NON-NEGATIVE MATRIX FACTORIZATION

Unlike LDA, **Non-Negative Matrix Factorization** (**NMF**) is not a probabilistic model. instead, it is, as the name implies, an approach involving linear algebra. Using matrix factorization as an approach to topic modeling was introduced by Daniel D. Lee and H. Sebastian Seung in 1999. The approach falls into the decomposition family of models that includes PCA, the modeling technique introduced in *Chapter 4, Introduction to Dimensionality Reduction and PCA*.

The major differences between PCA and NMF are that PCA requires components to be perpendicular while allowing them to be either positive or negative. NMF requires that matrix components be non-negative, which should make sense if you think of this requirement in the context of the data. Topics cannot be negatively related to documents, and words cannot be negatively related to topics.

If you are not convinced, try to interpret a negative weight associating a topic with a document. It would be something like, topic T makes up -30% of document D; but what does that even mean? It is nonsensical, so NMF has non-negative requirements for every part of the matrix factorization.

Let's define the matrix to be factorized, X, as a term-document matrix where the rows are words and the columns are documents. Each element of matrix X is either the number of occurrences of word *i* (the row) in document *j* (the column) or some other quantification of the relationship between word *i* and document *j*. The matrix, X, is naturally a sparse matrix as most elements in the term-document matrix will be zero, since each document only contains a limited number of words. There will be more on creating this matrix and deriving the quantifications later.

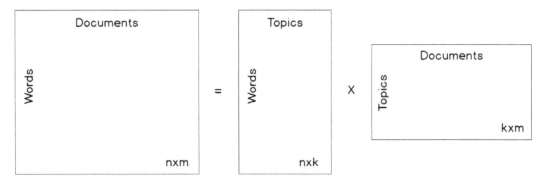

Figure 7.41: The matrix factorization

The matrix factorization takes the form $X_{n \times m} \approx W_{n \times k} H_{k \times m}$, where the two component matrices, W and H, represent the topics as collections of words and the topic weights for each document, respectively. More specifically, $W_{n \times k}$ is a word by topic matrix, while $H_{k \times m}$ is a topic by document matrix and, as stated earlier, $X_{n \times m}$ is a word by document matrix.

A nice way to think of this factorization is as a weighted sum of word groupings defining abstract topics. The equivalency symbol in the formula for the matrix factorization is an indicator that the factorization WH is an approximation, and thus, the product of those two matrices will not reproduce the original term-document matrix exactly.

The goal, as it was with LDA, is to find the approximation that is closest to the original matrix. Like X, both W and H are sparse matrices as each topic is only related to a few words, and each document is a mixture of only a small number of topics—one topic in many cases.

THE FROBENIUS NORM

The goal of solving NMF is the same as that of LDA: find the best approximation. To measure the distance between the input matrix and the approximation, NMF can use virtually any distance measure, but the standard is the Frobenius norm, also known as the Euclidean norm. The Frobenius norm is the sum of the element-wise squared errors mathematically expressed as $\|A\|_F = \sqrt{\sum_{i=1}^{m}\sum_{j=1}^{n}|a_{i,j}|^2}$.

With the measure of distance selected, the next step is to define the objective function. The minimization of the Frobenius norm will return the best approximation of the original term-document matrix and, thus, the most reasonable topics. Note that the objective function is minimized with respect to W and H so that both matrices are non-negative. It is expressed as $\min_{W \geq 0, H \geq 0} |X - WH|_F^2$.

THE MULTIPLICATIVE UPDATE ALGORITHM

The optimization algorithm used to solve NMF by Lee and Seung in their 1999 paper is the Multiplicative Update algorithm, and it is still one of the most commonly used solutions. It will be implemented in the exercises and activities later in the chapter.

The update rules, for both W and H, are derived by expanding the objective function and taking the partial derivatives with respect to W and H. The derivatives are not difficult but do, require fairly extensive linear algebra knowledge, and are time-consuming, so let's skip the derivatives and just state the updates. Note that, in the update rules, *i* is the current iteration and *T* means the transpose of the matrix. The first update rule is as follows:

$$H^{i+1} \leftarrow H^i \frac{(w^i)^T X}{(w^i)^T W^i H^i}$$

Figure 7.42: First update rule

The second update rule is as follows:

$$w^{i+1} \leftarrow w^i \frac{X(H^{i+1})^T}{W^i H^{i+1}(H^{i+1})^T}$$

Figure 7.43: Second update rule

W and H are updated iteratively until the algorithm converges. The objective function can also be shown to be non-increasing; that is, with each iterative update of W and H, the objective function gets closer to the minimum. Note that the multiplicative update optimizer, if the update rules are reorganized, is a rescaled gradient descent algorithm.

The final component of building a successful NMF algorithm is initializing the W and H component matrices so that the multiplicative update works quickly. A popular approach to initializing matrices is **Singular Value Decomposition** (**SVD**), which is a generalization of Eigen decomposition.

In the implementation of NMF undertaken in the forthcoming exercises, the matrices are initialized via non-negative Double Singular Value Decomposition, which is basically a more advanced version of SVD that is strictly non-negative. The full details of these initialization algorithms are not important for understanding NMF. Just note that initialization algorithms are used as a starting point for the optimization algorithms and can drastically speed up convergence.

EXERCISE 7.11: NON-NEGATIVE MATRIX FACTORIZATION

In this exercise, we'll fit the NMF algorithm and output the same two result tables we previously did with LDA. Those tables are the word-topic table, which shows the top 10 words associated with each topic, and the document-topic table, which shows the top 10 documents associated with each topic.

There are two additional parameters in the NMF algorithm function that we have not previously discussed, which are `alpha` and `l1_ratio`. If an overfit model is of concern, these parameters control how (`l1_ratio`) and the extent to which (`alpha`) regularization is applied to the objective function:

> **NOTE**
>
> More details can be found in the documentation for the scikit-learn library (https://scikit-learn.org/stable/modules/generated/sklearn.decomposition.NMF.html).

1. Define the NMF model and call the **fit** function using the output of the TF-IDF vectorizer:

   ```
   nmf = sklearn.decomposition.NMF(n_components=4, \
                                   init="nndsvda", \
                                   solver="mu", \
                                   beta_loss="frobenius", \
                                   random_state=0, \
                                   alpha=0.1, \
                                   l1_ratio=0.5)
   nmf.fit(clean_vec2)
   ```

   The output is as follows:

   ```
   NMF(alpha=0.1, beta_loss='frobenius', init='nndsvda', l1_ratio=0.5,
       max_iter=200, n_components=4, random_state=0, shuffle=False, solver='mu',
       tol=0.0001, verbose=0)
   ```

   Figure 7.44: Defining the NMF model

2. Run the **get_topics** functions to produce the two output tables:

   ```
   W_df, H_df = get_topics(mod=nmf, \
                           vec=clean_vec2, \
                           names=feature_names_vec2, \
                           docs=raw, \
                           ndocs=number_docs, \
                           nwords=number_words)
   ```

3. Print the **W** table:

   ```
   print(W_df)
   ```

The output is as follows:

```
           Topic0                  Topic1              Topic2  \
Word0  (0.0696, obama)    (0.0628, economy)    (0.087, microsoft)
Word1  (0.0645, president) (0.0212, economic)  (0.0305, windows)
Word2  (0.0484, barack)   (0.0179, growth)     (0.0196, company)
Word3  (0.0157, washington) (0.0144, global)   (0.0162, announce)
Word4  (0.0149, house)    (0.0128, china)      (0.0124, microsofts)
Word5  (0.0144, white)    (0.0111, percent)    (0.0118, update)
Word6  (0.0127, obamas)   (0.0109, world)      (0.0106, release)
Word7  (0.0109, state)    (0.0097, quarter)    (0.01, today)
Word8  (0.0096, administration) (0.0093, market) (0.0096, surface)
Word9  (0.0081, first)    (0.0086, country)    (0.0085, cloud)

           Topic3
Word0  (0.0881, palestine)
Word1  (0.0765, palestinian)
Word2  (0.0309, israeli)
Word3  (0.0279, israel)
Word4  (0.0171, state)
Word5  (0.0094, international)
Word6  (0.0092, ramallah)
Word7  (0.0089, minister)
Word8  (0.0079, unite)
Word9  (0.0078, force)
```

Figure 7.45: The word-topic table containing probabilities

4. Print the **H** table:

```
print(H_df)
```

The output is as follows:

```
                                                            Topic0  \
Doc0    (0.0844, A makeshift racetrack in the Palestin...
Doc1    (0.0844, (CNN) -- President Barack Obama will ...
Doc2    (0.0844, America is emerging as a top tax have...
Doc3    (0.0844, Microsoft wants its software to run e...
Doc4    (0.0844, The process began in Obama's first ye...
Doc5    (0.0844, Negative Interest Rates Benefit Globa...
Doc6    (0.0844, South Africa's economy shrank sharply...
Doc7    (0.0844, Conservatives mock President Barack O...
Doc8    (0.0844, AFP ISTANBUL (AFP-Jiji) """ Six suici...
Doc9    (0.0844, Japanese officials have raided the he...

                                                            Topic1  \
Doc0    (0.0677, The super PAC backing New Jersey Gov....
Doc1    (0.0677, President Obama speaks about the Nati...
Doc2    (0.0677, Windows 10 is being offered by Micros...
Doc3    (0.0677, President Obama is a man of many tale...
Doc4    (0.0677, New Zealand's finance minister said o...
Doc5    (0.0677, Each year nearly $4 billion new dolla...
Doc6    (0.0677, Even those who disagree with Presiden...
Doc7    (0.0677, Central banks' ultra-loose monetary p...
Doc8    (0.0677, New Zealand's economy grew at a faste...
Doc9    (0.0677, Venezuela's economy contracted 4.5 pe...

                                                            Topic2  \
Doc0    (0.0836, But the argument that Microsoft is wi...
Doc1    (0.0836, Oil spills and other forms of polluti...
Doc2    (0.0836, President Obama promised he would run...
Doc3    (0.0836, President Obama will propose in his u...
Doc4    (0.0836, Join the Palestine Public Library at ...
Doc5    (0.0836, Colorado's economy had the fourth str...
Doc6    (0.0836, An international law firm has withdra...
Doc7    (0.0836, Certainly, we are all on the shoulder...
Doc8    (0.0836, (Photo : Getty Images) Seems like Red...
Doc9    (0.0836, Thousands March During Obama Argentin...

                                                            Topic3
Doc0    (0.1078, Full details of Barack Obama's histor...
Doc1    (0.1078, It's been half a year since we propos...
Doc2    (0.0857, (Reuters) """ On Monday, U.S. Distric...
Doc3    (0.0853, Yeah, it seemed radical at the time""...
Doc4    (0.0842, U.S. President Barack Obama, left, an...
Doc5    (0.0828, Senate Majority Leader Mitch McConnel...
Doc6    (0.0815, American multinational technology com...
Doc7    (0.0815, Former Florida Gov. Jeb Bush (R) said...
Doc8    (0.0815, (Reuters) - Microsoft Corp executives...
Doc9    (0.0815, President Obama's crusade to bury the...
```

Figure 7.46: The document-topic table containing probabilities

The word-topic table contains word groupings that suggest the same abstract topics that the four-topic LDA model produced in *Exercise 7.09, Trying Four Topics*. However, the interesting part of the comparison is that some of the individual words contained in these groupings are new or in a new place in the grouping. This is not surprising given that the methodologies are distinct. Given the alignment with the topics specified in the original dataset, we have shown that both of these methodologies are effective tools for extracting the underlying topic structure of the corpus.

As we did with our previously fit LDA model, we will visualize the results of our NMF model.

> **NOTE**
>
> To access the source code for this specific section, please refer to https://packt.live/34gLGKa.
>
> You can also run this example online at https://packt.live/3fbWQES.
>
> You must execute the entire Notebook in order to get the desired result.

EXERCISE 7.12: VISUALIZING NMF

The purpose of this exercise is to visualize the results of NMF. Visualizing the results gives insight into the distinctness of the topics and the prevalence of each topic in the corpus. In this exercise, we'll do the visualizing using t-SNE, which was discussed fully in *Chapter 6, t-Distributed Stochastic Neighbor Embedding*:

1. Run **transform** on the cleaned data to get the topic-document allocations. Print both the shape and an example of the data:

    ```
    nmf_transform = nmf.transform(clean_vec2)
    print(nmf_transform.shape)
    print(nmf_transform)
    ```

The output is as follows:

```
(92946, 4)
[[5.12653315e-02 3.60582233e-15 3.19729419e-34 8.17267206e-16]
 [7.43734737e-04 2.04138105e-02 6.85552731e-15 2.11679327e-03]
 [2.92397552e-15 1.94083984e-02 4.76691813e-21 1.24269313e-18]
 ...
 [9.83404082e-06 3.41225477e-03 6.14009658e-04 3.23919592e-02]
 [6.51294966e-07 1.32359509e-07 3.32509174e-08 6.14671536e-02]
 [4.53925928e-05 1.16401194e-04 1.84755839e-02 2.00616344e-03]]
```

2. Run the **plot_tsne** function to fit a t-SNE model and plot the results:

```
plot_tsne(data=nmf_transform, threshold=0)
```

The plot appears as follows:

```
LENGTH:
92946

COUNTS:
[[    0 28973]
 [    1 32951]
 [    2 22148]
 [    3  8874]]
```

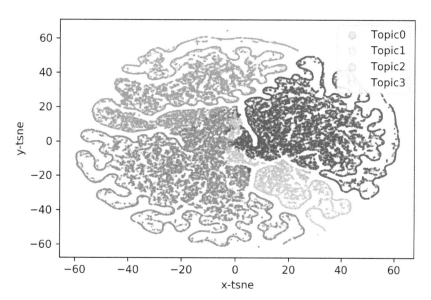

Figure 7.47: t-SNE plot with metrics summarizing the topic distribution across the corpus

> **NOTE**
>
> The results can differ slightly because of the optimization algorithms that support both LDA and NMF. Many of the functions do not have a seed setting capability.
>
> To access the source code for this specific section, please refer to https://packt.live/34gLGKa.
>
> You can also run this example online at https://packt.live/3fbWQES.
>
> You must execute the entire Notebook in order to get the desired result.

The t-SNE plot, with no threshold specified, shows some topic overlap and a clear discrepancy in the topic frequency across the corpus. These two facts explain why, when using perplexity, the optimal number of topics was three. There seems to be some correlation between topics that the model can't fully accommodate. Even with the correlation between topics, the model is finding the topics it should when the number of topics is set to four.

To recap, NMF is a non-probabilistic topic model that seeks to answer the same question LDA is trying to answer. It uses a popular concept of linear algebra known as matrix factorization, which is the process of breaking a large and intractable matrix down into smaller and more easily interpretable matrices that can be leveraged to answer many questions about the data. Remember that the non-negative requirement is not rooted in mathematics, but in the data itself. It does not make sense for the components of any document to be negative.

In many cases, NMF does not perform as well as LDA, because LDA incorporates prior distributions that add an extra layer of information to help inform the topic word groupings. However, we know that there are cases, especially when the topics are highly correlated, when NMF is the better performer. One of those cases was the headline data on which all the exercises were based.

Let's now try to apply our new knowledge of NMF to the Twitter dataset used in the previous activities.

ACTIVITY 7.03: NON-NEGATIVE MATRIX FACTORIZATION

This activity is the summation of the topic modeling analysis done on the health Twitter data loaded and cleaned in *Activity 7.01, Loading and Cleaning Twitter Data*, and on which LDA was done in *Activity 7.02, LDA and Health Tweets*. The execution of NMF is straightforward and requires limited coding. We can take this opportunity to play with the parameters of the model while thinking about the limitations and benefits of NMF.

Here are the steps to complete the activity:

1. Create the appropriate bag-of-words model and output the feature names as another variable.

2. Define and fit the NMF algorithm using the number of topics (**n_components**) value from *Activity 7.02, LDA and Health Tweets*.

3. Get the topic-document and word-topic tables. Take a few minutes to explore the word groupings and try to define the abstract topics. Can you quantify the meanings of the word groupings? Do the word groupings make sense? Are the results similar to those produced using LDA?

4. Adjust the model parameters and rerun *Step 3* and *Step 4*. How do the results change?

    The output will be as follows:

	Topic0	Topic1
Word0	(0.3764, study)	(0.5933, latfit)
Word1	(0.0258, cancer)	(0.049, steps)
Word2	(0.0208, people)	(0.0449, today)
Word3	(0.0184, obesity)	(0.0405, exercise)
Word4	(0.0184, health)	(0.0274, healthtips)
Word5	(0.0181, brain)	(0.0258, workout)
Word6	(0.0173, suggest)	(0.0204, getting)
Word7	(0.0168, weight)	(0.0193, fitness)
Word8	(0.0153, woman)	(0.0143, great)
Word9	(0.0131, death)	(0.0132, morning)

    Figure 7.48: The word-topic table with probabilities

> **NOTE**
>
> The solution to this activity can be found on page 487.

SUMMARY

When faced with the task of extracting information from an as yet unseen large collection of documents, topic modeling is a great approach, as it provides insights into the underlying structure of the documents. That is, topic models find word groupings using proximity, not context.

In this chapter, we have learned how to apply two of the most common and most effective topic modeling algorithms: latent Dirichlet allocation and non-negative matrix factorization. You should now feel comfortable cleaning raw text documents using several different techniques; techniques that can be utilized in many other modeling scenarios. We continued by learning how to convert the cleaned corpus into the appropriate data structure of per-document raw word counts or word weights by applying bag-of-words models.

The main focus of the chapter was fitting the two topic models, including optimizing the number of topics, converting the output to easy-to-interpret tables, and visualizing the results. With this information, you should be able to apply fully functioning topic models to derive value and insights for your business.

In the next chapter, we will change direction entirely. We will deep dive into market basket analysis.

8
MARKET BASKET ANALYSIS

OVERVIEW

In this chapter, we will explore market basket analysis, which is an algorithm originally designed to help retailers understand and improve their businesses. It is not, however, exclusive to retail, as we will discuss throughout the chapter. Market basket analysis unlocks the underlying relationships between the items that customers purchase. By the end of this chapter, you should have a solid grasp of transaction data, the basic metrics that define the relationship between two items, the Apriori algorithm, and association rules.

INTRODUCTION

Most data science practitioners would agree that natural language processing, including topic modeling, is toward the cutting edge of data science and is an active research area. We now understand that topic models can, and should, be leveraged wherever text data could potentially drive insights or growth, including in social media analyses, recommendation engines, and news filtering. The preceding chapter featured an exploration of the fundamental features of topic models and two of the major algorithms. In this chapter, we are going to change direction entirely.

This chapter takes us into the retail space to explore a foundational and reliable algorithm for analyzing transaction data. While this algorithm might not be on the cutting edge or in the catalog of the most popular machine learning algorithms, it is ubiquitous and undeniably impactful in the retail space. The insights it drives are easily interpretable, immediately actionable, and instructive for determining analytical next steps. If you work in the retail space or with transaction data, you would be well-served to dive deep into market basket analysis. Market basket analysis is important because it provides insight into why people buy certain items together and whether those item combinations can be leveraged to hasten growth and or increase profitability.

MARKET BASKET ANALYSIS

Imagine you work for a retailer that sells dozens of products and your boss comes to you and asks the following questions:

- What products are purchased together most frequently?
- How should the products be organized and positioned in the store?
- How do we identify the best products to discount via coupons?

You might reasonably respond with complete bewilderment, as those questions are very diverse and do not immediately seem answerable using a single algorithm and dataset. However, the answer to all those questions and many more is **market basket analysis**. The general idea behind market basket analysis is to identify and quantify which items, or groups of items, are purchased together frequently enough to drive insight into customer behavior and product relationships.

Before we dive into the analytics, it is worth defining the term *market basket*. A market basket is a permanent set of products in an economic system. Permanent does not necessarily mean permanent in the traditional sense. It means that until such time as the product is taken out of the catalog, it will consistently be available for purchase. The product referenced in the preceding definition is any good, service, or element of a group, including a bicycle, having your house painted, or a website. Lastly, an economic system could be a company, a collection of activities, or a country. The easiest example of a market basket is a grocery store, which is a system made up of a collection of food and drink items:

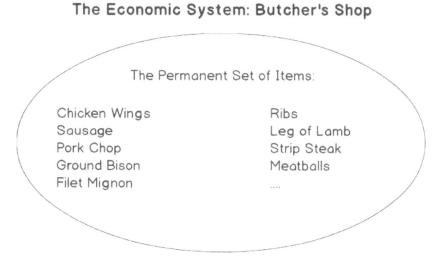

Figure 8.1: An example market basket

Even without using any models or analyses, certain product relationships are obvious. Let's take the relationship between meat and vegetables. Typically, market basket analysis models return relationships more specific than meat and vegetables, but, for argument's sake, we will generalize to meat and vegetables. Okay, there is a relationship between meat and vegetables. So what? Well, we know these are staple items that are frequently purchased together. We can leverage this information by putting the vegetables and meats on opposite sides of the store, which you will notice is often the positioning of those two items, forcing customers to walk the full distance of the store, and thereby increasing the likelihood that they will buy additional items that they might not have bought had they not traversed the whole store.

One of the things retail companies struggle with is how to discount items effectively. Let's consider another obvious relationship: peanut butter and jelly. In the United States, peanut butter and jelly sandwiches are incredibly popular, especially among children. When peanut butter is in a shopping basket, the chance jelly is also there can be assumed to be quite high. Since we know peanut butter and jelly are purchased together, it does not make sense to discount them both. If we want customers to buy both items, we can just discount one of the items, knowing that if we can get the customers to buy the discounted item, they will probably buy the other item too, even if it is full price. Just like the topic models in the preceding chapter, market basket analysis is all about identifying frequently occurring groups. The following figure presents an example of such groups:

Figure 8.2: A visualization of market basket analysis

In market basket analysis, we are looking for frequently occurring groups of products, whereas in topic models, we were looking for frequently occurring groups of words. Thus, as it was to topic models, the word *clustering* could be applied to market basket analysis. The major differences are that the clusters in market basket analysis are micro – only a few products per cluster – and the order of the items in the cluster matters when it comes to computing probabilistic metrics. We will dive much deeper into these metrics and how they are calculated later in this chapter.

What has clearly been implied by the previous two examples is that, in market basket analysis, retailers can discover the relationships – obvious and surprising – between the products that customers buy. Once uncovered, the relationships can be used to inform and improve the decision-making process. A great aspect of market basket analysis is that while this analysis was developed in relation to, discussed in terms of, and mostly applied to, the retail world, it can be applied to many different types of businesses.

The only requirement for performing this type of analysis is that the data is a list of collections of items. In the retail case, this would be a list of transactions where each transaction is a group of purchased products. One example of an alternative application is analyzing website traffic. With website traffic, we consider the products to be websites, so each element of the list is the collection of websites visited by an individual over a specified time period. Needless to say, the applications of market basket analysis extend well beyond the initial retail application.

USE CASES

There are three principal use cases in the traditional retail application: pricing enhancement, coupon and discount recommendations, and store layout. As was briefly mentioned previously, by using the product associations uncovered by the model, retailers can strategically place products in their stores to get customers to buy more items and thus spend more money. If any relationship between two or more products is sufficiently strong, meaning the product grouping occurs often in the dataset and the individual products in the grouping appear separate from the group infrequently, then the products could be placed far away from one another in the store without significantly jeopardizing the odds of the customer purchasing both products. By forcing the customer to traverse the whole store to get both products, the retailer increases the chances that the customer will notice and purchase additional products. Likewise, retailers can increase the chances of customers purchasing two weakly related or non-staple products by placing the two items next to each other. Obviously, there are a lot of factors that drive store layout, but market basket analysis is definitely one of those factors:

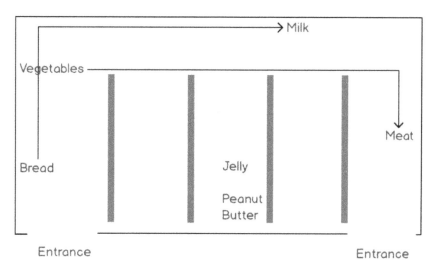

Figure 8.3: Product associations that can help inform efficient and lucrative store layouts

Pricing enhancement and coupon and discount recommendations are two sides of the same coin. They can simply be interpreted as where to raise and where to lower prices. Consider the case of two strongly related items. These two items are most likely going to be purchased in the same transaction, so one way to increase the profitability of that transaction would be to increase the price of one of the items. If the association between the two items is sufficiently strong, the price increase can be made with little to no risk of the customer not purchasing both items. In a similar way, retailers can encourage customers to purchase an item weakly associated with another through discounting or couponing.

For example, retailers could compare the purchase history of individual customers with the results of market basket analysis done on all transactions and find where some of the items certain customers are purchasing are weakly associated with items those customers are not currently purchasing. Using this comparison, retailers could offer discounts to the customers for the as-yet-unpurchased items the model suggested were related to the items previously purchased by those customers. If you have ever had coupons print out with your receipt at the end of a transaction, the chances are high that those items were found to be related to the items involved in your just-completed transaction.

A non-traditional, but viable, use of market basket analysis would be to enhance online advertising and search engine optimization. Imagine we had access to lists of websites visited by individuals. Using market basket analysis, we could find relationships between websites and use those relationships to both strategically order and group the websites resulting from a search engine query. In many ways, this is similar to the store layout use case.

With a general sense of what market basket analysis is all about and a clear understanding of its use cases, let's dig into the data used in these models.

IMPORTANT PROBABILISTIC METRICS

Market basket analysis is built upon the computation of several probabilistic metrics. The five major metrics covered here are **support**, **confidence**, **lift**, **leverage**, and **conviction**. Before digging into transaction data and the specific market basket analysis models, including the **Apriori algorithm** and **association rules**, we should spend some time defining and exploring these metrics using a small, made-up set of transactions. We begin by making up some data to use.

EXERCISE 8.01: CREATING SAMPLE TRANSACTION DATA

Since this is the first exercise of the chapter, let's set the environment. This chapter will use the same environment requirements that were used in *Chapter 7*, *Topic Modeling*. If any of the packages do not load, as happened in the preceding chapter, use **pip** to install them via the command line. One of the libraries we will use is **mlxtend**, which may be unfamiliar to you. It is a machine learning extensions library that contains useful supplemental tools, including ensembling, stacking, and of course, market basket analysis models. This exercise does not have any real output. We will simply create a sample transaction dataset for use in subsequent exercises:

1. Open a Jupyter notebook with Python 3.

2. Install the following libraries: **matplotlib.pyplot**, which is used to plot the results of the models; **mlxtend.frequent_patterns**, which is used to run the models; **mlxtend.preprocessing**, which is used to encode and prep the data for the models; **numpy**, which is used to work with arrays; and **pandas**, which is used to work with DataFrames.

> **NOTE**
>
> To install **mlxtend**, go to the Anaconda prompt and execute **pip install mlxtend**.

The code is as follows:

```
import matplotlib.pyplot as plt
import mlxtend.frequent_patterns
import mlxtend.preprocessing
import numpy
import pandas
```

3. Create 10 fake transactions featuring grocery store items, and then print out the transactions. The data will take the form of a list of lists, a data structure that will be relevant later when discussing formatting transaction data for the models:

```
example = [['milk', 'bread', 'apples', 'cereal', 'jelly', \
            'cookies', 'salad', 'tomatoes'], \
           ['beer', 'milk', 'chips', 'salsa', 'grapes', \
            'wine', 'potatoes', 'eggs', 'carrots'], \
           ['diapers', 'baby formula', 'milk', 'bread', \
            'chicken', 'asparagus', 'cookies'], \
```

```
                ['milk', 'cookies', 'chicken', 'asparagus', \
                 'broccoli', 'cereal', 'orange juice'], \
                ['steak', 'asparagus', 'broccoli', 'chips', \
                 'salsa', 'ketchup', 'potatoes', 'salad'], \
                ['beer', 'salsa', 'asparagus', 'wine', 'cheese', \
                 'crackers', 'strawberries', 'cookies'],\
                ['chocolate cake', 'strawberries', 'wine', 'cheese', \
                 'beer', 'milk', 'orange juice'],\
                ['chicken', 'peas', 'broccoli', 'milk', 'bread', \
                 'eggs', 'potatoes', 'ketchup', 'crackers'],\
                ['eggs', 'bread', 'cheese', 'turkey', 'salad', \
                 'tomatoes', 'wine', 'steak', 'carrots'],\
                ['bread', 'milk', 'tomatoes', 'cereal', 'chicken', \
                 'turkey', 'chips', 'salsa', 'diapers']]
print(example)
```

The output is as follows:

```
[['milk', 'bread', 'apples', 'cereal', 'jelly', 'cookies', 'salad', 'tomatoes'], ['beer', 'milk',
'chips', 'salsa', 'grapes', 'wine', 'potatoes', 'eggs', 'carrots'], ['diapers', 'baby formula', 'm
ilk', 'bread', 'chicken', 'asparagus', 'cookies'], ['milk', 'cookies', 'chicken', 'asparagus', 'br
occoli', 'cereal', 'orange juice'], ['steak', 'asparagus', 'broccoli', 'chips', 'salsa', 'ketchu
p', 'potatoes', 'salad'], ['beer', 'salsa', 'asparagus', 'wine', 'cheese', 'crackers', 'strawberri
es', 'cookies'], ['chocolate cake', 'strawberries', 'wine', 'cheese', 'beer', 'milk', 'orange juic
e'], ['chicken', 'peas', 'broccoli', 'milk', 'bread', 'eggs', 'potatoes', 'ketchup', 'crackers'],
['eggs', 'bread', 'cheese', 'turkey', 'salad', 'tomatoes', 'wine', 'steak', 'carrots'], ['bread',
'milk', 'tomatoes', 'cereal', 'chicken', 'turkey', 'chips', 'salsa', 'diapers']]
```

Figure 8.4: Printing the transactions

Now that we have created our dataset, we will explore several probabilistic metrics that quantify the relationship between pairs of items.

> **NOTE**
>
> To access the source code for this specific section, please refer to https://packt.live/3fhf9bS.
>
> You can also run this example online at https://packt.live/303vIBJ.
>
> You must execute the entire Notebook in order to get the desired result.

SUPPORT

Support is simply the probability that a given item set appears in the data, which can be calculated by counting the number of transactions in which the item set appears and dividing that count by the total number of transactions.

> **NOTE**
>
> An item set can be a single item or a group of items.

Support is an important metric, despite being very simple, as it is one of the primary metrics used to determine the believability and strength of association between groups of items. For example, it is possible to have two items that only occur with each other, suggesting that their association is very strong, but in a dataset containing 100 transactions, only appearing twice is not very impressive. Because the item set appears in only 2% of the transactions, and 2% is small in terms of the raw number of appearances, the association cannot be considered significant and, therefore, is probably unusable in decision making.

Note that since support is a probability, it will fall in the range [0,1]. The formula takes the following form if the item set is two items, X and Y, and N is the total number of transactions:

$$Support(X \Rightarrow Y) = Support(X,Y) = P(X,Y) = \frac{Frequency(X,Y)}{N}$$

Figure 8.5: Formula for support

While working with market basket analysis, if the support for an item or group of items is lower than a pre-defined threshold, then the purchase of that item or group of items is rare enough to not be actionable. Let's return momentarily to the made-up data from *Exercise 8.01, Creating Sample Transaction Data*, and define an item set as being milk and bread. We can easily look through the 10 transactions and count the number of transactions in which this milk and bread item set occurs – that would be 4 times. Given that there are 10 transactions, the support of milk and bread is 4 divided by 10, or 0.4. Whether this is large enough support depends on the dataset itself, which we will get into in a later section.

CONFIDENCE

The **confidence** metric can be thought of in terms of conditional probability, as it is basically the probability that product B is purchased given the purchase of product A. Confidence is typically notated as A ⇒ B, and expressed as the proportion of transactions containing A that also contain B. Hence, confidence is found by filtering the full set of transactions down to those containing A, and then computing the proportion of those transactions that contain B. Like support, confidence is a probability, so its range is [0,1]. Using the same variable definitions as in the *Support* section, the following is the formula for confidence:

$$Confidence\ (X \Rightarrow Y) = P(Y|X) = \frac{Support(X,Y)}{P(X)} = \frac{\frac{Frequency(X,Y)}{N}}{\frac{Frequency(X)}{N}}$$

Figure 8.6: Formula for confidence

To demonstrate confidence, we will use the items beer and wine. Specifically, let's compute the confidence of Beer ⇒ Wine. To begin, we need to identify the transactions that contain beer. There are three of them, and they are transactions 2, 6, and 7. Now, of those transactions, how many contain wine? The answer is all of them. Thus, the confidence of Beer ⇒ Wine is 1. Every time a customer bought beer, they also bought wine. It might be obvious, but for identifying actionable associations, higher confidence values are better.

LIFT AND LEVERAGE

We will discuss the next two metrics, lift and leverage, simultaneously, since despite being calculated differently, both seek to answer the same question. Like confidence, **lift** and **leverage** are notated as A ⇒ B. The question to which we seek an answer is, can one item, say A, be used to determine anything about another item, say B? Stated another way, if product A is bought by an individual, can we say anything about whether they will or will not purchase product B with some level of confidence? These questions are answered by comparing the support of A and B under the standard case when A and B are not assumed to be independent with the case where the two products are assumed to be independent. Lift calculates the ratio of these two cases, so its range is [0, Infinity]. When lift equals one, the two products are independent and, hence, no conclusions can be made about product B when product A is purchased:

$$Lift(X \Rightarrow Y) = \frac{Support(X,Y)}{Support(X) * Support(Y)} = \frac{P(X,Y)}{P(X)*P(Y)}$$

Figure 8.7: Formula for lift

Leverage calculates the difference between the two cases, so its range is [-1, 1]. Leverage equaling zero can be interpreted the same way as lift equaling one:

$$Leverage(X \Rightarrow Y) = Support(X,Y) - (Support(X) * Support(Y)) = P(X,Y) - (P(X)*P(Y))$$

Figure 8.8: Formula for leverage

The values of the metrics measure the degree and orientation (in other words, positive or negative) of the relationship between the items. A value of lift other than 1 means that some dependency exists between the items. When the value is greater than 1, the second item is more likely to be purchased if the first item is purchased. Likewise, when the value is less than 1, the second item is less likely to be purchased if the first item is purchased. If the lift value is 0.1, we could say that the relationship between the two items is strong in the negative direction. That is, it could be said that when one product is purchased, the chance the second product is purchased is diminished. A lift of 1 indicates that the products are independent of one another. In the case of leverage, a positive value implies a positive association, while a negative value indicates a negative association. The positive and negative associations are separated by the points of independence, which, as stated earlier, are 1 for lift and 0 for leverage, and the further away the value gets from these points, the stronger the association.

CONVICTION

The last metric to be discussed is conviction, which is a bit less intuitive than the other metrics. Conviction is the ratio of the expected frequency that X occurs without Y, given that X and Y are independent of the frequency of incorrect predictions. The frequency of incorrect predictions is defined as 1 minus the confidence of $X \Rightarrow Y$. Remember that confidence can be defined as $P(Y|X)$, which means $1 - P(Y|X) = P(Not\ Y|X)$. The numerator could also be thought of as $1 - P(Y|X) = P(Not\ Y|X)$. The only difference between the two is that the numerator has the assumption of independence between X and Y, while the denominator does not. A value greater than 1 is ideal because that means the association between products or item sets X and Y is incorrect more often if the association between X and Y is random (in other words, X and Y are independent). To reiterate, this stipulates that the association between X and Y is meaningful. A value of 1 applies independence, and a value of less than 1 signifies that the random chance X and Y relationship is correct more often than the X and Y relationship that has been defined as $X \Rightarrow Y$. Under this situation, the relationship might go the other way (in other words, $Y \Rightarrow X$). Conviction has the range [0, Infinity] and the following form:

$$Conviction(X \Rightarrow Y) = \frac{1 - Support(Y)}{1 - Confidence\ (X \Rightarrow Y)}$$

Figure 8.9: Formula for conviction

Let's again return to the products beer and wine, but for this explanation, we will consider the opposite association of Wine ⟹ Beer. Support(Y) or, in this case, Support(Beer) is 3/10, and Confidence X ⟹ Y, or, in this case, Confidence(Wine ⟹ Beer) is 3/4. Thus, the Conviction(Wine ⟹ Beer) is (1-3/10) / (1-3/4) = (7/10) * (4/1). We can conclude by saying that Wine ⟹ Beer would be incorrect 2.8 times as often if wine and beer were independent. Thus, the previously articulated association between wine and beer is legitimate.

EXERCISE 8.02: COMPUTING METRICS

In this exercise, we'll use the fake data from *Exercise 8.01, Creating Sample Transaction Data*, to compute the five previously described metrics, which we will use again in the covering of the Apriori algorithm and association rules. The association on which these metrics will be evaluated is Milk ⟹ Bread:

> **NOTE**
>
> All exercises in this chapter need to be performed in the same Jupyter notebook.

1. Define and print the frequencies that are the basis of all five metrics, which would be Frequency(Milk), Frequency(Bread), and Frequency(Milk, Bread). Also, define **N** as the total number of transactions in the dataset:

```
# the number of transactions
N = len(example)
# the frequency of milk
f_x = sum(['milk' in i for i in example])
# the frequency of bread
f_y = sum(['bread' in i for i in example])
# the frequency of milk and bread
f_x_y = sum([all(w in i for w in ['milk', 'bread']) \
            for i in example])
# print out the metrics computed above
print("N = {}\n".format(N) + "Freq(x) = {}\n".format(f_x) \
      + "Freq(y) = {}\n".format(f_y) \
      + "Freq(x, y) = {}".format(f_x_y))
```

The output is as follows:

```
N = 10
Freq(x) = 7
Freq(y) = 5
Freq(x, y) = 4
```

2. Calculate and print Support(Milk ⟹ Bread):

```
support = f_x_y / N
print("Support = {}".format(round(support, 4)))
```

The support of **x** to **y** is **0.4**. From experience, if we were working with a full transaction dataset, this support value would be considered very large in many cases.

3. Calculate and print Confidence(Milk ⟹ Bread):

```
confidence = support / (f_x / N)
print("Confidence = {}".format(round(confidence, 4)))
```

The confidence of **x** to **y** is **0.5714**. This means that the probability of Y being purchased given that **x** was purchased is just slightly higher than 50%.

4. Calculate and print Lift(Milk ⟹ Bread):

```
lift = confidence / (f_y / N)
print("Lift = {}".format(round(lift, 4)))
```

The lift of **x** to **y** is **1.1429**.

5. Calculate and print Leverage(Milk ⟹ Bread):

```
leverage = support - ((f_x / N) * (f_y / N))
print("Leverage = {}".format(round(leverage, 4)))
```

The leverage of **x** to **y** is **0.05**. Both lift and leverage can be used to say that the association **x** to **y** is positive (in other words, **x** implies **y**) but weak. The values for lift and leverage are close to 1 and 0, respectively.

6. Calculate and print Conviction(Milk ⇒ Bread):

```
conviction = (1 - (f_y / N)) / (1 - confidence)
print("Conviction = {}".format(round(conviction, 4)))
```

The conviction value of **1.1667** can be interpreted by saying the Milk ⇒ Bread association would be incorrect **1.1667** times as often if milk and bread were independent.

In this exercise, we explored a series of probabilistic metrics designed to quantify the relationship between two items. The five metrics are support, confidence, lift, leverage, and conviction. We will use these metrics again as they are the foundation of both the Apriori algorithm and association rules.

> **NOTE**
>
> To access the source code for this specific section, please refer to https://packt.live/3fhf9bS.
>
> You can also run this example online at https://packt.live/303vlBJ.
>
> You must execute the entire Notebook in order to get the desired result.

Before diving into the Apriori algorithm and association rule learning on actual data, we will explore transaction data and get some retail data loaded and prepped for modeling.

CHARACTERISTICS OF TRANSACTION DATA

The data used in market basket analysis is transaction data or any type of data that resembles transaction data. In its most basic form, transaction data has some sort of transaction identifier, such as an invoice or transaction number, and a list of products associated with said identifier. It just so happens that these two base elements are all that is needed to perform market basket analysis. However, transaction data rarely – it is probably even safe to say never – comes in this basic form. Transaction data typically includes pricing information, dates and times, and customer identifiers, among many other things. Here is how each product is mapped to multiple invoices:

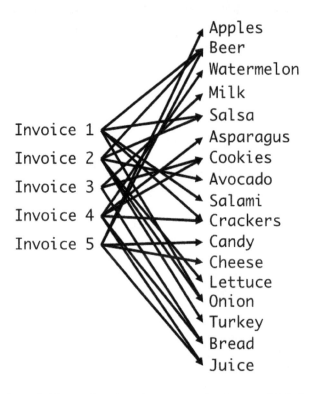

Figure 8.10: Each available product is going to map back to multiple invoice numbers

6. Calculate and print Conviction(Milk ⟹ Bread):

    ```
    conviction = (1 - (f_y / N)) / (1 - confidence)
    print("Conviction = {}".format(round(conviction, 4)))
    ```

    The conviction value of **1.1667** can be interpreted by saying the Milk ⟹ Bread association would be incorrect **1.1667** times as often if milk and bread were independent.

    In this exercise, we explored a series of probabilistic metrics designed to quantify the relationship between two items. The five metrics are support, confidence, lift, leverage, and conviction. We will use these metrics again as they are the foundation of both the Apriori algorithm and association rules.

    > **NOTE**
    >
    > To access the source code for this specific section, please refer to https://packt.live/3fhf9bS.
    >
    > You can also run this example online at https://packt.live/303vIBJ.
    >
    > You must execute the entire Notebook in order to get the desired result.

Before diving into the Apriori algorithm and association rule learning on actual data, we will explore transaction data and get some retail data loaded and prepped for modeling.

328 | Market Basket Analysis

CHARACTERISTICS OF TRANSACTION DATA

The data used in market basket analysis is transaction data or any type of data that resembles transaction data. In its most basic form, transaction data has some sort of transaction identifier, such as an invoice or transaction number, and a list of products associated with said identifier. It just so happens that these two base elements are all that is needed to perform market basket analysis. However, transaction data rarely – it is probably even safe to say never – comes in this basic form. Transaction data typically includes pricing information, dates and times, and customer identifiers, among many other things. Here is how each product is mapped to multiple invoices:

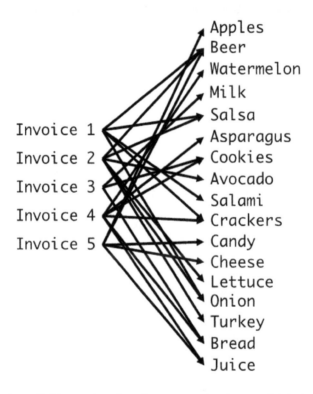

Figure 8.10: Each available product is going to map back to multiple invoice numbers

Due to the complexity of transaction data, data cleaning is crucial. The goal of data cleaning in the context of market basket analysis is to filter out all the unnecessary information, which includes removing variables in the data that are not relevant and filtering out problematic transactions. The techniques used to complete these two cleaning steps vary, depending on the particular transaction data file. In an attempt to not get bogged down in data cleaning, the exercises from here on out will use a subset of an online retail dataset from the UCI Machine Learning Repository, and the activities will use the entire dataset. This both limits the data cleaning discussion, but also gives us an opportunity to discuss how the results change when the size of the dataset changes. This is important because if you work for a retailer and run market basket analysis, it will be important to understand and be able to clearly articulate the fact that, as more data is received, product relationships can, and most likely will, shift. Before discussing the specific cleaning process required for this dataset, let's load the online retail dataset.

> **NOTE**
>
> In all subsequent exercises and activities, there could be slight differences in the output from what is shown in the following. This is due to one of two things: issues with data loading (in other words, rows getting shuffled) or the fact that `mlxtend` does not have a seed setting option to guarantee consistency across executions.

EXERCISE 8.03: LOADING DATA

In this exercise, we will load and view an example online retail dataset. This dataset is originally sourced from the UCI Machine Learning Repository and can be found at https://packt.live/2XeT6ft. Once you have downloaded the dataset, save it and note the path. Now, let's proceed with the exercise. The output of this exercise is the transaction data that will be used in future modeling exercises and some exploratory figures to help us better understand the data with which we are working.

> **NOTE**
>
> This dataset is sourced from http://archive.ics.uci.edu/ml/datasets/online+retail# (UCI Machine Learning Repository [http://archive.ics.uci.edu/ml]. Irvine, CA: University of California, School of Information and Computer Science).
>
> Citation: This is a subset of the online retail dataset obtained from the UCI Machine Learning repository. Daqing Chen, Sai Liang Sain, and Kun Guo, *Data mining for the online retail industry: A case study of RFM model-based customer segmentation using data mining*, Journal of Database Marketing and Customer Strategy Management, Vol. 19, No. 3, pp. 197-208, 2012.
>
> It can be downloaded from https://packt.live/2XeT6ft.

Perform the following steps:

1. Using the **read_excel** function from **pandas**, load the data. Note that we can define the first row as containing column names by adding **header=0** to the **read_excel** function:

```
online = pandas.read_excel(io="./Online Retail.xlsx", \
                           sheet_name="Online Retail", \
                           header=0)
```

> **NOTE**
>
> The path to **Online Retail.xlsx** should be changed as per the location of the file on your system.

2. Print out the first 10 rows of the DataFrame. Notice that the data contains some columns that will not be relevant to market basket analysis:

```
online.head(10)
```

The output is as follows:

	InvoiceNo	StockCode	Description	Quantity	InvoiceDate	UnitPrice	CustomerID	Country
0	536365	85123A	WHITE HANGING HEART T-LIGHT HOLDER	6	2010-12-01 08:26:00	2.55	17850.0	United Kingdom
1	536365	71053	WHITE METAL LANTERN	6	2010-12-01 08:26:00	3.39	17850.0	United Kingdom
2	536365	84406B	CREAM CUPID HEARTS COAT HANGER	8	2010-12-01 08:26:00	2.75	17850.0	United Kingdom
3	536365	84029G	KNITTED UNION FLAG HOT WATER BOTTLE	6	2010-12-01 08:26:00	3.39	17850.0	United Kingdom
4	536365	84029E	RED WOOLLY HOTTIE WHITE HEART.	6	2010-12-01 08:26:00	3.39	17850.0	United Kingdom
5	536365	22752	SET 7 BABUSHKA NESTING BOXES	2	2010-12-01 08:26:00	7.65	17850.0	United Kingdom
6	536365	21730	GLASS STAR FROSTED T-LIGHT HOLDER	6	2010-12-01 08:26:00	4.25	17850.0	United Kingdom
7	536366	22633	HAND WARMER UNION JACK	6	2010-12-01 08:28:00	1.85	17850.0	United Kingdom
8	536366	22632	HAND WARMER RED POLKA DOT	6	2010-12-01 08:28:00	1.85	17850.0	United Kingdom
9	536367	84879	ASSORTED COLOUR BIRD ORNAMENT	32	2010-12-01 08:34:00	1.69	13047.0	United Kingdom

Figure 8.11: The raw online retail data

3. Print out the data type for each column in the DataFrame. This information will come in handy when trying to perform specific cleaning tasks. Columns need to be of the correct type in order for filtering and computing to execute as expected:

```
online.dtypes
```

The output is as follows:

```
InvoiceNo               object
StockCode               object
Description             object
Quantity                 int64
InvoiceDate     datetime64[ns]
UnitPrice              float64
CustomerID             float64
Country                 object
dtype: object
```

4. Get the dimensions of the DataFrame, as well as the number of unique invoice numbers and customer identifications:

```
print("Data dimension (row count, col count): {dim}" \
      .format(dim=online.shape))
print("Count of unique invoice numbers: {cnt}" \
      .format(cnt=online.InvoiceNo.nunique()))
print("Count of unique customer ids: {cnt}" \
      .format(cnt=online.CustomerID.nunique()))
```

The output is as follows:

```
Data dimension (row count, col count): (541909, 8)
Count of unique invoice numbers: 25900
Count of unique customer ids: 4372
```

From the preceding output, we can say that we successfully loaded the data and obtained some key information which will be further used as we progress with exercises in this chapter.

> **NOTE**
>
> To access the source code for this specific section, please refer to https://packt.live/3fhf9bS.
>
> You can also run this example online at https://packt.live/303vIBJ.
>
> You must execute the entire Notebook in order to get the desired result.

DATA CLEANING AND FORMATTING

With the dataset now loaded, let's delve into the specific data cleaning processes to be performed. Since we are going to filter the data down to just the invoice numbers and items, we focus the data cleaning on these two columns of the dataset. Remember that market basket analysis looks to identify associations between the items purchased by all customers over time. As such, the main focus of data cleaning involves removing transactions with a non-positive number of items. This could happen when the transaction involves voiding another transaction, when items are returned, or when the transaction is an administrative task. These types of transactions will be filtered out in two ways. The first is that canceled transactions have invoice numbers that are prefixed with "C," so we will identify those specific invoice numbers and remove them from the data. The other approach is to remove all transactions with either zero or a negative number of items. After performing these two steps, the data will be subset down to just the invoice number and item description columns, and any row of the now two-column dataset with at least one missing value is removed.

The next stage of the data cleaning exercise involves putting the data in the appropriate format for modeling. In this and subsequent exercises, we will use a subset of the full data. The subset will be created by taking the first 5,000 unique invoice numbers. Once we have cut the data down to the first 5,000 unique invoice numbers, we change the data structure to the structure needed to run the models. Note that the data currently features one item per row, so transactions with multiple items take up multiple rows. The desired format is a list of lists, like the made-up data from earlier in the chapter. Each subset list represents a unique invoice number, so in this case, the outer list should contain 5,000 sub-lists. The elements of the sub-lists are all the items belonging to the invoice number that that sub-list represents. With the cleaning process described, let's proceed to the exercise.

EXERCISE 8.04: DATA CLEANING AND FORMATTING

In this exercise, we will perform the cleaning steps described previously. As we work through the process, the evolution of the data will be monitored by printing out the current state of the data and computing some basic summary metrics. Be sure to perform data cleaning in the same notebook in which the data is loaded:

1. Create an indicator column stipulating whether the invoice number begins with "**C**":

```
# create new column called IsCPresent
online['IsCPresent']  = (# looking for C in InvoiceNo column \
                        .astype(str)\
                        .apply(lambda x: 1 if x.find('C') \
                               != -1 else 0))
```

2. Filter out all transactions having either zero or a negative number of items (in other words, items were returned), remove all invoice numbers starting with "**C**" using the column created in step one, subset the DataFrame down to **InvoiceNo** and **Description**, and lastly, drop all rows with at least one missing value. Rename the DataFrame **online1**:

```
online1 = (online\
           # filter out non-positive quantity values\
           .loc[online["Quantity"] > 0]\
           # remove InvoiceNos starting with C\
           .loc[online['IsCPresent'] != 1]\
           # column filtering\
           .loc[:, ["InvoiceNo", "Description"]]\
           # dropping all rows with at least one missing value\
           .dropna())
```

3. Print out the first 10 rows of the filtered DataFrame, `online1`:

   ```
   online1.head(10)
   ```

   The output is as follows:

	InvoiceNo	Description
0	536365	WHITE HANGING HEART T-LIGHT HOLDER
1	536365	WHITE METAL LANTERN
2	536365	CREAM CUPID HEARTS COAT HANGER
3	536365	KNITTED UNION FLAG HOT WATER BOTTLE
4	536365	RED WOOLLY HOTTIE WHITE HEART.
5	536365	SET 7 BABUSHKA NESTING BOXES
6	536365	GLASS STAR FROSTED T-LIGHT HOLDER
7	536366	HAND WARMER UNION JACK
8	536366	HAND WARMER RED POLKA DOT
9	536367	ASSORTED COLOUR BIRD ORNAMENT

   Figure 8.12: The cleaned online retail dataset

4. Print out the dimensions of the cleaned DataFrame and the number of unique invoice numbers using the **nunique** function, which counts the number of unique values in a DataFrame column:

   ```
   print("Data dimension (row count, col count): {dim}"\
         .format(dim=online1.shape)\
   )
   print("Count of unique invoice numbers: {cnt}"\
         .format(cnt=online1.InvoiceNo.nunique())\
   )
   ```

   The output is as follows:

   ```
   Data dimension (row count, col count): (530693, 2)
   Count of unique invoice numbers: 20136
   ```

Notice that we have already removed approximately 10,000 rows and 5,800 invoice numbers.

5. Extract the invoice numbers from the DataFrame as a list. Remove duplicate elements to create a list of unique invoice numbers. Confirm that the process was successful by printing the length of the list of unique invoice numbers. Compare with the output of *Step 4*:

```
invoice_no_list = online1.InvoiceNo.tolist()
invoice_no_list = list(set(invoice_no_list))
print("Length of list of invoice numbers: {ln}" \
    .format(ln=len(invoice_no_list)))
```

The output is as follows:

```
Length of list of invoice numbers: 20136
```

6. Take the list from *Step 5* and cut it to only include the first 5,000 elements. Print out the length of the new list to confirm that it is, in fact, the expected length of 5,000:

```
subset_invoice_no_list = invoice_no_list[0:5000]
print("Length of subset list of invoice numbers: {ln}"\
    .format(ln=len(subset_invoice_no_list)))
```

The output is as follows:

```
Length of subset list of invoice numbers: 5000
```

7. Filter the **online1** DataFrame down by only keeping the invoice numbers in the list from the preceding step:

```
online1 = online1.loc[online1["InvoiceNo"]\
                    .isin(subset_invoice_no_list)]
```

8. Print out the first 10 rows of **online1**:

```
online1.head(10)
```

The output is as follows:

	InvoiceNo	Description
229435	557056	SET OF 4 KNICK KNACK TINS DOILEY
229436	557057	RED POLKADOT BEAKER
229437	557057	BLUE POLKADOT BEAKER
229438	557057	DAIRY MAID TOASTRACK
229439	557057	BLUE EGG SPOON
229440	557057	RED EGG SPOON
229441	557057	MODERN FLORAL STATIONERY SET
229442	557057	FLORAL FOLK STATIONERY SET
229443	557057	CERAMIC BOWL WITH LOVE HEART DESIGN
229444	557057	WOOD STAMP SET THANK YOU

Figure 8.13: The cleaned dataset with only 5,000 unique invoice numbers

9. Print out the dimensions of the DataFrame and the number of unique invoice numbers to confirm that the filtering and cleaning process was successful:

```
print("Data dimension (row count, col count): {dim}"\
    .format(dim=online1.shape))
print("Count of unique invoice numbers: {cnt}"\
    .format(cnt=online1.InvoiceNo.nunique()))
```

The output is as follows:

```
Data dimension (row count, col count): (133315, 2)
Count of unique invoice numbers: 5000
```

10. Transform the data in **online1** into the aforementioned list of lists called **invoice_item_list**. The process for doing this is to iterate over the unique invoice numbers and, at each iteration, extract the item descriptions as a list and append that list to the larger **invoice_item_list** list. Print out elements one through four of the list:

```
invoice_item_list = []
for num in list(set(online1.InvoiceNo.tolist())):
    # filter dataset down to one invoice number
    tmp_df = online1.loc[online1['InvoiceNo'] == num]
    # extract item descriptions and convert to list
```

```python
        tmp_items = tmp_df.Description.tolist()
        # append list invoice_item_list
        invoice_item_list.append(tmp_items)
print(invoice_item_list[1:5])
```

The output is as follows:

```
[['RED POLKADOT BEAKER ', 'BLUE POLKADOT BEAKER ', 'DAIRY MAID TOASTRACK', 'BLUE EGG  SPOON', 'RED
EGG  SPOON', 'MODERN FLORAL STATIONERY SET', 'FLORAL FOLK STATIONERY SET', 'CERAMIC BOWL WITH LOVE
HEART DESIGN', 'WOOD STAMP SET THANK YOU', 'WOOD STAMP SET HAPPY BIRTHDAY', 'PENS ASSORTED SPACEBA
LL', 'PENS ASSORTED FUNNY FACE', 'PENS ASSORTED FUNKY JEWELED ', 'SCOTTIE DOGS BABY BIB', 'CHARLIE
AND LOLA TABLE TINS', 'CHARLIE & LOLA WASTEPAPER BIN FLORA', 'CHARLIE & LOLA WASTEPAPER BIN BLUE',
'CHARLIE AND LOLA FIGURES TINS', 'TV DINNER TRAY DOLLY GIRL', 'SET/20 RED RETROSPOT PAPER NAPKINS
', 'MINT KITCHEN SCALES', 'RED KITCHEN SCALES', '36 FOIL HEART CAKE CASES', '36 FOIL STAR CAKE CAS
ES ', 'ILLUSTRATED CAT BOWL ', 'POTTING SHED TEA MUG', 'CERAMIC STRAWBERRY DESIGN MUG', 'RED RETRO
SPOT SHOPPER BAG', 'BUTTON BOX ', 'MINI CAKE STAND  HANGING STRAWBERY', 'LUNCH BAG DOILEY PATTERN
', 'JUMBO BAG STRAWBERRY', 'STRAWBERRY SHOPPER BAG', 'SUKI  SHOULDER BAG', 'JUMBO BAG ALPHABET',
'SKULL SHOULDER BAG', 'LUNCH BAG  BLACK SKULL.', 'TRADITIONAL WOODEN CATCH CUP GAME ', '10 COLOUR
SPACEBOY PEN', 'JUMBO BAG SPACEBOY DESIGN', 'LUNCH BAG SPACEBOY DESIGN ', "CHILDREN'S APRON DOLLY
GIRL ", 'LUNCH BAG DOLLY GIRL DESIGN', 'TEATIME ROUND PENCIL SHARPENER ', 'SILVER HEARTS TABLE DEC
ORATION', 'PARISIENNE KEY CABINET ', 'PARISIENNE JEWELLERY DRAWER ', 'BUNDLE OF 3 SCHOOL EXERCISE
BOOKS  ', 'JUMBO BAG DOILEY PATTERNS', 'DOILEY STORAGE TIN', 'SET OF 4 KNICK KNACK TINS POPPIES',
'SET OF 4 KNICK KNACK TINS DOILEY ', 'SET OF 3 REGENCY CAKE TINS', 'SET OF 3 WOODEN HEART DECORATI
ONS', 'SPACEBOY CHILDRENS BOWL', 'DOLLY GIRL CHILDRENS CUP', 'DOLLY GIRL CHILDRENS BOWL', 'SPACE B
OY CHILDRENS CUP', 'GARDENERS KNEELING PAD CUP OF TEA ', 'GARDENERS KNEELING PAD KEEP CALM ', 'CAR
TOON  PENCIL SHARPENERS', 'POPART RECT PENCIL SHARPENER ASST', 'PIECE OF CAMO STATIONERY SET', 'PO
PART WOODEN PENCILS ASST', 'ORIGAMI VANILLA INCENSE/CANDLE SET ', 'ORIGAMI JASMINE INCENSE/CANDLE
SET', 'FRENCH FLORAL CUSHION COVER ', 'FRENCH LATTICE CUSHION COVER '], ['SET OF TEA COFFEE SUGAR
TINS PANTRY', 'SET OF 3 CAKE TINS PANTRY DESIGN '], ['JUMBO BAG PINK VINTAGE PAISLEY', 'JUMBO  BAG
BAROQUE BLACK WHITE', 'RIBBON REEL STRIPES DESIGN ', 'RIBBON REEL LACE DESIGN ', 'RIBBON REEL POLK
ADOTS ', 'TRAVEL CARD WALLET TRANSPORT', 'TRAVEL CARD WALLET FLOWER MEADOW', 'TRAVEL CARD WALLET V
INTAGE LEAF', 'TRAVEL CARD WALLET VINTAGE TICKET', 'VINTAGE  2 METER FOLDING RULER', 'IVORY WICKER
HEART LARGE', 'BUNDLE OF 3 ALPHABET EXERCISE BOOKS', 'BUNDLE OF 3 RETRO NOTE BOOKS', '20 DOLLY PEG
S RETROSPOT', 'CLOTHES PEGS RETROSPOT PACK 24 ', 'VICTORIAN  METAL POSTCARD SPRING', 'ROLL WRAP VI
NTAGE CHRISTMAS', 'ROLL WRAP VINTAGE SPOT ', 'ENAMEL MEASURING JUG CREAM', 'JUMBO BAG VINTAGE CHRI
STMAS ', "JUMBO BAG 50'S CHRISTMAS ", 'SET OF 4 KNICK KNACK TINS DOILY ', 'SET OF 4 KNICK KNACK TI
NS POPPIES', 'IVORY WICKER HEART LARGE', 'JINGLE BELL HEART ANTIQUE GOLD', 'SET OF 4 NAPKIN CHARMS
CUTLERY', 'SET OF 4 NAPKIN CHARMS HEARTS', 'SET OF 4 KNICK KNACK TINS LEAF', 'MADRAS NOTEBOOK MEDI
UM ', 'SET OF 3 WOODEN HEART DECORATIONS', 'FAMILY ALBUM WHITE PICTURE FRAME', 'REX CASH+CARRY JUMB
O SHOPPER'], ['COFFEE MUG PEARS  DESIGN', 'TRAVEL CARD WALLET VINTAGE TICKET', 'AIRLINE BAG VINTAG
E JET SET RED', 'AIRLINE BAG VINTAGE JET SET WHITE', 'GREY HEART HOT WATER BOTTLE', 'LOVE HOT WATE
R BOTTLE', 'TRAVEL CARD WALLET I LOVE LONDON', 'KNITTED UNION FLAG HOT WATER BOTTLE', 'HOT WATER B
OTTLE I AM SO POORLY', 'AIRLINE BAG VINTAGE WORLD CHAMPION ', 'AIRLINE BAG VINTAGE TOKYO 78', 'HOT
WATER BOTTLE TEA AND SYMPATHY', 'BLUE PAISLEY POCKET BOOK', 'ABSTRACT CIRCLES POCKET BOOK', 'HAND
WARMER RED RETROSPOT', 'PLASTERS IN TIN WOODLAND ANIMALS', 'PLASTERS IN TIN VINTAGE PAISLEY ', 'HA
ND WARMER SCOTTY DOG DESIGN', 'HAND WARMER BIRD DESIGN', 'PLASTERS IN TIN STRONGMAN', 'PLASTERS IN
TIN CIRCUS PARADE ']]
```

Figure 8.14: Four elements of the list of lists

In the preceding list of lists, each sub-list contains all the items belonging to an individual invoice.

> **NOTE**
>
> This step can take some minutes to complete.

In this exercise, the DataFrame was filtered and subset to only the needed columns and the relevant rows. We then cut the full dataset down to the first 5,000 unique invoice numbers. The full dataset will be used in the forthcoming activities. The last step converted the DataFrame to a list of lists, which is the format the data needs to be in for the encoder that will be discussed next.

> **NOTE**
>
> To access the source code for this specific section, please refer to https://packt.live/3fhf9bS.
>
> You can also run this example online at https://packt.live/303vIBJ.
>
> You must execute the entire Notebook in order to get the desired result.

DATA ENCODING

While cleaning the data is crucial, the most important part of the data preparation process is molding the data into the correct form. Before running the models, the data, currently in the list of lists form, needs to be encoded and recast as a DataFrame. To do this, we will leverage **TransactionEncoder** from the **preprocessing** module of **mlxtend**. The output from the encoder is a multidimensional array, where each row is the length of the total number of unique items in the transaction dataset and the elements are Boolean variables, indicating whether that particular item is linked to the invoice number that row represents. With the data encoded, we can recast it as a DataFrame where the rows are the invoice numbers and the columns are the unique items in the transaction dataset.

In the following exercise, the data encoding will be done using **mlxtend**, but it is very easy to encode the data without using a package. The first step is to unlist the list of lists and return one list with every value from the original list of lists. Next, the duplicate products are filtered out and, if preferred, the data is sorted in alphabetical order. Before doing the actual encoding, we initialize the final DataFrame by having all elements equal to false, the number of rows equal to the number of invoice numbers in the dataset, and column names equal to the non-duplicated list of product names.

In this case, we have 5,000 transactions and over 3,100 unique products. Thus, the DataFrame has over 15,000,000 elements. The actual encoding is done by looping over each transaction and each item in each transaction. Change the row i and column j cell values in the initialized dataset from **false** to **true** if the ith transaction contains the jth product. This double loop is not fast as we need to iterate over 15,000,000 cells. There are ways to improve performance, including some that have been implemented in **mlxtend**, but to better understand the process, it is helpful to work through the double loop methodology. The following is an example function to do the encoding from scratch without the assistance of a package other than **pandas**:

```
def manual_encoding(ll):
    # unlist the list of lists input
    # result is one list with all the elements of the sublists
    list_dup_unsort_items = \
    [element for sub in ll for element in sub]
    # two cleaning steps:
    """
    1. remove duplicate items, only want one of each item in list
    """
    #    2. sort items in alphabetical order
    list_nondup_sort_items = \
    sorted(list(set(list_dup_unsort_items)))
    # initialize DataFrame with all elements having False value
    # name the columns the elements of list_dup_unsort_items
    manual_df = pandas.DataFrame(False, \
                                 index=range(len(ll)), \
                                 columns=list_dup_unsort_items)
```

```
    """
    change False to True if element is
    in individual transaction list
    """
    """
    each row is represents the contains of an individual transaction
    """
    # (sublist from the original list of lists)
    for i in range(len(ll)):
        for j in ll[i]:
            manual_df.loc[i, j] = True
    # return the True/False DataFrame
    return manual_df
```

EXERCISE 8.05: DATA ENCODING

In this exercise, we'll continue the data preparation process by taking the list of lists generated in the preceding exercise and encoding the data in the specific way required to run the models:

1. Initialize and fit the transaction encoder. Print out an example of the resulting data:

    ```
    online_encoder = mlxtend.preprocessing.TransactionEncoder()
    online_encoder_array = online_encoder\
                            .fit_transform(invoice_item_list)
    print(online_encoder_array)
    ```

    The output is as follows:

    ```
    [[False False False ... False False False]
     [False False False ... False False False]
     [False False False ... False False False]
     ...
     [False False False ... False False False]
     [False False False ... False False False]
     [False False False ... False False False]]
    ```

    The preceding array contains the Boolean variables indicating the product presence in each transaction.

2. Recast the encoded array as a DataFrame named **online_encoder_df**. Print the predefined subset of the DataFrame that features both **True** and **False** values:

```
online_encoder_df = pandas.DataFrame(online_encoder_array, \
                                     columns=online_encoder\
                                     .columns_)
"""
this is a very big table, so for more
easy viewing only a subset is printed
"""
online_encoder_df.loc[4970:4979, \
                      online_encoder_df.columns.tolist()[0:8]]
```

The output will be similar to the following:

	4 PURPLE FLOCK DINNER CANDLES	50'S CHRISTMAS GIFT BAG LARGE	DOLLY GIRL BEAKER	I LOVE LONDON MINI BACKPACK	NINE DRAWER OFFICE TIDY	OVAL WALL MIRROR DIAMANTE	RED SPOT GIFT BAG LARGE	SET 2 TEA TOWELS I LOVE LONDON
4970	False	False	False	False	False	False	False	False
4971	False	False	True	False	False	False	False	False
4972	False	False	False	False	False	False	False	False
4973	False	False	False	False	False	False	False	False
4974	False	False	False	False	False	False	False	False
4975	False	False	False	False	False	False	False	False
4976	False	False	False	False	False	False	False	False
4977	False	False	False	False	False	False	False	False
4978	False	False	False	False	False	False	False	False
4979	False	False	False	False	False	False	False	False

Figure 8.15: A small section of the encoded data recast as a DataFrame

3. Print out the dimensions of the encoded DataFrame. It should have 5,000 rows because the data used to generate it was previously filtered down to 5,000 unique invoice numbers:

```
print("Data dimension (row count, col count): {dim}"\
      .format(dim=online_encoder_df.shape))
```

The output will be similar to the following:

```
Data dimension (row count, col count): (5000, 3135)
```

The data is now prepared for modeling, which we will perform in *Exercise 8.06, Executing the Apriori Algorithm*.

> **NOTE**
>
> To access the source code for this specific section, please refer to https://packt.live/3fhf9bS.
>
> You can also run this example online at https://packt.live/303vIBJ.
>
> You must execute the entire Notebook in order to get the desired result.

ACTIVITY 8.01: LOADING AND PREPARING FULL ONLINE RETAIL DATA

In this activity, we are charged with loading and preparing a large transaction dataset for modeling. The final output will be an appropriately encoded dataset that has one row for each unique transaction in the dataset, and one column for each unique item in the dataset. If an item appears in an individual transaction, that element of the DataFrame will be marked **true**.

This activity will largely repeat the last few exercises but will use the complete online retail dataset file. No new downloads need to be executed, but you will need the path to the file downloaded previously. Perform this activity in a separate Jupyter notebook.

The following steps will help you to complete the activity:

1. Load the online retail dataset file.

> **NOTE**
>
> This dataset is sourced from http://archive.ics.uci.edu/ml/datasets/online+retail# (UCI Machine Learning Repository [http://archive.ics.uci.edu/ml]. Irvine, CA: University of California, School of Information and Computer Science).
>
> Citation: Daqing Chen, Sai Liang Sain, and Kun Guo, *Data mining for the online retail industry: A case study of RFM model-based customer segmentation using data mining*, Journal of Database Marketing and Customer Strategy Management, Vol. 19, No. 3, pp. 197-208, 2012.
>
> It can be downloaded from https://packt.live/39Nx3iQ.

2. Clean and prepare the data for modeling, including turning the cleaned data into a list of lists.

3. Encode the data and recast it as a DataFrame.

> **NOTE**
> The solution to this activity can be found on page 490.

The output will be similar to the following:

	6 CHOCOLATE LOVE HEART T-LIGHTS	6 EGG HOUSE PAINTED WOOD	6 GIFT TAGS 50'S CHRISTMAS	6 GIFT TAGS VINTAGE CHRISTMAS	6 RIBBONS ELEGANT CHRISTMAS	6 RIBBONS EMPIRE	6 RIBBONS RUSTIC CHARM	6 RIBBONS SHIMMERING PINKS	6 ROCKET BALLOONS	60 CAKE CASES DOLLY GIRL DESIGN
20125	False	False	False	False	False	False	False	False	False	False
20126	False	False	False	False	False	False	False	False	False	False
20127	False	False	False	False	False	False	False	False	False	False
20128	False	False	False	False	False	False	False	False	False	False
20129	False	False	False	False	False	False	False	False	False	False
20130	False	False	False	False	False	False	False	False	False	False
20131	False	False	False	False	False	False	False	False	False	False
20132	False	False	False	False	False	False	False	False	False	False
20133	False	False	False	False	False	False	False	False	False	False
20134	False	False	False	False	False	False	False	False	False	False
20135	False	False	False	False	False	False	False	False	False	False

Figure 8.16: A subset of the cleaned, encoded, and recast DataFrame built from the complete online retail dataset

THE APRIORI ALGORITHM

The **Apriori** algorithm is a data mining methodology for identifying and quantifying frequent item sets in transaction data and is the foundational component of association rule learning. Extending the results of the Apriori algorithm to association rule learning will be discussed in the next section. The threshold for an item set being frequent is an input (in other words, a hyperparameter) of the model and, as such, is adjustable. Frequency is quantified here as support, so the value input into the model is the minimum support acceptable for the analysis being done. The model then identifies all item sets whose support is greater than, or equal to, the minimum support provided to the model.

> **NOTE**
>
> The minimum support hyperparameter is not a value that can be optimized via grid search because there is no evaluation metric for the Apriori algorithm. Instead, the minimum support parameter is set based on the data, the use case, and domain expertise.

The main idea behind the Apriori algorithm is the Apriori principle: any subset of a frequent item set must itself be frequent.

Another aspect worth mentioning is the corollary: no superset of an infrequent item set can be frequent.

Let's take some examples. If the item set {hammer, saw, and nail} is frequent, then, according to the Apriori principle, and what is hopefully obvious, any less complex item set derived from it, say {hammer, saw}, is also frequent. On the contrary, if that same item set, {hammer, saw, nail}, is infrequent, then adding complexity, such as incorporating wood into the item set {hammer, saw, nail, wood}, is not going to result in the item set becoming frequent.

It might seem straightforward to calculate the support value for every item set in a transactional database and only return those item sets whose support is greater than or equal to the pre-specified minimum support threshold, but it is not because of the number of computations that need to happen. For example, take an item set with 10 unique items. This would result in 1,023 individual item sets for which support would need to be calculated. Now, try to extrapolate out to our working dataset that has 3,135 unique items. That is going to be an enormous number of item sets for which we need to compute a support value. Computational efficiency is a major issue:

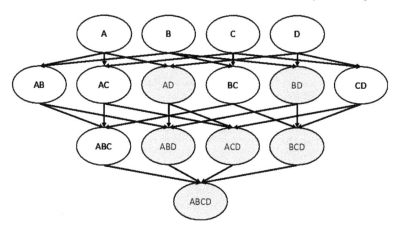

Figure 8.17: Representation of the computational efficiency issue

346 | Market Basket Analysis

The preceding diagram shows a mapping of how item sets are built and how the Apriori principle can greatly decrease the computational requirements (all the grayed-out nodes are infrequent).

In order to address the computational demands, the Apriori algorithm is defined as a bottom-up model that has two steps. These steps involve generating candidate item sets by adding items to already existing frequent item sets and testing these candidate item sets against the dataset to determine whether these candidate item sets are also frequent. No support value is computed for item sets that contain infrequent item sets. This process repeats until no further candidate item sets exist:

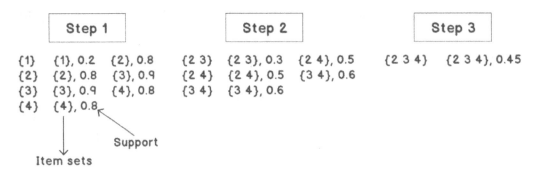

Figure 8.18: General Apriori algorithm structure

Assuming a minimum support threshold of 0.4, the preceding diagram shows the general Apriori algorithm structure.

Figure 8.20 includes establishing an item set, computing support values, filtering out infrequent item sets, creating new item sets, and repeating the process.

There is a clear tree-like structure that serves as the path for identifying candidate item sets. The specific search technique used, which was built for traversing tree-like data structures, is called a breadth-first search, which means that each step of the search process focuses on completely searching one level of the tree before moving on instead of searching branch by branch.

The high-level steps of the algorithm are designed to do the following:

1. Define the set of frequent items (in other words, choose only those items that have support greater than the pre-defined threshold). To start, this is typically the set of individual items.

2. Derive candidate item sets by combining frequent item sets. Move up in size one item at a time. That is, go from item sets with one item to two, two to three, and so on.

3. Compute the support value for each candidate item set.

4. Create a new frequent item set made up of the candidate item sets whose support value exceeds the specified threshold.

Repeat *Steps 1* to *4* until there are no more frequent item sets; that is, until we have worked through all the combinations.

The pseudo code for the Apriori algorithm is as follows:

```
L₁ = {frequent items}
k = 1
L = {}
while Lₖ.Length is not an empty set do
    Cₖ₊₁ = candidate item sets derived from Lₖ
    For each transaction t in the dataset do
        Increment the count of the candidates \
            in Cₖ₊₁ that appear in t
    Compute the support for the candidates in Cₖ₊₁ \
        using the appearance counts
    Lₖ₊₁ = the candidates in Cₖ₊₁ meeting \
        the minimum support requirement
    L.append(Lₖ)
    k = k + 1
End
Return L = all frequent item sets with corresponding support values
```

Despite the Apriori principle, this algorithm can still face significant computational challenges depending on the size of the transaction dataset. There are several strategies currently accepted to further reduce the computational demands.

COMPUTATIONAL FIXES

Transaction reduction is an easy way to reduce the computational load. Note that after each candidate set of items is generated, the entirety of the transaction data needs to be scanned in order to count the number of appearances of each candidate item set. If we could shrink the size of the transaction dataset, the size of the dataset scan would decrease dramatically. The shrinking of the transaction dataset is done by realizing that any transaction containing no frequent item sets in the ith iteration is not going to contain any frequent item sets in subsequent iterations. Therefore, once each transaction contains no frequent item sets, it can be removed from the transaction dataset used for future scans.

348 | Market Basket Analysis

Sampling the transaction dataset and testing each candidate item set against it is another approach to reducing the computational requirements associated with scanning the transaction dataset to calculate the support of each item set. When this approach is implemented, it is important to lower the minimum support requirement to guarantee that no item sets that should be present in the final data are left out. Given that the sampled transaction dataset will naturally cause the support values to be smaller, leaving the minimum support at its original value will incorrectly remove what should be frequent item sets from the output of the model.

A similar approach is partitioning. In this case, the dataset is randomly partitioned into several individual datasets on which the evaluation of each candidate item set is executed. Item sets are deemed frequent in the full transaction dataset if frequent in one of the partitions. Each partition is scanned consecutively until the frequency for an item set is established. Like sampling, partitioning is just another way to avoid testing each item set on the full dataset, which could be very computationally expensive if the full dataset is really big. If the frequency is established on the first partition, then we have established it for the whole dataset without testing it against a majority of the partitions.

Regardless of whether or not one of these techniques is employed, the computational requirements are always going to be fairly substantial when it comes to the Apriori algorithm. As should now be clear, the essence of the algorithm, the computation of support, is not as complex as other models discussed in this text.

EXERCISE 8.06: EXECUTING THE APRIORI ALGORITHM

The execution of the Apriori algorithm is made easy with `mlxtend`. As a result, this exercise will focus on how to manipulate the output dataset and to interpret the results. You will recall that the cleaned and encoded transaction data was defined as `online_encoder_df`:

> **NOTE**
>
> Perform this exercise in the same notebook that all previous exercises were run in as we will continue using the environment, data, and results already established in that notebook. (So, you should be using the notebook that contains the reduced dataset of 5,000 invoices, not the full dataset as was used in the activity.)

1. Run the Apriori algorithm using **mlxtend** without changing any of the default parameter values:

    ```
    mod = mlxtend.frequent_patterns.apriori(online_encoder_df)
    mod
    ```

    The output is an empty DataFrame. The default minimum support value is set to 0.5, so since an empty DataFrame was returned, we know that all item sets have a support value of less than 0.5. Depending on the number of transactions and the diversity of available items, having no item set with a plus 0.5 support value is not unusual.

2. Rerun the Apriori algorithm, but with the minimum support set to 0.01:

    ```
    mod_minsupport = mlxtend.frequent_patterns\
                    .apriori(online_encoder_df, \
                    min_support=0.01)
    mod_minsupport.loc[0:6]
    ```

    This minimum support value is the same as saying that when analyzing 5,000 transactions, we need an item set to appear 50 times to be considered frequent. As mentioned previously, the minimum support can be set to any value in the range [0,1]. There is no best minimum support value; the setting of this value is entirely subjective. Many businesses have their own specific thresholds for significance, but there is no industry standard or method for optimizing this value.

    The output will be similar to the following:

	support	itemsets
0	0.0168	(2)
1	0.0150	(10)
2	0.0116	(15)
3	0.0144	(18)
4	0.0210	(19)
5	0.0144	(20)
6	0.0138	(21)

    Figure 8.19: Basic output of the Apriori algorithm run using mlxtend

Notice that the item sets are designated numerically in the output, which makes the results hard to interpret.

3. Rerun the Apriori algorithm with the same minimum support as in *Step 2*, but this time set **use_colnames** to **True**. This will replace the numerical designations with the actual item names:

```
mod_colnames_minsupport = mlxtend.frequent_patterns\
                    .apriori(online_encoder_df, \
                    min_support=0.01, \
                    use_colnames=True)
mod_colnames_minsupport.loc[0:6]
```

The output will be similar to the following:

	support	itemsets
0	0.0168	(DOLLY GIRL BEAKER)
1	0.0150	(10 COLOUR SPACEBOY PEN)
2	0.0116	(12 MESSAGE CARDS WITH ENVELOPES)
3	0.0144	(12 PENCILS SMALL TUBE SKULL)
4	0.0210	(12 PENCILS TALL TUBE POSY)
5	0.0144	(12 PENCILS TALL TUBE RED RETROSPOT)
6	0.0138	(12 PENCILS TALL TUBE SKULLS)

Figure 8.20: The output of the Apriori algorithm with the actual item names instead of numerical designations

This DataFrame contains every item set whose support value is greater than the specified minimum support value. That is, these item sets occur with sufficient frequency to potentially be meaningful and therefore actionable.

4. Add an additional column to the output of *Step 3* that contains the size of the item set (in other words, how many items are in the set), which will help with filtering and further analysis:

```
mod_colnames_minsupport['length'] = \
(mod_colnames_minsupport['itemsets'].apply(lambda x: len(x)))
mod_colnames_minsupport.loc[0:6]
```

The output will be similar to the following:

	support	itemsets	length
0	0.0168	(DOLLY GIRL BEAKER)	1
1	0.0150	(10 COLOUR SPACEBOY PEN)	1
2	0.0116	(12 MESSAGE CARDS WITH ENVELOPES)	1
3	0.0144	(12 PENCILS SMALL TUBE SKULL)	1
4	0.0210	(12 PENCILS TALL TUBE POSY)	1
5	0.0144	(12 PENCILS TALL TUBE RED RETROSPOT)	1
6	0.0138	(12 PENCILS TALL TUBE SKULLS)	1

Figure 8.21: The Apriori algorithm output plus an additional column containing the lengths of the item sets

5. Find the support of the item set containing **10 COLOUR SPACEBOY PEN**:

```
mod_colnames_minsupport[mod_colnames_minsupport['itemsets'] \
                    == frozenset({'10 COLOUR SPACEBOY PEN'})]
```

The output is as follows:

	support	itemsets	length
1	0.015	(10 COLOUR SPACEBOY PEN)	1

Figure 8.22: The output DataFrame filtered down to a single item set

This single-row DataFrame gives us the support value for this specific item set that contains one item. The support value says that this specific item set appears in 1.78% of the transactions.

6. Return all item sets of length 2 whose support is in the range [0.02, 0.021]:

```
mod_colnames_minsupport[(mod_colnames_minsupport['length'] == 2) \
                    & (mod_colnames_minsupport\
                       ['support'] >= 0.02) \
                    & (mod_colnames_minsupport\
                       ['support'] < 0.021)]
```

The output will be similar to the following:

	support	itemsets	length
837	0.0202	(ALARM CLOCK BAKELIKE IVORY, ALARM CLOCK BAKEL...	2
956	0.0202	(CHARLOTTE BAG APPLES DESIGN, LUNCH BAG APPLE ...	2
994	0.0200	(CHARLOTTE BAG PINK POLKADOT, LUNCH BAG PINK P...	2
1026	0.0206	(CHARLOTTE BAG SUKI DESIGN, LUNCH BAG BLACK S...	2
1032	0.0206	(CHARLOTTE BAG SUKI DESIGN, LUNCH BAG RED RETR...	2
1131	0.0200	(JUMBO SHOPPER VINTAGE RED PAISLEY, DOTCOM POS...	2
1298	0.0208	(HEART OF WICKER LARGE, HEART OF WICKER SMALL)	2
1305	0.0200	(SMALL WHITE HEART OF WICKER, HEART OF WICKER ...	2
1316	0.0204	(JAM MAKING SET WITH JARS, JAM MAKING SET PRIN...	2
1349	0.0208	(JAM MAKING SET PRINTED, SET OF 3 REGENCY CAKE...	2
1440	0.0200	(JUMBO BAG ALPHABET, LUNCH BAG ALPHABET DESIGN)	2
1464	0.0206	(JUMBO BAG APPLES, JUMBO BAG DOILEY PATTERNS)	2
1471	0.0202	(JUMBO BAG APPLES, JUMBO BAG SCANDINAVIAN BLUE...	2
1472	0.0202	(JUMBO BAG APPLES, JUMBO BAG SPACEBOY DESIGN)	2
1479	0.0204	(JUMBO BAG APPLES, JUMBO STORAGE BAG SKULLS)	2
1575	0.0200	(JUMBO BAG OWLS, JUMBO BAG PINK POLKADOT)	2
1583	0.0208	(JUMBO BAG WOODLAND ANIMALS, JUMBO BAG OWLS)	2

Figure 8.23: The Apriori algorithm output DataFrame filtered by length and support

This DataFrame contains all the item sets (pairs of items bought together) whose support value is in the range specified at the start of the step. Each of these item sets appears in between 2.0% and 2.1% of transactions.

Note that when filtering on `support`, it is wise to specify a range instead of a specific value since it is quite possible to pick a value for which there are no item sets. The preceding output has 32 item sets; only a subset is shown. Keep note of the particular items in the item sets because we will be running this same filter when we scale up to the full data and we will want to execute a comparison.

7. Plot the support values. Note that this plot will have no support values less than **0.01** because that was the value used as the minimum support in *Step 2*:

   ```
   mod_colnames_minsupport.hist("support", grid=False, bins=30)
   plt.xlabel("Support of item")
   plt.ylabel("Number of items")
   plt.title("Frequency distribution of Support")
   plt.show()
   ```

   The output will be similar to the following plot:

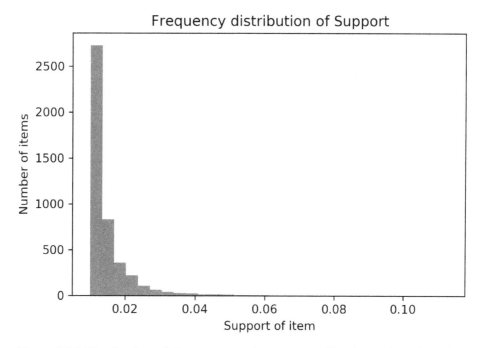

Figure 8.24: Distribution of the support values returned by the Apriori algorithm

The maximum support value is approximately 0.14, which is approximately 700 transactions. What might appear to be a small value may not be, given the number of products available. Larger numbers of products tend to result in lower support values because the variability of item combinations increases.

> **NOTE**
>
> To access the source code for this specific section, please refer to https://packt.live/3fhf9bS.
>
> You can also run this example online at https://packt.live/303vIBJ.
>
> You must execute the entire Notebook in order to get the desired result.

Hopefully, you can think of more ways in which this data could be used. We will generate even more useful information in the next section by using the Apriori algorithm results to generate association rules.

ACTIVITY 8.02: RUNNING THE APRIORI ALGORITHM ON THE COMPLETE ONLINE RETAIL DATASET

Imagine you work for an online retailer. You are given all the transaction data from the last month and told to find all the item sets appearing in at least 1% of the transactions. Once the qualifying item sets are identified, you are subsequently told to identify the distribution of the support values. The distribution of support values will tell all interested parties whether groups of items exist that are purchased together with high probability as well as the mean of the support values. Let's collect all the information for the company's leadership and strategists.

In this activity, you will run the Apriori algorithm on the full online retail dataset.

> **NOTE**
>
> This dataset is sourced from http://archive.ics.uci.edu/ml/datasets/online+retail# (UCI Machine Learning Repository [http://archive.ics.uci.edu/ml]. Irvine, CA: University of California, School of Information and Computer Science).
>
> Citation: Daqing Chen, Sai Liang Sain, and Kun Guo, *Data mining for the online retail industry: A case study of RFM model-based customer segmentation using data mining*, Journal of Database Marketing and Customer Strategy Management, Vol. 19, No. 3, pp. 197-208, 2012.
>
> It can be downloaded from https://packt.live/39Nx3iQ.

Ensure that you complete this activity in the same notebook as the preceding activity (in other words, the notebook that uses the full dataset, not the notebook that uses the subset of 5,000 invoices that you're using for the exercises).

This will also provide you with an opportunity to compare the results with those generated using only 5,000 transactions. This is an interesting activity, as it provides some insight into the ways in which the data may change as more data is collected, as well as some insight into how support values change when the partitioning technique is employed. Note that what was done in the exercises is not a perfect representation of the partitioning technique because 5,000 was an arbitrary number of transactions to sample.

> **NOTE**
> All the activities in this chapter need to be performed in the same notebook.

The following steps will help you to complete the activity:

1. Run the Apriori algorithm on the full data with reasonable parameter settings.

2. Filter the results down to the item set containing **10 COLOUR SPACEBOY PEN**. Compare the support value to that of *Exercise 8.06, Executing the Apriori Algorithm*.

3. Add another column containing the item set length. Then, filter down to those item sets whose length is two and whose support is in the range [0.02, 0.021]. Compare this to the result from *Exercise 8.06, Executing the Apriori Algorithm*.

4. Plot the **support** values.

> **NOTE**
> The solution to this activity can be found on page 492.

The output of this activity will be similar to the following:

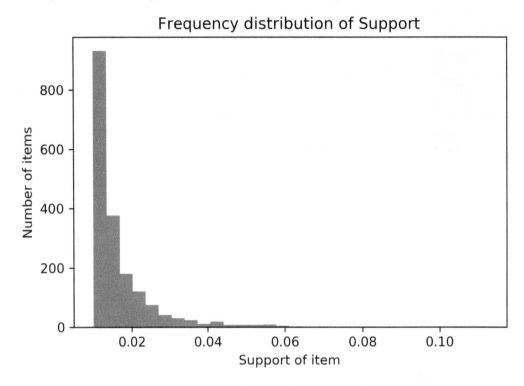

Figure 8.25: Distribution of support values

ASSOCIATION RULES

Association rule learning is a machine learning model that seeks to unearth the hidden patterns (in other words, relationships) in transaction data that describe the shopping habits of the customers of any retailer. The definition of an association rule was hinted at when the common probabilistic metrics were defined and explained earlier in the chapter.

Consider the imaginary frequent item set {Milk, Bread}. Two association rules can be formed from that item set: Milk $\Rightarrow$ Bread and Bread $\Rightarrow$ Milk. For simplicity, the first item set in the association rule is referred to as the antecedent, while the second item set in the association rule is referred to as the consequent. Once the association rules have been identified, all the previously discussed metrics can be computed to evaluate the validity of the association rules, determining whether or not the rules can be leveraged in the decision-making process.

The establishment of an association rule is based on support and confidence. Support, as we discussed in the last section, identifies which item sets are frequent, while confidence measures the frequency of truthfulness for a particular rule. Confidence is typically referred to as one of the measures of interestingness, as it is one of the metrics that determines whether an association should be formed. Thus, the establishment of an association rule is a two-step process. Identify frequent datasets and then evaluate the confidence of a candidate association rule and, if that confidence value exceeds some arbitrary threshold, the result is an association rule.

A major issue of association rule learning is the discovery of spurious associations, which are highly likely given the huge numbers of potential rules. Spurious associations are defined as associations that occur with surprising regularity in the data, given that the association occurs entirely by chance. To clearly articulate the idea, assume we are in a situation where we have 100 candidate rules. If we run a statistical test for independence at the 0.05 significance level, we are still faced with a 5% chance that an association is found when no association exists. Let's further assume that all 100 candidate rules are not valid associations. Given the 5% chance, we should still expect to find 5 valid association rules. Now, scale the imaginary candidate rule list up to millions or billions, so that 5% amounts to an enormous number of associations. This problem is not unlike the issue of statistical significance and error faced by virtually every model. It is worth calling out that some techniques exist to combat the spurious association issue, but they are neither consistently incorporated into the frequently used association rule libraries nor in the scope of this chapter.

Let's now apply our working knowledge of association rule learning to the online retail dataset.

EXERCISE 8.07: DERIVING ASSOCIATION RULES

In this exercise, we will derive association rules for the online retail dataset and explore the associated metrics.

> **NOTE**
>
> Ensure that you complete this exercise in the same notebook as the previous exercises (in other words, the notebook that uses the 5,000-invoice subset, not the full dataset from the activities).

1. Use the **mlxtend** library to derive association rules for the online retail dataset. Use confidence as the measure of interestingness, set the minimum threshold to **0.6**, and return all the metrics, not just support:

```
rules = mlxtend.frequent_patterns\
    .association_rules(mod_colnames_minsupport, \
    metric="confidence", \
    min_threshold=0.6, \
    support_only=False)
rules.loc[0:6]
```

The output is similar to the following:

	antecedents	consequents	antecedent support	consequent support	support	confidence	lift	leverage	conviction
0	(SPACEBOY BEAKER)	(DOLLY GIRL BEAKER)	0.0172	0.0168	0.0126	0.732558	43.604651	0.012311	3.676313
1	(DOLLY GIRL BEAKER)	(SPACEBOY BEAKER)	0.0168	0.0172	0.0126	0.750000	43.604651	0.012311	3.931200
2	(ALARM CLOCK BAKELIKE CHOCOLATE)	(ALARM CLOCK BAKELIKE GREEN)	0.0208	0.0580	0.0160	0.769231	13.262599	0.014794	4.082000
3	(ALARM CLOCK BAKELIKE CHOCOLATE)	(ALARM CLOCK BAKELIKE RED)	0.0208	0.0498	0.0142	0.682692	13.708681	0.013164	2.994570
4	(ALARM CLOCK BAKELIKE IVORY)	(ALARM CLOCK BAKELIKE GREEN)	0.0302	0.0580	0.0202	0.668874	11.532313	0.018448	2.844840
5	(ALARM CLOCK BAKELIKE ORANGE)	(ALARM CLOCK BAKELIKE GREEN)	0.0282	0.0580	0.0212	0.751773	12.961604	0.019564	3.794914
6	(ALARM CLOCK BAKELIKE PINK)	(ALARM CLOCK BAKELIKE GREEN)	0.0380	0.0580	0.0254	0.668421	11.524501	0.023196	2.840952

Figure 8.26: The first seven rows of the association rules generated using only 5,000 transactions

2. Print the number of associations:

   ```
   print("Number of Associations: {}".format(rules.shape[0]))
   ```

   1,064 association rules were found.

   > **NOTE**
   >
   > The number of association rules may differ.

3. Try running another version of the model. Choose any minimum threshold and any measure of interestingness. Explore the returned rules:

   ```
   rules2 = mlxtend.frequent_patterns\
           .association_rules(mod_colnames_minsupport, \
           metric="lift", \
           min_threshold=50,\
           support_only=False)
   rules2.loc[0:6]
   ```

   The output is as follows:

	antecedents	consequents	antecedent support	consequent support	support	confidence	lift	leverage	conviction
0	(POPPY'S PLAYHOUSE BEDROOM , POPPY'S PLAYHOUSE...)	(POPPY'S PLAYHOUSE LIVINGROOM)	0.0136	0.0148	0.0102	0.750000	50.675676	0.009999	3.940800
1	(POPPY'S PLAYHOUSE LIVINGROOM)	(POPPY'S PLAYHOUSE BEDROOM , POPPY'S PLAYHOUSE...)	0.0148	0.0136	0.0102	0.689189	50.675676	0.009999	3.173635
2	(SPACEBOY CHILDRENS BOWL, DOLLY GIRL CHILDRENS...)	(SPACEBOY CHILDRENS CUP, DOLLY GIRL CHILDRENS ...)	0.0140	0.0136	0.0120	0.857143	63.025210	0.011810	6.904800
3	(SPACEBOY CHILDRENS CUP, DOLLY GIRL CHILDRENS ...)	(SPACEBOY CHILDRENS BOWL, DOLLY GIRL CHILDRENS...)	0.0136	0.0140	0.0120	0.882353	63.025210	0.011810	8.381000
4	(GREEN REGENCY TEACUP AND SAUCER, REGENCY TEA ...)	(PINK REGENCY TEACUP AND SAUCER, REGENCY TEA P...)	0.0160	0.0138	0.0112	0.700000	50.724638	0.010979	3.287333
5	(PINK REGENCY TEACUP AND SAUCER, REGENCY TEA P...)	(GREEN REGENCY TEACUP AND SAUCER, REGENCY TEA ...)	0.0138	0.0160	0.0112	0.811594	50.724638	0.010979	5.222769
6	(ROSES REGENCY TEACUP AND SAUCER , REGENCY TEA...)	(GREEN REGENCY TEACUP AND SAUCER, REGENCY TEA ...)	0.0166	0.0124	0.0106	0.638554	51.496308	0.010394	2.732360

   Figure 8.27: The first seven rows of the association rules

4. Print the number of associations:

   ```
   print("Number of Associations: {}".format(rules2.shape[0]))
   ```

   The number of association rules found using the lift metric and the minimum threshold value of **50** is **176**, which is significantly lower than in *Step 2*. We will see in a future step that **50** is quite a high threshold value, so it is not surprising that we returned fewer association rules.

5. Plot confidence against support and identify specific trends in the data:

```
rules.plot.scatter("support", "confidence", \
                alpha=0.5, marker="*")
plt.xlabel("Support")
plt.ylabel("Confidence")
plt.title("Association Rules")
plt.show()
```

The output is as follows:

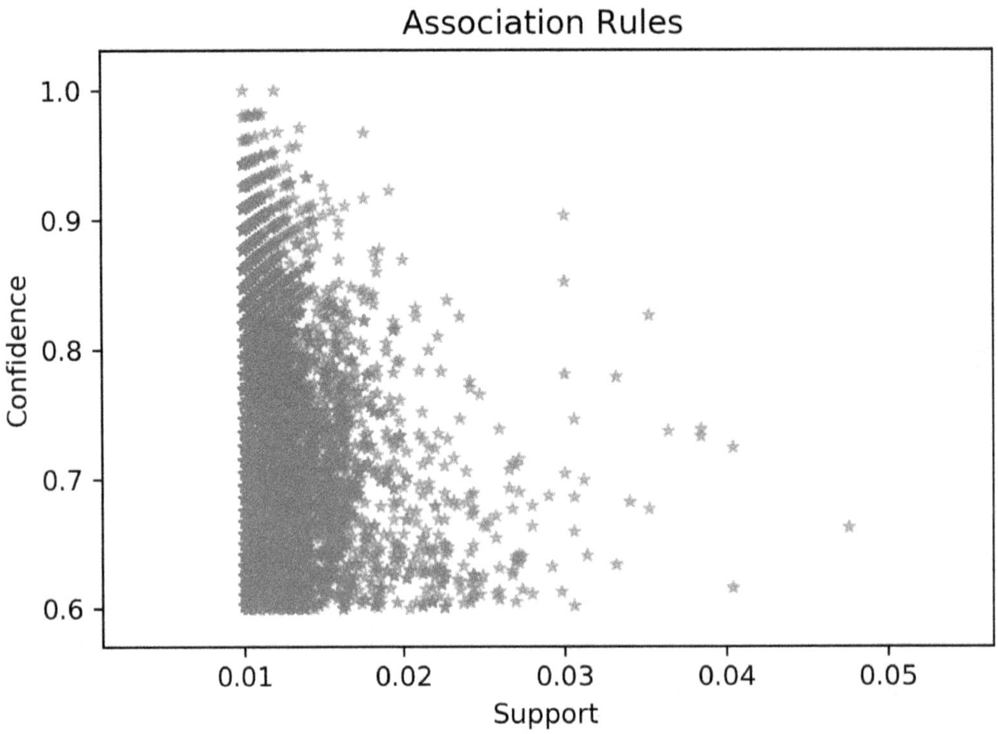

Figure 8.28: A plot of confidence against support

Notice that there are no association rules with both extremely high confidence and extremely high support. This should hopefully make sense. If an item set has high support, the items are likely to appear with many other items, making the chances of high confidence very low.

6. Look at the distribution of confidence:

   ```
   rules.hist("confidence", grid=False, bins=30)
   plt.xlabel("Confidence of item")
   plt.ylabel("Number of items")
   plt.title("Frequency distribution of Confidence")
   plt.show()
   ```

   The output is as follows:

   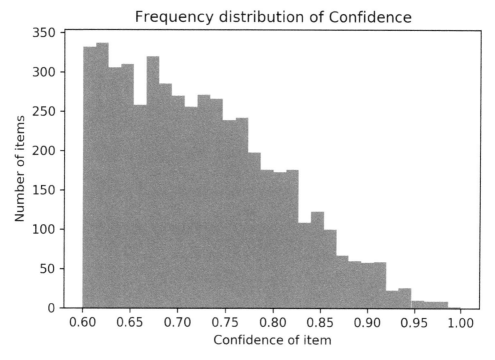

   Figure 8.29: The distribution of confidence values

7. Now, look at the distribution of lift:

   ```
   rules.hist("lift", grid=False, bins=30)
   plt.xlabel("Lift of item")
   plt.ylabel("Number of items")
   plt.title("Frequency distribution of Lift")
   plt.show()
   ```

The output is as follows:

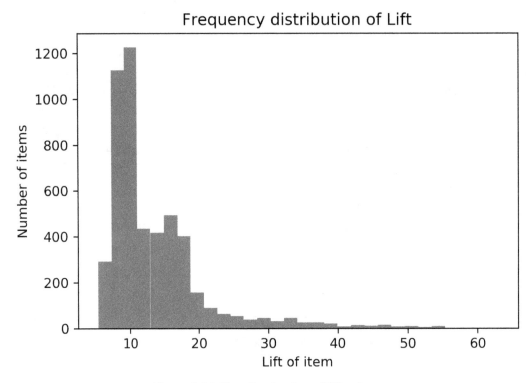

Figure 8.30: The distribution of lift values

As mentioned previously, this plot shows that **50** is a high threshold value in that there are not many points above that value.

8. Now, look at the distribution of leverage:

```
rules.hist("leverage", grid=False, bins=30)
plt.xlabel("Leverage of item")
plt.ylabel("Number of items")
plt.title("Frequency distribution of Leverage")
plt.show()
```

The output is as follows:

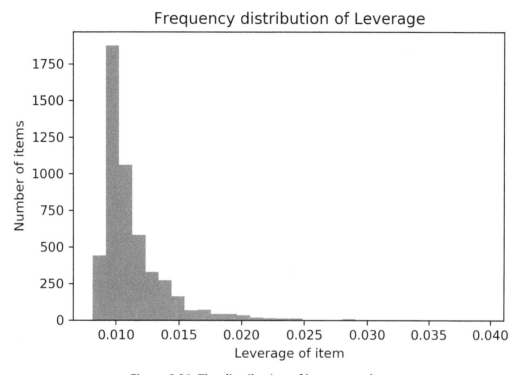

Figure 8.31: The distribution of leverage values

9. Now, look at the distribution of conviction:

```
plt.hist(rules[numpy.isfinite(rules['conviction'])]\
                            .conviction.values, bins = 30)
plt.xlabel("Coviction of item")
plt.ylabel("Number of items")
plt.title("Frequency distribution of Conviction")
plt.show()
```

The output is as follows:

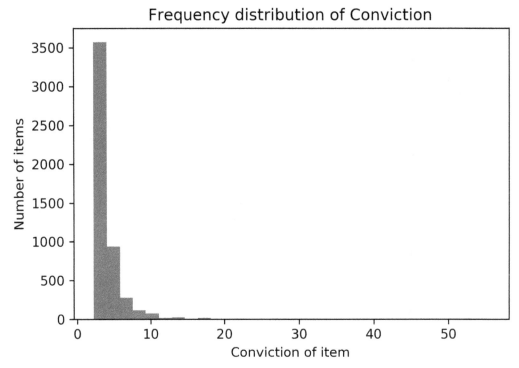

Figure 8.32: The distribution of conviction values

What is interesting about the four distributions is that spikes of varying sizes appear at the upper ends of the plots, implying that there are a few very strong association rules. The distribution of confidence tails off as the confidence values get larger, but at the very end, around the highest values, the distribution jumps up a little. The lift distribution has the most obvious spike. The conviction distribution plot shows a small spike, perhaps more accurately described as a bump, around 50. Lastly, the leverage distribution does not really show any spike in the higher values, but it does feature a long tail with some very high leverage values.

Take some time to explore the association rules found by the model. Do the product pairings make sense to you? What happened to the number of association rules when you changed the model parameter values? Do you appreciate the impact that these rules would have when attempting to improve any retail business?

In the preceding exercise, we built and plotted association rules. Association rules can be difficult to interpret and are heavily dependent on the thresholds and metrics used to create them. The questions in the preceding paragraph are meant to get you thinking creatively about how the algorithm works and how the rules can be used. Let's go through the questions one by one. There are obviously many rules, so the question regarding whether or not the rules make sense is hard to answer as a whole. Spot-checking the pairs seems to suggest that the rules are reasonable. For example, three of the pairings include children's cups and bowls, teacups and plates, and a playhouse kitchen and living room, all of which make sense. When the metrics and parameters changed, so did the results. As is the case in almost all modeling exercises, the ideal course of action is to look at the results under various circumstances and leverage all the findings to make the best decisions.

> **NOTE**
>
> To access the source code for this specific section, please refer to https://packt.live/3fhf9bS.
>
> You can also run this example online at https://packt.live/303vIBJ.
>
> You must execute the entire Notebook in order to get the desired result.

ACTIVITY 8.03: FINDING THE ASSOCIATION RULES ON THE COMPLETE ONLINE RETAIL DATASET

Let's pick up the scenario set out in *Activity 8.02, Running the Apriori Algorithm on the Complete Online Retail Dataset*. The company leadership comes back to you and says it is great that we know how frequently each item set occurs in the dataset, but which item sets can we act upon? Which item sets can we use to change the store layout or adjust pricing? To find these answers, we derive the full association rules.

In this activity, let's derive association rules from the complete online retail transaction dataset. Ensure that you complete this activity in the notebook that uses the full dataset (in other words, the notebook with the complete retail dataset, not the notebook from the exercises that use the 5,000-invoice subset).

366 | Market Basket Analysis

These steps will help us to perform the activity:

1. Fit the association rule model on the full dataset. Use the confidence metric and a minimum threshold of **0.6**.

2. Count the number of association rules. Is the number different from that found in *Step 1* of *Exercise 8.07, Deriving Association Rules*?

3. Plot confidence against support.

4. Look at the distributions of confidence, lift, leverage, and conviction.

Expected association rules output the following:

	antecedents	consequents	antecedent support	consequent support	support	confidence	lift	leverage	conviction
0	(ALARM CLOCK BAKELIKE CHOCOLATE)	(ALARM CLOCK BAKELIKE GREEN)	0.021255	0.048669	0.013756	0.647196	13.297902	0.012722	2.696488
1	(ALARM CLOCK BAKELIKE CHOCOLATE)	(ALARM CLOCK BAKELIKE RED)	0.021255	0.052195	0.014501	0.682243	13.071023	0.013392	2.982798
2	(ALARM CLOCK BAKELIKE ORANGE)	(ALARM CLOCK BAKELIKE GREEN)	0.022100	0.048669	0.013558	0.613483	12.605201	0.012482	2.461292
3	(ALARM CLOCK BAKELIKE GREEN)	(ALARM CLOCK BAKELIKE RED)	0.048669	0.052195	0.031784	0.653061	12.511932	0.029244	2.731908
4	(ALARM CLOCK BAKELIKE RED)	(ALARM CLOCK BAKELIKE GREEN)	0.052195	0.048669	0.031784	0.608944	12.511932	0.029244	2.432722
5	(ALARM CLOCK BAKELIKE IVORY)	(ALARM CLOCK BAKELIKE RED)	0.028308	0.052195	0.018524	0.654386	12.537313	0.017047	2.742380
6	(ALARM CLOCK BAKELIKE ORANGE)	(ALARM CLOCK BAKELIKE RED)	0.022100	0.052195	0.014998	0.678652	13.002217	0.013845	2.949463

Figure 8.33: Expected association rules

> **NOTE**
>
> The solution to this activity can be found on page 496.

By the end of this activity, you will have plots of lift, leverage, and conviction.

SUMMARY

Market basket analysis is used to analyze and extract insights from transaction or transaction-like data that can be used to help drive growth in many industries, most famously the retail industry. These decisions can include how to lay out the retail space, what products to discount, and how to price products. One of the central pillars of market basket analysis is the establishment of association rules. Association rule learning is a machine learning approach to uncovering the associations between the products individuals purchase that are strong enough to be leveraged for business decisions. Association rule learning relies on the Apriori algorithm to find frequent item sets in a computationally efficient way. These models are atypical of machine learning models because no prediction is being done, the results cannot really be evaluated using any one metric, and the parameter values are selected not by grid search, but by domain requirements specific to the question of interest. That being said, the goal of pattern extraction that is at the heart of all machine learning models is most definitely present here.

At the conclusion of this chapter, you should feel comfortable evaluating and interpreting probabilistic metrics, be able to run and adjust the Apriori algorithm and association rule learning models using `mlxtend` and know how these models are applied in business. Know that there is a decent chance that the positioning and pricing of items in your neighborhood grocery store were chosen based on the past actions made by you and many other customers in that store!

In the next chapter, we will explore hotspot analysis using kernel density estimation, arguably one of the most frequently used algorithms in all of statistics and machine learning.

9
HOTSPOT ANALYSIS

OVERVIEW

In this chapter, we will perform hotspot analysis. We will also visualize the results of hotspot analysis. We will use kernel density estimation, which is the most popular algorithm for building distributions using a collection of observations. We will build kernel density estimation models. We will describe the fundamentals behind probability density functions. By the end of the chapter, you should be able to leverage Python libraries to build multi-dimensional density estimation models and work with geo-spatial data.

INTRODUCTION

In the preceding chapter, we explored market basket analysis. Market basket analysis, as you hopefully recall, is an algorithm that seeks to understand the relationships between all the items and groups of items in transaction data. These relationships are then leveraged to help retailers optimize store layouts, more accurately order inventory, and adjust prices without shrinking the number of items in each transaction. We now change directions to explore hotspot modeling.

Let's consider an imaginary scenario: a new disease has begun spreading through numerous communities in the country that you live in and the government is trying to figure out how to confront this health emergency. Critical to any plan to confront this health emergency is epidemiological knowledge, including where the patients are located and how the disease is moving. The ability to locate and quantify problem areas (which are classically referred to as hotspots) can help health professionals, policy makers, and emergency response teams craft the most effective and efficient strategies for combating the disease. This scenario highlights one of the many applications of hotspot modeling.

Hotspot modeling is an approach that is used to identify how a population is distributed across a geographical area; for example, how the population of individuals infected with the previously mentioned disease is spread across the country. The creation of this distribution relies on the availability of representative sample data. Note that the population can be anything definable in geographical terms, which includes, but is not limited to, crime, disease-infected individuals, people with certain demographic characteristics, or hurricanes:

Figure 9.1: A fabricated example of fire location data showing some potential hotspots

Hotspot analysis is incredibly popular, and this is mainly because of how easy it is to visualize and interpret the results. Newspapers, websites, blogs, and TV shows all leverage hotspot analysis to support the arguments, chapters, and topics included in or on them. While it might not be as well-known as the most popular machine learning models, the main hotspot analysis algorithm, known as **kernel density estimation**, is arguably one of the most widely used analytical techniques. People even perform kernel density estimation mentally on a daily basis without knowing it. Kernel density estimation is a hotspot analysis technique that is used to estimate the true population distribution of specific geographical events. Before getting into the algorithm itself, we need to briefly review spatial statistics and probability density functions.

SPATIAL STATISTICS

Spatial statistics is the branch of statistics that focuses on the analysis of data that has spatial properties, including geographic or topological coordinates. It is similar to time series analysis in that the goal is to analyze data that changes across some dimension. In the case of time series analysis, the dimension across which the data changes is time, whereas in the spatial statistics case, the data changes across the spatial dimension. There are a number of techniques that are included under the spatial statistics umbrella, but the technique we are concerned with here is kernel density estimation. As is the goal of most statistical analyses, in spatial statistics, we are trying to take samples of geographic data and use them to generate insights and make predictions. The analysis of earthquakes is one arena in which spatial statistical analyses are commonly deployed. By collecting earthquake location data, maps that identify areas of high and low earthquake likelihood can be generated, which can help scientists determine both where future earthquakes are likely to occur and what to expect in terms of intensity.

PROBABILITY DENSITY FUNCTIONS

Kernel density estimation uses the idea of the **Probability Density Function (PDF)**, which is one of the foundational concepts in statistics. The probability density function is a function that describes the behavior of a continuous **random variable**. That is, it expresses the likelihood, or probability, that the random variable takes on some range of values. Consider the heights of males in the United States as an example. Using the probability density function of the heights of males in the United States, we could calculate the probability that some United States-based male is between 1.9 and 1.95 meters tall.

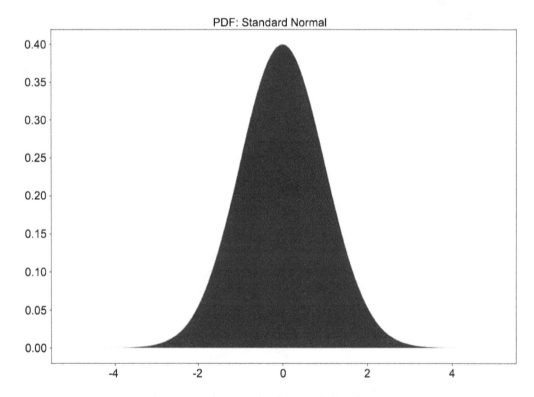

Figure 9.2: The standard normal distribution

Possibly the most popular density function in statistics is the standard normal distribution, which is simply the normal distribution centered at zero with a standard deviation equal to one.

Instead of the density function, what is typically available to statisticians or data scientists are randomly collected sample values coming from a population distribution that is unknown. This is where kernel density estimation comes in; it is a technique that is used for estimating the unknown probability density function of a random variable using sample data. The following figure represents a simple, but somewhat more reasonable example of a distribution that we would want to estimate with kernel density estimation. We would take some number of observations (sample data points) and use those observations to create a smooth distribution that mimics the true underlying distribution that is unknown to us.

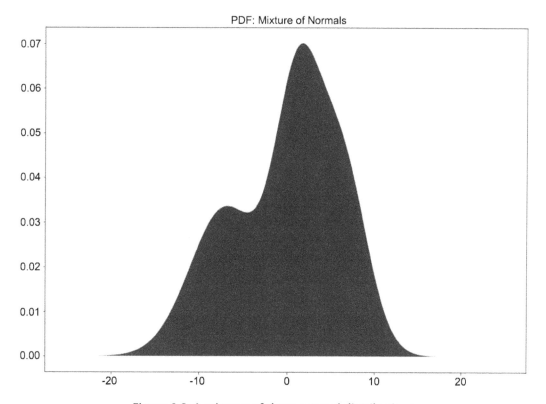

Figure 9.3: A mixture of three normal distributions

USING HOTSPOT ANALYSIS IN BUSINESS

We have already mentioned some of the ways in which hotspot modeling can be leveraged to meaningfully impact industry. When reporting on infectious diseases, health organizations and media companies typically use hotspot analysis to communicate where the diseases are located and the likelihood of contracting the disease based on geographic location. Using hotspot analysis, this information could be reliably computed and disseminated. Hotspot analysis is great for dealing with health data because the visualizations are very straightforward. This means that the chances of data being misinterpreted either intentionally or unintentionally are relatively low.

Hotspot analysis can also be used to predict where certain events are likely to occur geographically. One research area that is leveraging the predictive capabilities of hotspot analysis more and more are the environmental sciences, which includes the study of natural disasters and extreme weather events. Earthquakes, for example, are notorious for being difficult to predict, because the time between significant earthquakes can be large, and the machinery needed to track and measure earthquakes to the degree required to make these predictions is relatively new.

In terms of public policy and resource deployment, hotspot analysis can be very impactful when dealing with the analysis of population demographics. Determining where resources, both monetary and personnel, should be deployed can be challenging; however, given that resources are often demographic-specific, hotspot analysis is a useful technique since it can be used to determine the distribution of certain demographic characteristics. By demographic characteristics we mean that we could find the geographic distribution of high school graduates, immigrants from a specific global region, or individuals making $100,000 or more annually.

The number of applications of hotspot modeling are virtually endless. We have here only discussed three of the major ones.

KERNEL DENSITY ESTIMATION

One of the main methodological approaches to hotspot analysis is kernel density estimation. Kernel density estimation builds an estimated density using sample data and two parameters known as the **kernel function** and the **bandwidth value**. The estimated density is, like any distribution, essentially a guideline for the behavior of a random variable. Here, we mean how frequently the random variable takes on any specific value, $\{x_1, ..., x_n\}$. When dealing with hotspot analysis where the data is typically geographic, the estimated density answers the question *How frequently do specific longitude and latitude pairs appear for a given event?* If a specific longitude and latitude pair $\{x_{longitude}, x_{latitude}\}$ and other nearby pairs occur with high frequency, then the estimated density built using the sample data will show that the area around the aforementioned longitude and latitude pair occurs with high likelihood.

Kernel density estimation is referred to as a smoothing algorithm because a smooth curve is drawn over the sample data, which, if the data is a representative sample, can be a good estimate of the true population density function. Stated another way, when it is done correctly, kernel density estimation aims to remove the noise that is inherent in sampled data, but is not a feature of the total population. The only assumption of the model is that the data truly belongs to some interpretable and meaningful density from which insights can be derived and acted upon. That is, there exists a true underlying distribution. We assume that the sample data contains clusters of data points and that these clusters align to regions of high likelihood in the true population. A benefit of creating a quality estimate of the true population density is that the estimated density can then be used to sample more data from the population.

Following this brief introduction, you probably have the following two questions:

- What is the bandwidth value?
- What is the kernel function?

We answer both of these questions next.

THE BANDWIDTH VALUE

The most crucial parameter in kernel density estimation is called the **bandwidth value** and its impact on the quality of the estimate cannot be overestimated. A high-level definition of the bandwidth value is that it is a value that determines the degree of smoothing. If the bandwidth value is low, then the estimated density will feature limited smoothing, which means that the density will capture all the noise in the sample data. If the bandwidth value is high, then the estimated density will be very smooth. An overly smooth density will remove characteristics of the true density from the estimated density, which are legitimate and not noise.

In more statistical parlance, the bandwidth parameter controls the bias-variance trade-off. That is, high variance is the result of low bandwidth values because the density is sensitive to the variance of the sample data. Low bandwidth values limit any ability the model may have had to adapt to and work around gaps in the sample data that are not present in the population. A density estimated using a low bandwidth value will tend to overfit the data (this is also known as an under-smoothed density). When high bandwidth values are used, then the resulting density is underfit and the estimated density has a high bias (this is also known as an over-smoothed density).

> **NOTE**
>
> In all the subsequent exercises and activities, the output could vary slightly from what is shown below. This is because of the following: differences in sampled data can lead to slightly different output and the `sklearn` and `seaborn` libraries have some non-deterministic elements that could cause the results to change from run to run.

EXERCISE 9.01: THE EFFECT OF THE BANDWIDTH VALUE

In this exercise, we will fit nine different models with nine different bandwidth values to sample data created in the exercise. The goal here is to solidify our understanding of the impact the bandwidth parameter can have and make clear that if an accurate estimated density is sought, then the bandwidth value needs to be selected with care. Note that finding an optimal bandwidth value will be the topic of the next section. All exercises will be done in a Jupyter notebook utilizing Python 3; ensure that all package installation is done using `pip`. The easiest way to install the `basemap` module from `mpl_toolkits` is by using *Anaconda*. Instructions for downloading and installing *Anaconda* can be found at the beginning of this book:

1. Load all of the libraries that are needed for the exercises in this chapter. The **basemap** library is used to create graphics involving location data. All the other libraries have been used previously in this title.

```
get_ipython().run_line_magic('matplotlib', 'inline')
import matplotlib.pyplot as plt
import mpl_toolkits.basemap
import numpy
import pandas
import scipy.stats
import seaborn
import sklearn.model_selection
import sklearn.neighbors
seaborn.set()
```

2. Create some sample data (**vals**) by mixing three normal distributions. In addition to the sample data, define the true density curve (**true_density**) and the range over which the data will be plotted (**x_vec**):

```
x_vec = numpy.linspace(-30, 30, 10000)[:, numpy.newaxis]
numpy.random.seed(42)
vals = numpy.concatenate(( \
        numpy.random.normal(loc=1, scale=2.5, size=500), \
        numpy.random.normal(loc=10, scale=4, size=500), \
        numpy.random.normal(loc=-12, scale=5, size=500) \
))[:, numpy.newaxis]
true_density = ((1 / 3) * scipy.stats.norm(1, 2.5)\
                        .pdf(x_vec[:, 0]) \
                + (1 / 3) * scipy.stats.norm(10, 4)\
                        .pdf(x_vec[:, 0]) \
                + (1 / 3) * scipy.stats.norm(-12, 5)\
                        .pdf(x_vec[:, 0]))
```

3. Define a list of tuples that will guide the creation of the multiplot graphic. Each tuple contains the row and column indices of the specific subplot, and the bandwidth value used to create the estimated density in that particular subplot. Note that, for the sake of this exercise, the bandwidth values are picked randomly, but there is some strategy that goes into picking optimal bandwidth values. We will dig more into this in the next section.

```
position_bandwidth_vec = [(0, 0, 0.1), (0, 1, 0.4), (0, 2, 0.7), \
                         (1, 0, 1.0), (1, 1, 1.3), (1, 2, 1.6), \
                         (2, 0, 1.9), (2, 1, 2.5), (2, 2, 5.0)]
```

4. Create nine plots each using a different bandwidth value. The first plot, with the index of (0, 0), will have the lowest bandwidth value and the last plot, with the index of (2, 2), will have the highest. These values are not the absolute lowest or absolute highest bandwidth values, rather they are only the minimum and maximum of the list defined in the previous step:

```
fig, ax = plt.subplots(3, 3, sharex=True, \
                      sharey=True, figsize=(12, 9))
fig.suptitle('The Effect of the Bandwidth Value', fontsize=16)
for r, c, b in position_bandwidth_vec:
    kde = sklearn.neighbors.KernelDensity(bandwidth=b).fit(vals)
    log_density = kde.score_samples(x_vec)
    ax[r, c].hist(vals, bins=50, density=True, alpha=0.5)
    ax[r, c].plot(x_vec[:, 0], numpy.exp(log_density), \
                  '-', linewidth=2)
    ax[r, c].set_title('Bandwidth = {}'.format(b))
plt.show()
```

The output is as follows:

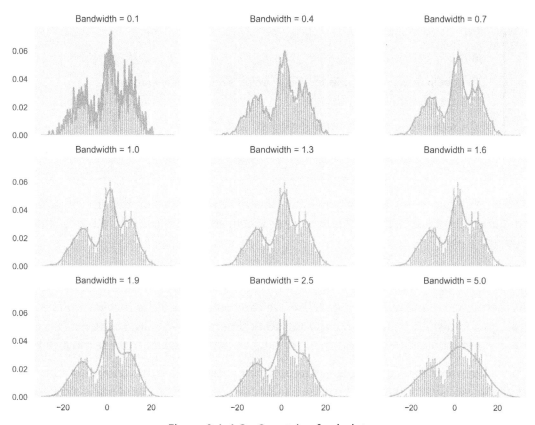

Figure 9.4: A 3 x 3 matrix of subplots

Notice that the estimated density curve in the ninth subplot (where the bandwidth is 5) clearly underfits the data. As the bandwidth values increase, the estimated density becomes smoother until it noticeably underfits the data. Visually, it looks like the optimal bandwidth may be around **1.6**.

> **NOTE**
>
> To access the source code for this specific section, please refer to https://packt.live/2UOHbTZ.
>
> You can also run this example online at https://packt.live/38DbmTo.
>
> You must execute the entire Notebook in order to get the desired result.

The next step is to design an algorithm to identify the optimal bandwidth value, so that the estimated density is the most reasonable and, therefore, the most reliable and actionable.

SELECTING THE OPTIMAL BANDWIDTH

As mentioned in the preceding exercise, we can come quite close to selecting the optimal bandwidth by simply comparing several densities visually. However, this is neither the most efficient method of selecting parameter values nor the most reliable.

There are two standard approaches to optimizing the bandwidth value, and both of these will appear in future exercises and activities. The first approach is a plug-in method (or a formulaic approach) that is deterministic and not optimized on the sample data. Plug-in methods are generally much faster to implement, simpler to code, and easier to explain. However, these methods have one big downside, which is that their accuracy tends to suffer compared to approaches that are optimized on the sample data. These methods also have distributional assumptions. The most popular plug-in methods are Silverman's Rule and Scott's Rule. Explaining these rules in detail is beyond the scope of this text, not necessary for fully understanding kernel density estimation, and would require some tricky mathematical work, so we will skip any further exploration here. That being said, if interested, there are a number of great sources publicly available that explain these rules at various levels of detail. By default, the **seaborn** package (which will be used in future exercises) uses Scott's Rule as the method to determine the bandwidth value.

The second, and arguably the more robust, approach to finding an optimal bandwidth value is by searching a predefined grid of bandwidth values. Grid search is an empirical approach that is used frequently in machine learning and predictive modeling to optimize model hyperparameters. The process starts by defining the bandwidth grid, which is simply the collection of bandwidth values to be evaluated. The bandwidth grid is chosen at random. Use each bandwidth value in the grid to create an estimated density; then, score the estimated density using the pseudo-log-likelihood value. The optimal bandwidth value is that which has the maximum pseudo-log-likelihood value. Think of the pseudo-log-likelihood value as the probability of getting data points where we got data points and the probability of not getting points where we did not get any data points. Ideally, both of these probabilities would be large. Consider the case where the probability of getting data points where we did get points is low. In this situation, the implication would be that the data points in the sample were anomalous because, under the true distribution, getting points where we did would not be expected with a high likelihood value. The pseudo-log-likelihood value is an evaluation metric that plays the same role as the accuracy score in classification problems and root mean squared error in regression problems.

Let's now implement the grid search approach to optimize the bandwidth value.

EXERCISE 9.02: SELECTING THE OPTIMAL BANDWIDTH USING GRID SEARCH

In this exercise, we will create an estimated density for the sample data created in *Exercise 9.01, The Effect of the Bandwidth Value* with an optimal bandwidth value identified using grid search and cross-validation. To run the grid search with cross-validation, we will leverage `sklearn`, which we have used throughout this book.

> **NOTE**
>
> This exercise is a continuation of *Exercise 9.01, The Effect of the Bandwidth Value* as we are using the same sample data and continuing our exploration of the bandwidth value.

1. Define a grid of bandwidth values and the grid search cross-validation model. Ideally, the leave-one-out approach to cross-validation should be used, but for the sake of having the model run in a reasonable amount of time, we will do a 10-fold cross-validation. Fit the model on the sample data, as follows:

```
# define a grid of 100 possible bandwidth values
bandwidths = 10 ** numpy.linspace(-1, 1, 100)
# define the grid search cross validation model
grid = sklearn.model_selection.GridSearchCV\
        (estimator=sklearn.neighbors.KernelDensity(),\
        param_grid={"bandwidth": bandwidths},\
        cv=10)
# run the model on the previously defined data
grid.fit(vals)
```

The output is as follows:

```
GridSearchCV(cv=10, error_score='raise-deprecating',
        estimator=KernelDensity(algorithm='auto', atol=0, bandwidth=1.0, breadth_first=True,
        kernel='gaussian', leaf_size=40, metric='euclidean',
        metric_params=None, rtol=0),
        fit_params=None, iid='warn', n_jobs=None,
        param_grid={'bandwidth': array([ 0.1    ,  0.10476, ...,  9.54548, 10.    ])},
        pre_dispatch='2*n_jobs', refit=True, return_train_score='warn',
        scoring=None, verbose=0)
```

Figure 9.5: Output of cross-validation model

2. Extract the optimal bandwidth value from the model. The **best_params_** function extracts from the model object the best performing parameters in the grid.

```
best_bandwidth = grid.best_params_["bandwidth"]
print("Best Bandwidth Value: {}" \
        .format(best_bandwidth))
```

The optimal bandwidth value should be approximately **1.6**. We can interpret the optimal bandwidth value as the bandwidth value producing the maximum pseudo-log-likelihood value. Note that depending on the values included in the grid, the optimal bandwidth value can change.

3. Plot the histogram of the sample data overlaid by both the true and estimated densities. In this case, the estimated density will be the optimal estimated density:

```
fig, ax = plt.subplots(figsize=(14, 10))
ax.hist(vals, bins=50, density=True, alpha=0.5, \
        label='Sampled Values')
ax.fill(x_vec[:, 0], true_density, \
        fc='black', alpha=0.3, label='True Distribution')
log_density = numpy.exp(grid.best_estimator_ \
                        .score_samples(x_vec))
ax.plot(x_vec[:, 0], log_density, \
        '-', linewidth=2, label='Kernel = Gaussian')
ax.legend(loc='upper right')
plt.show()
```

The output is as follows:

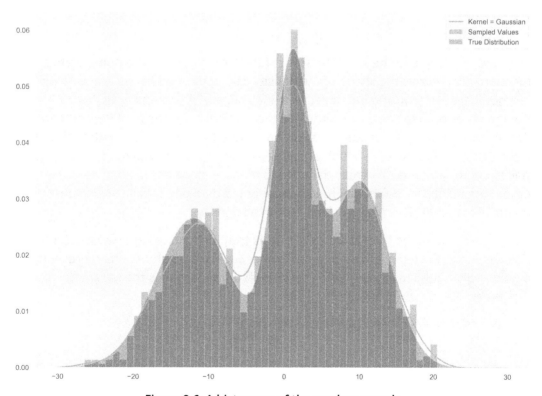

Figure 9.6: A histogram of the random sample

In this histogram, the true density and the optimal estimated density are overlaid. The estimated density is neither overfit or underfit to any noticeable degree and it definitely captures the three clusters. Arguably, it could map to the true density better, but this is just an estimated density generated by a model that is limited by the dataset provided.

> **NOTE**
>
> To access the source code for this specific section, please refer to https://packt.live/2UOHbTZ.
>
> You can also run this example online at https://packt.live/38DbmTo.
>
> You must execute the entire Notebook in order to get the desired result.

Let's now move onto the second question: what is the kernel function and what role does it play?

KERNEL FUNCTIONS

The other parameter to be set is the kernel function. The kernel is a non-negative function that controls the shape of the density. Like topic models, we are working in a non-negative environment because it does not make sense to have negative likelihoods or probabilities. The kernel function controls the shape of the estimated density by weighting the points in a systematic way. This systematic methodology for weighting is fairly simple; data points that are in close proximity to many other data points are up-weighted, whereas data points that are alone or far away from any other data points are down-weighted. Up-weighted data points will correspond to points of higher likelihood in the final estimated density.

Many functions can be used as kernels, but six frequent choices are Gaussian, Tophat, Epanechnikov, Exponential, Linear, and Cosine. Each of these functions represents a unique distributional shape. Note that in each of the formulas the parameter, h, represents the bandwidth value:

- Gaussian: Each observation has a bell-shaped weight.

$$K(x;h) \propto exp\left(-\frac{x^2}{2h^2}\right)$$

Figure 9.7: The formula for the Gaussian kernel

- Tophat: Each observation has a rectangular-shaped weight.

$$K(x;h) \propto \begin{cases} 0 & \text{if } |x| \geq h \\ \dfrac{1}{2h} & \text{if } |x| < h \end{cases}$$

Figure 9.8: The formula for the Tophat kernel

- Epanechnikov: Each observation has a mound-shaped weight.

$$K(x;h) \propto 1 - \dfrac{x^2}{h^2}$$

Figure 9.9: The formula for the Epanechnikov kernel

- Exponential: Each observation has a triangular-shaped weight. The sides of the triangle are concave.

$$K(x;h) \propto \exp\left(-\dfrac{|x|}{h}\right)$$

Figure 9.10: The formula for the Exponential kernel

- Linear: Each observation has a triangular-shaped weight.

$$K(x;h) \propto \begin{cases} 0 & \text{if } |x| \geq h \\ 1 - \dfrac{x}{h} & \text{if } |x| < h \end{cases}$$

Figure 9.11: The formula for the Linear kernel

- Cosine: Each observation has a mound-shaped weight. This mound-shape is narrower at the top than the Epanechnikov kernel.

$$K(x;h) \propto \begin{cases} 0 & \text{if } |x| \geq h \\ 1 - \cos\dfrac{\pi x}{2h} & \text{if } |x| < h \end{cases}$$

Figure 9.12: The formula for the Cosine kernel

Here are the distributional shapes of the six kernel functions:

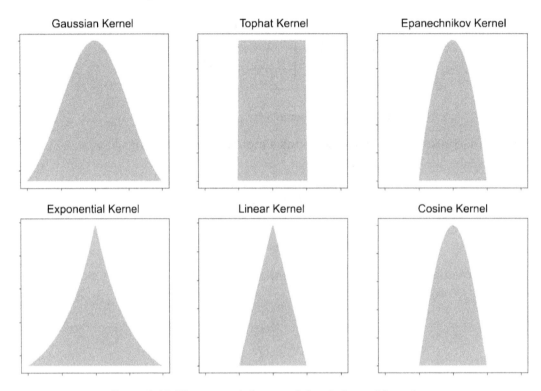

Figure 9.13: The general shapes of the six kernel functions

The choice of kernel function is not completely insignificant, but it is definitely not nearly as important as the choice of bandwidth value. A reasonable course of action would be to use the gaussian kernel for all density estimation problems, which is what we will do in the following exercises and activities.

EXERCISE 9.03: THE EFFECT OF THE KERNEL FUNCTION

We will demonstrate how the choice of kernel function affects the quality of the density estimate. Like we did when exploring the bandwidth value effect, we will hold all other parameters constant, use the same data generated in the first two exercises, and run six different kernel density estimation models using the six kernel functions previously specified. Clear differences should be noticeable between the six estimated densities, but these differences should be slightly less dramatic than the differences between the densities estimated using the different bandwidth values. Note that this exercise should be executed in the same notebook as the previous exercises.

1. Define a list of tuples along the same lines as the one defined previously. Each tuple includes the row and column indices of the subplot, and the kernel function to be used to create the density estimation:

   ```
   position_kernel_vec = [(0, 0, 'gaussian'), (0, 1, 'tophat'), \
                          (1, 0, 'epanechnikov'), \
                          (1, 1, 'exponential'), \
                          (2, 0, 'linear'), (2, 1, 'cosine'),]
   ```

2. Fit and plot six kernel density estimation models using a different kernel function for each. To truly understand the differences between the kernel functions, we will set the bandwidth value to the optimal bandwidth value found in *Exercise 9.02, Selecting the Optimal Bandwidth Using Grid Search* and not adjust it:

   ```
   fig, ax = plt.subplots(3, 2, sharex=True, \
                          sharey=True, figsize=(12, 9))
   fig.suptitle('The Effect of Different Kernels', fontsize=16)
   for r, c, k in position_kernel_vec:
       kde = sklearn.neighbors.KernelDensity(\
             kernel=k, bandwidth=best_bandwidth).fit(vals)
       log_density = kde.score_samples(x_vec)
       ax[r, c].hist(vals, bins=50, density=True, alpha=0.5)
       ax[r, c].plot(x_vec[:, 0], numpy.exp(log_density), \
                     '-', linewidth=2)
       ax[r, c].set_title('Kernel = {}'.format(k.capitalize()))
   plt.show()
   ```

The output is as follows:

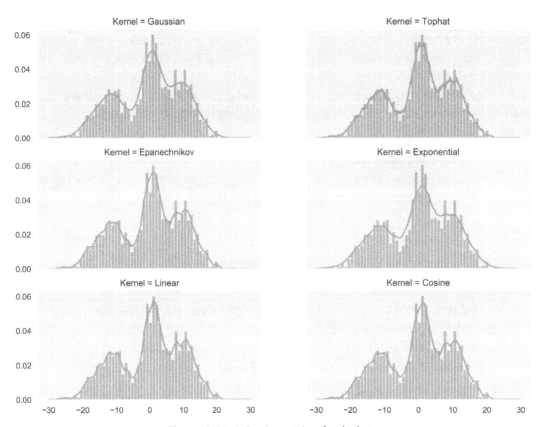

Figure 9.14: A 3 x 2 matrix of subplots

Out of the six kernel functions, the gaussian kernel produced the most reasonable estimated density. Beyond that, notice that the difference between the estimated densities with different kernels is less than the difference between the estimated densities with different bandwidth values. This goes to the previously made claim that the bandwidth value is the more important parameter and should be the focus during the model building process.

> **NOTE**
>
> To access the source code for this specific section, please refer to https://packt.live/2UOHbTZ.
>
> You can also run this example online at https://packt.live/38DbmTo.
>
> You must execute the entire Notebook in order to get the desired result.

With our understanding mostly formed, let's discuss the derivation of kernel density estimation at a high-level.

KERNEL DENSITY ESTIMATION DERIVATION

Let's skip the formal mathematical derivation in favor of the popular derivation by intuition. Kernel density estimation turns each data point in the sample into its own distribution whose width is controlled by the bandwidth value. The individual distributions are then summed to create the desired density estimate. This concept is fairly easy to demonstrate; however, before doing that in the next exercise, let's try to think through it in an abstract way. For geographic regions containing many sample data points, the individual densities will overlap and, through the process of summing those densities, will create points of higher likelihood in the estimated density. Similarly, for geographic regions containing few to no sample data points, the individual densities will not overlap and, therefore, will correspond to points of lower likelihood in the estimated density.

EXERCISE 9.04: SIMULATING THE DERIVATION OF KERNEL DENSITY ESTIMATION

The goal here is to demonstrate the concept of summing individual distributions to create an overall estimated density for a random variable. We will establish the concept incrementally by starting with one sample data point and then work up too many sample data points. Additionally, different bandwidth values will be applied, so our understanding of the effect of the bandwidth value on these individual densities will further solidify. Note that this exercise should be done in the same notebook as all the other exercises.

1. D Hotspot Analysis efine a function that will evaluate the normal distribution. The input values are the grid representing the range of the random variable, **x**, the sampled data point, **m**, and the bandwidth, **b**:

   ```
   def eval_gaussian(x, m, b):
       numerator = numpy.exp(-numpy.power(x - m, 2) \
                   / (2 * numpy.power(b, 2)))
   ```

```
            denominator = b * numpy.sqrt(2 * numpy.pi)
            return numerator / denominator
```

2. Plot a single sample data point as a histogram and as an individual density with varying bandwidth values:

```
m = numpy.array([5.1])
b_vec = [0.1, 0.35, 0.8]
x_vec = numpy.linspace(1, 10, 100)[:, None]
figOne, ax = plt.subplots(2, 3, sharex=True, \
                          sharey=True, figsize=(15, 10))
for i, b in enumerate(b_vec):
    ax[0, i].hist(m[:], bins=1, fc='#AAAAFF', density=True)
    ax[0, i].set_title("Histogram: Normed")
    evaluation = eval_gaussian(x_vec, m=m[0], b=b)
    ax[1, i].fill(x_vec, evaluation, '-k', fc='#AAAAFF')
    ax[1, i].set_title("Gaussian Dist: b={}".format(b))
plt.show()
```

The output is as follows:

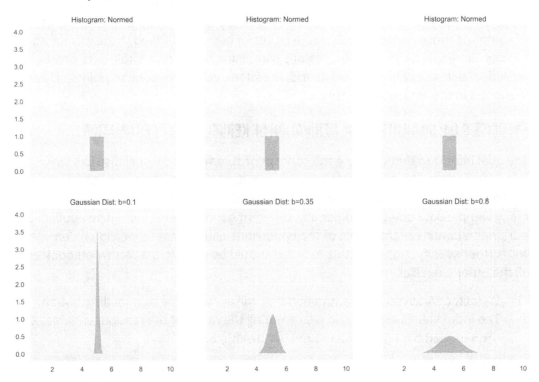

Figure 9.15: Showing one data point and its individual density at various bandwidth values

Here, we see what has already been established, which is that lower bandwidth values produce very narrow densities that tend to overfit the data.

3. Reproduce the work done in *Step 2*, but now scale up to 16 data points:

```
m = numpy.random.normal(4.7, 0.88, 16)
n = len(m)
b_vec = [0.1, 0.35, 1.1]
x_vec = numpy.linspace(-1, 11, 100)[:, None]
figMulti, ax = plt.subplots(2, 3, sharex=True, \
                            sharey=True, figsize=(15, 10))
for i, b in enumerate(b_vec):
    ax[0, i].hist(m[:], bins=n, fc='#AAAAFF', density=True)
    ax[0, i].set_title("Histogram: Normed")
    sum_evaluation = numpy.zeros(len(x_vec))
    for j in range(n):
        evaluation = eval_gaussian(x_vec, m=m[j], b=b) / n
        sum_evaluation += evaluation[:, 0]
        ax[1, i].plot(x_vec, evaluation, \
                      '-k', linestyle="dashed")
    ax[1, i].fill(x_vec, sum_evaluation, '-k', fc='#AAAAFF')
    ax[1, i].set_title("Gaussian Dist: b={}".format(b))
plt.show()
```

The output is as follows:

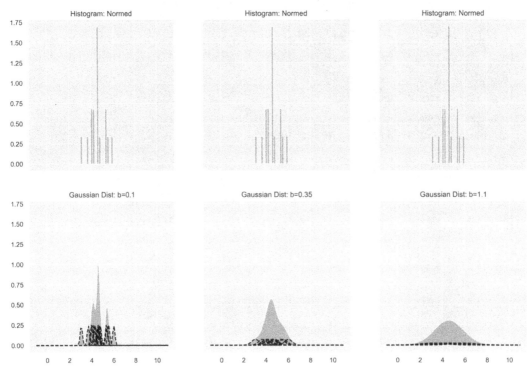

Figure 9.16: Plotting the data points

The preceding image shows 16 data points, their individual densities at various bandwidth values, and the sum of their individual densities.

Again, unsurprisingly, the plot utilizing the smallest bandwidth value features a wildly overfit estimated density. That is, the estimated density captures all the noise in the sample data. Of these three densities, the second one, where the bandwidth value was set to **0.35**, is the most reasonable.

> **NOTE**
>
> To access the source code for this specific section, please refer to https://packt.live/2UOHbTZ.
>
> You can also run this example online at https://packt.live/38DbmTo.
>
> You must execute the entire Notebook in order to get the desired result.

ACTIVITY 9.01: ESTIMATING DENSITY IN ONE DIMENSION

In this activity, we will be generating some fake sample data and estimating the density function using kernel density estimation. The bandwidth value will be optimized using grid search cross-validation. The goal is to solidify our understanding of this useful methodology by running the model in a simple one-dimensional case. We will once again leverage Jupyter notebooks to do our work.

Imagine that the sample data we will be creating describes the price of homes in a state in the United States. Momentarily ignore the values in the following sample data. The question is, *what does the distribution of home prices look like, and can we extract the probability of a house having a price that falls in some specific range?* These questions and more are answerable using kernel density estimation.

Here are the steps to complete the activity:

1. Open a new notebook and install all the necessary libraries.

2. Sample 1,000 data points from the standard normal distribution. Add 3.5 to each of the last 625 values of the sample (that is, the indices between 375 and 1,000). Set a random state of 100 using `numpy.random.RandomState` to guarantee the same sampled values, and then randomly generate the data points using the `rand.randn(1000)` call.

3. Plot the 1,000-point sample data as a histogram and add a scatterplot below it.

4. Define a grid of bandwidth values. Then, define and fit a grid search cross-validation algorithm.

5. Extract the optimal bandwidth value.

6. Replot the histogram from *Step 3* and overlay the estimated density.

The output will be as follows:

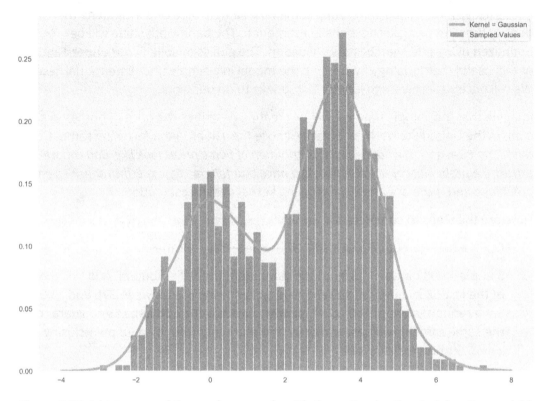

Figure 9.17: A histogram of the random sample with the optimal estimated density overlaid

> **NOTE**
> The solution for this activity can be found on page 501.

HOTSPOT ANALYSIS

To start, hotspots are areas of higher concentrations of data points, such as particular neighborhoods where the crime rate is abnormally high or swaths of the country that are impacted by an above-average number of tornadoes. Hotspot analysis is the process of finding these hotspots, should any exist, in a population using sampled data. This process is generally done by leveraging kernel density estimation.

Hotspot analysis can be described in four high-level steps:

1. **Collect the data**: The data should include the locations of the objects or events. As we have briefly mentioned, the amount of data needed to run and achieve actionable results is relatively flexible. The optimal state is to have a sample dataset that is representative of the population.

2. **Identify the base map**: The next step is to identify which base map would best suit the analytical and presentational needs of the project. On this base map, the results of the model will be overlaid, so that the locations of the hotspots can be easily articulated in much more digestible terms, such as city, neighborhood, or region.

3. **Execute the model**: In this step, you select and execute one or multiple methodologies of extracting spatial patterns to identify hotspots. For us, this method will be – no surprise – kernel density estimation.

4. **Create the visualization**: The hotspot maps are generated by overlaying the model results on the base map to support whatever business questions are outstanding.

One of the principal issues with hotspot analysis from a usability standpoint is that the statistical significance of a hotspot is not particularly easy to ascertain. Most questions about statistical significance revolve around the existence of the hotspots. That is, do the fluctuations in likelihood of occurrence actually amount to statistically significant fluctuations? It is important to note that statistical significance is not required to perform kernel density estimation and that we will not be dealing with significance at all going forward.

While the term hotspot is traditionally reserved to describe a cluster of location data points, it is not limited to location data. Any data type can have hotspots regardless of whether or not they are referred to as hotspots. In one of the following exercises, we will model some non-location data to find hotspots, which will be regions of the feature space having a high or low likelihood of occurrence.

EXERCISE 9.05: LOADING DATA AND MODELING WITH SEABORN

In this exercise, we will work with the **seaborn** library to fit and visualize kernel density estimation models. This is done on both location and non-location data. Before getting into the modeling, we load the data, which is the California housing dataset that comes with **sklearn**, that has been provided in csv form. The file needs to be downloaded from the GitHub repository and saved on your local machine. Taken from the United States census in 1990, this dataset describes the housing situation in California during that time. One row of data describes one census block group. The definition of a census block group is irrelevant to this exercise, so we will bypass the definition here in favor of more hands-on coding and modeling. It is important to mention that all the variables are aggregated to the census block. For example, **MedInc** is the median income of households in each census block. Additional information on this dataset is available at https://scikit-learn.org/stable/datasets/index.html#california-housing-dataset.

1. Load the California housing dataset using **california_housing.csv**. Print the first five rows of the data frame:

    ```
    df = pandas.read_csv('./california_housing.csv', header=0)
    df.head()
    ```

    NOTE

    The path of the file depends on the location of the file on your system.

    The output is as follows:

	MedInc	HouseAge	AveRooms	AveBedrms	Population	AveOccup	Latitude	Longitude
0	8.3252	41.0	6.984127	1.023810	322.0	2.555556	37.88	-122.23
1	8.3014	21.0	6.238137	0.971880	2401.0	2.109842	37.86	-122.22
2	7.2574	52.0	8.288136	1.073446	496.0	2.802260	37.85	-122.24
3	5.6431	52.0	5.817352	1.073059	558.0	2.547945	37.85	-122.25
4	3.8462	52.0	6.281853	1.081081	565.0	2.181467	37.85	-122.25

    Figure 9.18: The first five rows of the California housing dataset from sklearn

2. Filter the data frame on the **HouseAge** feature, which is the median home age of each census block. Keep only the rows with **HouseAge** less than or equal to 15 and name the data frame **dfLess15**. Print out the first five rows of the data frame, then reduce the data frame down to just the longitude and latitude features:

```
dfLess15 = df[df['HouseAge'] <= 15.0]
dfLess15 = dfLess15[['Latitude', 'Longitude']]
dfLess15.head()
```

The output is as follows:

	Latitude	Longitude
59	37.82	-122.29
87	37.81	-122.27
88	37.80	-122.27
391	37.90	-122.30
437	37.87	-122.30

Figure 9.19: The first five rows of the filtered dataset

3. Use **seaborn** to fit and visualize the kernel density estimation model built on the longitude and latitude data points. There are four inputs to the model, which are the names of the two columns over which the estimated density is sought (that is, the longitude and latitude), the data frame to which those columns belong, and the method of density estimation (that is, the **kde** or kernel density estimation):

```
seaborn.jointplot("Longitude", "Latitude", dfLess15, kind="kde")
```

The output is as follows:

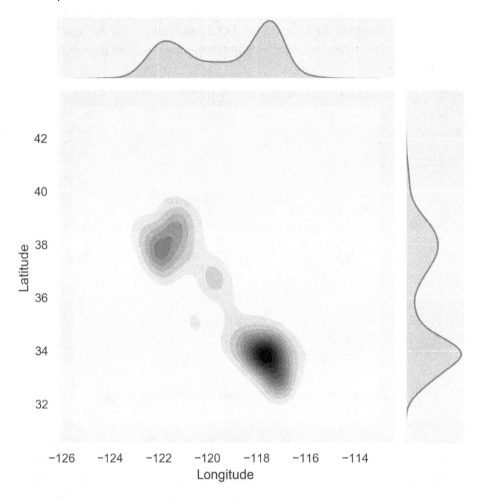

Figure 9.20: A joint plot

> **NOTE**
> The graph might differ as the estimation is not exactly the same every time.

This joint plot contains both the two-dimensional estimated density plus the marginal densities for the dfLess15 dataset.

If we overlay these results on a map of California, we will see that the hotspots are southern California, including Los Angeles and San Diego, the bay area, including San Francisco, and to a small degree the area known as the central valley. A benefit of this **seaborn** graphic is that we get the two-dimensional estimated density and the marginal densities for both longitude and latitude.

4. Create another filtered data frame based on the **HouseAge** feature; this time keep only the rows with **HouseAge** greater than 40 and name the data frame **dfMore40**. Additionally, remove all the columns other than longitude and latitude. Then, print the first five rows of the data frame:

```
dfMore40 = df[df['HouseAge'] > 40.0]
dfMore40 = dfMore40[['Latitude', 'Longitude']]
dfMore40.head()
```

The output is as follows:

	Latitude	Longitude
0	37.88	-122.23
2	37.85	-122.24
3	37.85	-122.25
4	37.85	-122.25
5	37.85	-122.25

Figure 9.21: The top of the dataset filtered

5. Repeat the process from *Step 3*, but using this new filtered data frame:

```
seaborn.jointplot("Longitude", "Latitude", dfMore40, kind="kde")
```

The output is as follows:

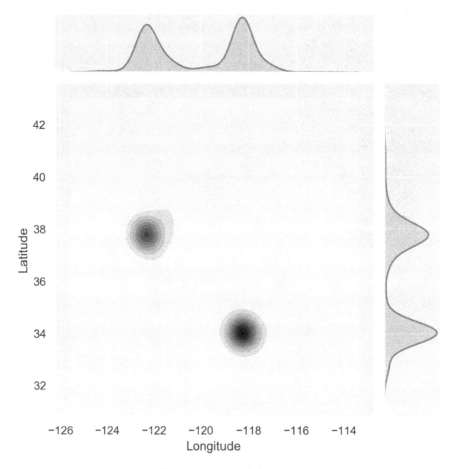

Figure 9.22: The joint plot

This joint plot contains both the two-dimensional estimated density plus the marginal densities for the dfMore40 dataset.

This estimated density is much more compact in that the data is clustered almost entirely in two areas. Those areas are Los Angeles and the bay area. Comparing this to the plot in *Step 3*, we notice that housing development has spread out across the state. Additionally, newer housing developments occur with much higher frequencies in a larger number of census blocks.

6. Let's again create another filtered data frame. This time only keeping rows where **HouseAge** is less than or equal to five and name the data frame **dfLess5**. Plot **Population** and **MedInc** as a scatterplot, as follows:

```
dfLess5 = df[df['HouseAge'] <= 5]
x_vals = dfLess5.Population.values
y_vals = dfLess5.MedInc.values
fig = plt.figure(figsize=(10, 10))
plt.scatter(x_vals, y_vals, c='black')
plt.xlabel('Population', fontsize=18)
plt.ylabel('Median Income', fontsize=16)
```

The output is as follows:

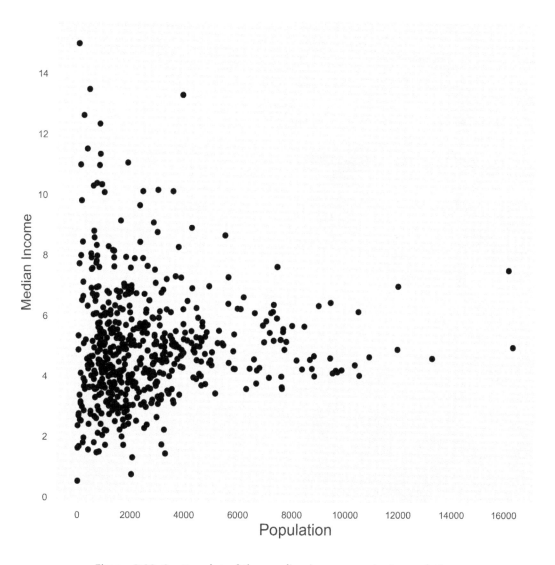

Figure 9.23: Scatterplot of the median income against population

This is the scatterplot of the median income against population for values of five or less in the **HouseAge** column.

7. Use yet another **seaborn** function to fit a kernel density estimation model. The optimal bandwidth is found using Scott's Rule. Replot the histogram and overlay the estimated density, as follows:

```
fig = plt.figure(figsize=(10, 10))
ax = seaborn.kdeplot(x_vals, \
                     y_vals, \
                     kernel='gau', \
                     cmap='Blues', \
                     shade=True, \
                     shade_lowest=False)
plt.scatter(x_vals, y_vals, c='black', alpha=0.05)
plt.xlabel('Population', fontsize=18)
plt.ylabel('Median Income', fontsize=18)
plt.title('Density Estimation With Scatterplot Overlay', size=18)
```

The output is as follows:

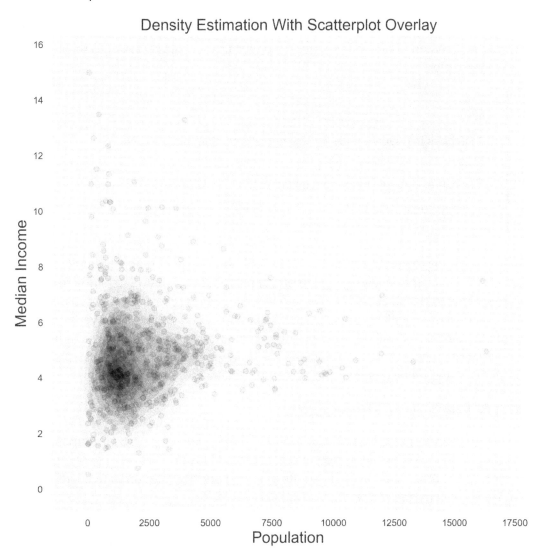

Figure 9.24: The same scatterplot as created in step 6 with the estimated density overlaid

Here, the estimated density shows that census blocks with smaller populations have lower median incomes at higher likelihoods than they have high median incomes. The point of this step is to showcase how kernel density estimation can be used on non-location data. A plot like this is typically referred to as a contour plot.

> **NOTE**
>
> To access the source code for this specific section, please refer to https://packt.live/2UOHbTZ.
>
> You can also run this example online at https://packt.live/38DbmTo.
>
> You must execute the entire Notebook in order to get the desired result.

When presenting the results of hotspot analysis, some type of map should be involved since hotspot analysis is generally done on location data. Acquiring maps on which estimated densities can be overlaid is not an easy process. Due to copyright issues, we will use very basic maps, called basemaps, on which we can overlay our estimated densities. It will be left to you to extend the knowledge you acquire in this chapter to fancier and more detailed maps. Mapping environments can also be complicated and time-consuming to download and install.

EXERCISE 9.06: WORKING WITH BASEMAPS

This exercise leverages the **basemap** module of **mpl_toolkits**. **basemap** is a mapping library, which can be used to create basic maps or outlines of geographic regions. These maps can have the results of kernel density estimation overlaid, so that we can clearly see where the hotspots are located.

First, check whether **basemap** is installed by running **import mpl_toolkits. basemap** in a Jupyter notebook. If it loads without error, then you are ready and need to take no further action. If the call fails, then install **basemap** using **pip** by running **python3 -m pip install basemap**. You should be good to go after restarting any already-open notebooks. Note that the **pip** installation will only work if Anaconda is installed.

The goal of this exercise is to remodel and replot the location data from *Exercise 9.05, Loading Data and Modeling with Seaborn*, using the kernel density estimation functions of **sklearn** and the mapping capabilities of **basemap**. Extract the longitude and latitude values from the filtered data frame called **dfLess15** to work through the steps. Note that this exercise should be done in the same notebook as all the other exercises.

1. Form the grid of locations over which the estimated density will be laid. The grid of locations is the two-dimensional location equivalent of the one-dimensional vector defining the range of the random variable in *Exercise 9.01, The Effect of the Bandwidth Value*:

```
xgrid15 = numpy.sort(list(dfLess15['Longitude']))
ygrid15 = numpy.sort(list(dfLess15['Latitude']))
x15, y15 = numpy.meshgrid(xgrid15, ygrid15)
print("X Grid Component:\n{}\n".format(x15))
print("Y Grid Component:\n{}\n".format(y15))
xy15 = numpy.vstack([y15.ravel(), x15.ravel()]).T
```

The output is as follows:

```
X Grid Component:
[[-124.23 -124.19 -124.17 ... -114.63 -114.57 -114.31]
 [-124.23 -124.19 -124.17 ... -114.63 -114.57 -114.31]
 [-124.23 -124.19 -124.17 ... -114.63 -114.57 -114.31]
 ...
 [-124.23 -124.19 -124.17 ... -114.63 -114.57 -114.31]
 [-124.23 -124.19 -124.17 ... -114.63 -114.57 -114.31]
 [-124.23 -124.19 -124.17 ... -114.63 -114.57 -114.31]]

Y Grid Component:
[[32.54 32.54 32.54 ... 32.54 32.54 32.54]
 [32.55 32.55 32.55 ... 32.55 32.55 32.55]
 [32.55 32.55 32.55 ... 32.55 32.55 32.55]
 ...
 [41.74 41.74 41.74 ... 41.74 41.74 41.74]
 [41.75 41.75 41.75 ... 41.75 41.75 41.75]
 [41.78 41.78 41.78 ... 41.78 41.78 41.78]]
```

Figure 9.25: The x and y components of the grid representing the dfLess15 dataset

2. Define and fit a kernel density estimation model. Set the bandwidth value to 0.05. Then create likelihood values for each point on the location grid:

   ```
   kde15 = sklearn.neighbors.KernelDensity(bandwidth=0.05, \
                                           metric='minkowski', \
                                           kernel='gaussian', \
                                           algorithm='ball_tree')
   kde15.fit(dfLess15.values)
   ```

   The output of kernel density estimation model is as follows:

   ```
   KernelDensity(algorithm='ball_tree', atol=0, bandwidth=0.05, \
                 breadth_first=True, kernel='gaussian', \
                 leaf_size=40, metric='minkowski', \
                 metric_params=None, rtol=0)
   ```

3. Fit the trained model on the **xy** grid and print the shape as follows:

   ```
   log_density = kde15.score_samples(xy15)
   density = numpy.exp(log_density)
   density = density.reshape(x15.shape)
   print("Shape of Density Values:\n{}\n".format(density.shape))
   ```

   The output is as follows:

   ```
   Shape of Density Values:
   (3287, 3287)
   ```

   Notice that if you print out the shape of the likelihood values, it is 3,287 rows by 3,287 columns, which is 10,804,369 likelihood values. This is the same number of values in the preestablished longitude and latitude grid, called **xy15**.

4. Create an outline of California and overlay the estimated density computed in *Step 2*:

Exercise9.01-Exercise9.06.ipynb

```
fig15 = plt.figure(figsize=(10, 10))
fig15.suptitle(\
    """
    Density Estimation:
    Location of Housing Blocks
    Where the Median Home Age <= 15 Years
    """,\
    fontsize=16)
```

The complete code for this step can be found at https://packt.live/38DbmTo.

The output is as follows:

Figure 9.26: The estimated density of dfLess15 overlaid onto an outline of California

The **0.05** value was set to purposefully overfit the data slightly. You'll notice that instead of the larger clusters that made up the density in *Exercise 9.05, Loading Data and Modeling with Seaborn* the estimated density here is made up of much smaller clusters. This slightly overfit density might be a bit more helpful than the previous version of the density because it gives you a clearer view of where the high likelihood census blocks are truly located. One of the high-likelihood areas in the previous density was southern California, but southern California is a huge area with an enormous population and many municipalities. Bear in mind that when using the results for business decisions, certain levels of specificity might be required and should be provided if the sample data can support results with that level of specificity or granularity.

5. Repeat *Step 1*, but with the **dfMore40** data frame:

   ```
   xgrid40 = numpy.sort(list(dfMore40['Longitude']))
   ygrid40 = numpy.sort(list(dfMore40['Latitude']))
   x40, y40 = numpy.meshgrid(xgrid40, ygrid40)
   print("X Grid Component:\n{}\n".format(x40))
   print("Y Grid Component:\n{}\n".format(y40))

   xy40 = numpy.vstack([y40.ravel(), x40.ravel()]).T
   ```

The output is as follows:

```
X Grid Component:
[[-124.35 -124.26 -124.23 ... -114.61 -114.6  -114.59]
 [-124.35 -124.26 -124.23 ... -114.61 -114.6  -114.59]
 [-124.35 -124.26 -124.23 ... -114.61 -114.6  -114.59]
 ...
 [-124.35 -124.26 -124.23 ... -114.61 -114.6  -114.59]
 [-124.35 -124.26 -124.23 ... -114.61 -114.6  -114.59]
 [-124.35 -124.26 -124.23 ... -114.61 -114.6  -114.59]]

Y Grid Component:
[[32.64 32.64 32.64 ... 32.64 32.64 32.64]
 [32.66 32.66 32.66 ... 32.66 32.66 32.66]
 [32.66 32.66 32.66 ... 32.66 32.66 32.66]
 ...
 [41.43 41.43 41.43 ... 41.43 41.43 41.43]
 [41.73 41.73 41.73 ... 41.73 41.73 41.73]
 [41.78 41.78 41.78 ... 41.78 41.78 41.78]]
```

Figure 9.27: The x and y components of the grid representing the dfMore40 dataset

6. Repeat *Step 2* using the grid established in *Step 4*:

```
kde40 = sklearn.neighbors.KernelDensity(bandwidth=0.05, \
                                        metric='minkowski', \
                                        kernel='gaussian', \
                                        algorithm='ball_tree')
kde40.fit(dfMore40.values)
log_density = kde40.score_samples(xy40)
density = numpy.exp(log_density)
density = density.reshape(x40.shape)
print("Shape of Density Values:\n{}\n".format(density.shape))
```

7. Repeat *Step 3* using the estimated density computed in *Step 5*:

Exercise9.01-Exercise9.06.ipynb

```
fig40 = plt.figure(figsize=(10, 10))
fig40.suptitle(\
    """
    Density Estimation:
    Location of Housing Blocks
    Where the Median Home Age > 40 Years
    """, \
    fontsize=16)
```

The complete code for this step can be found at https://packt.live/38DbmTo.

The output is as follows:

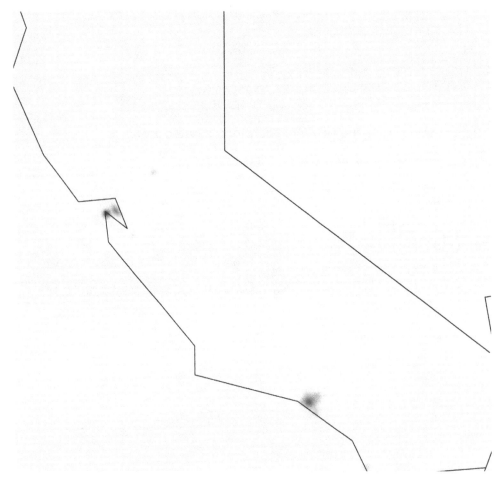

Figure 9.28: The estimated density of dfMore40 overlaid onto an outline of California

> **NOTE**
>
> To access the source code for this specific section, please refer to https://packt.live/2UOHbTZ.
>
> You can also run this example online at https://packt.live/38DbmTo. You must execute the entire Notebook in order to get the desired result.

This estimated density is again a redo of the one that we did in *Exercise 9.05, Loading Data and Modeling with Seaborn*. While the density from *Step 3* will provide more detail for a person interested in real estate or the census, this density does not actually look that different from its corollary density in *Exercise 9.05, Loading Data and Modeling with Seaborn*. The clusters are primarily around Los Angeles and San Francisco with almost no points anywhere else.

ACTIVITY 9.02: ANALYZING CRIME IN LONDON

In this activity, we will perform hotspot analysis with kernel density estimation on London crime data from https://data.police.uk/data/. Due to the difficulties of working with map data, we will visualize the results of the analysis using `seaborn`. However, if you feel brave and were able to run all the plots in *Exercise 9.06, Working with Basemaps* you are encouraged to try using maps.

The motivation for performing hotspot analysis on this crime data is two-fold. We are asked first to determine where certain types of crimes are occurring in high likelihood, so that police resources can be allocated for maximum impact. Then, as a follow up, we are asked to ascertain whether the hotspots for certain types of crime are changing over time. Both of these questions are answerable using kernel density estimation.

> **NOTE**
>
> This dataset is sourced from https://data.police.uk/data/. It contains public sector information licensed under the Open Government License v3.0.
>
> You can also download it from the Packt GitHub at https://packt.live/2JlWs2z.
>
> Alternatively, to download the data directly from the source, go to the preceding police website, check the box for `Metropolitan Police Service`, and then set the date range to `July 2018` to `Dec 2018`. Next, click `Generate file` followed by `Download now` and name the downloaded file `metro-jul18-dec18`. Make sure that you know how or can retrieve the path to the downloaded directory.
>
> This dataset contains public sector information licensed under the Open Government License v3.0.

Here are the steps to complete the activity:

1. Load the crime data. Use the path where you saved the downloaded directory, create a list of the year-month tags, use the **read_csv** command to load the individual files iteratively, and then concatenate these files together.

2. Print diagnostics of the complete (six months) and concatenated dataset.

3. Subset the data frame down to four variables (**Longitude**, **Latitude**, **Month**, and **Crime type**).

4. Using the **jointplot** function from **seaborn**, fit and visualize three kernel density estimation models for bicycle theft in July, September, and December 2018.

5. Repeat *Step 4*; this time, use shoplifting crimes for the months of August, October, and November 2018.

6. Repeat *Step 5*; this time, use burglary crimes for the months of July, October, and December 2018.

Hotspot Analysis | 413

The output from the last part of *Step 6* will be as follows:

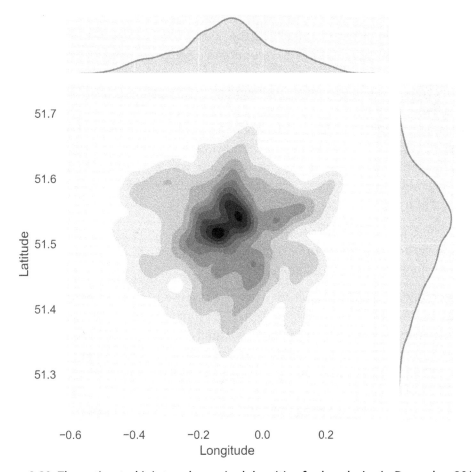

Figure 9.29: The estimated joint and marginal densities for burglaries in December 2018

To clarify one more time, the densities found in this activity should have been overlaid on maps so that we could see exactly what areas these densities cover. Attempting to overlay the results on maps on your own would be encouraged if you have the appropriate mapping platforms at your disposal. If not, you could go to the mapping services available online and use the longitude and latitude pairs to gain insight into the specific locations.

> **NOTE**
>
> The solution for this activity can be found on page 505.

SUMMARY

Kernel density estimation is a classic statistical technique that is in the same family of techniques as the histogram. It allows the user to extrapolate out from sample data to make insights and predictions about the population of particular objects or events. This extrapolation comes in the form of a probability density function, which is nice because the results read as likelihoods or probabilities. The quality of this model is dependent on two parameters: the bandwidth value and the kernel function. As discussed, the most crucial component of leveraging kernel density estimation successfully is the setting of an optimal bandwidth. Optimal bandwidths are most frequently identified using grid search cross-validation with pseudo-log-likelihood as the scoring metric. What makes kernel density estimation great is both its simplicity and its applicability to so many fields.

It is routine to find kernel density estimation models in criminology, epidemiology, meteorology, and real estate to only name a few. Regardless of your area of business, kernel density estimation should be applicable.

Between supervised and unsupervised learning, unsupervised learning is undoubtedly the least used and appreciated learning category. But this should not be the case. Supervised learning techniques are limited and most of the available data does not fit well with regression and classification. Expanding your skill set to include unsupervised learning techniques means that you will be able to leverage different datasets, answer business problems in more creative ways, and even enhance your existing supervised learning models. This text by no means exhausts all the unsupervised learning algorithms, but serves as a good start to pique your interest and drive continued learnings.

APPENDIX

CHAPTER 01: INTRODUCTION TO CLUSTERING

ACTIVITY 1.01: IMPLEMENTING K-MEANS CLUSTERING

Solution:

1. Import the required libraries:

   ```
   from sklearn.datasets import make_blobs
   from sklearn.cluster import KMeans
   from sklearn.metrics import accuracy_score, silhouette_score
   import matplotlib.pyplot as plt
   import pandas as pd
   import numpy as np
   from scipy.spatial.distance import cdist
   import math
   np.random.seed(0)

   %matplotlib inline
   ```

2. Load the seeds data file using **pandas**:

   ```
   seeds = pd.read_csv('Seed_Data.csv')
   ```

3. Return the first five rows of the dataset, as follows:

   ```
   seeds.head()
   ```

   The output is as follows:

	A	P	C	LK	WK	A_Coef	LKG	target
0	15.26	14.84	0.8710	5.763	3.312	2.221	5.220	0
1	14.88	14.57	0.8811	5.554	3.333	1.018	4.956	0
2	14.29	14.09	0.9050	5.291	3.337	2.699	4.825	0
3	13.84	13.94	0.8955	5.324	3.379	2.259	4.805	0
4	16.14	14.99	0.9034	5.658	3.562	1.355	5.175	0

   Figure 1.25: Displaying the first five rows of the dataset

4. Separate the **X** features as follows:

```
X = seeds[['A','P','C','LK','WK','A_Coef','LKG']]
y = seeds['target']
```

5. Check the features as follows:

```
X.head()
```

The output is as follows:

	A	P	C	LK	WK	A_Coef	LKG
0	15.26	14.84	0.8710	5.763	3.312	2.221	5.220
1	14.88	14.57	0.8811	5.554	3.333	1.018	4.956
2	14.29	14.09	0.9050	5.291	3.337	2.699	4.825
3	13.84	13.94	0.8955	5.324	3.379	2.259	4.805
4	16.14	14.99	0.9034	5.658	3.562	1.355	5.175

Figure 1.26: Printing the features

6. Define the **k_means** function as follows and initialize the k-centroids randomly. Repeat this process until the difference between the new/old **centroids** equals **0**, using the **while** loop:

Activity 1.01.ipynb

```
def k_means(X, K):
    # Keep track of history so you can see K-Means in action
    centroids_history = []
    labels_history = []

    # Randomly initialize Kcentroids
    rand_index = np.random.choice(X.shape[0], K)
    centroids = X[rand_index]
    centroids_history.append(centroids)
```

The complete code for this step can be found at https://packt.live/2JPZ4M8.

7. Convert the pandas DataFrame into a NumPy matrix:

```
X_mat = X.values
```

8. Run our seeds matrix through the **k_means** function we created earlier:

```
centroids, labels, centroids_history, labels_history = \
k_means(X_mat, 3)
```

9. Print the labels:

```
print(labels)
```

The output is as follows:

```
[1 1 1 1 1 1 1 1 1 1 1 1 1 1 1 1 0 1 1 0 1 1 1 0 1 1 0 0 1 1 1 1 1 1 1 1 1
 2 1 0 1 1 1 1 1 1 1 1 1 1 1 1 1 1 1 1 1 1 1 0 0 0 0 0 1 1 1 1 1 0 2 2 2 2
 2 2 2 2 2 2 2 2 2 2 2 2 2 2 2 2 2 2 2 2 2 2 2 2 2 2 1 2 2 2 2 2 2 2 2 2 2
 2 2 2 2 2 2 2 2 2 2 2 1 2 1 2 2 2 2 2 2 2 1 1 1 1 2 1 1 1 0 0 0 0 0 0 0 0
 0 0 0 0 0 0 0 0 0 0 0 0 0 0 0 0 0 0 0 0 0 0 0 0 0 0 0 0 0 0 0 0 0 0 0 0 0
 0 0 0 0 0 0 0 0 0 0 0 0 0 0 0 0 0 0 0 0 0 0 0 0 0]
```

Figure 1.27: Printing the labels

10. Plot the coordinates as follows:

```
plt.scatter(X['A'], X['LK'])
plt.title('Wheat Seeds - Area vs Length of Kernel')
plt.show()

plt.scatter(X['A'], X['LK'], c=labels, cmap='tab20b')
plt.title('Wheat Seeds - Area vs Length of Kernel')
plt.show()
```

The output is as follows:

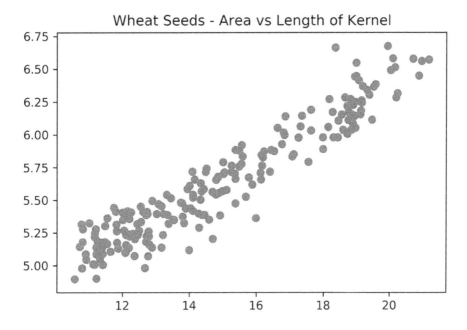

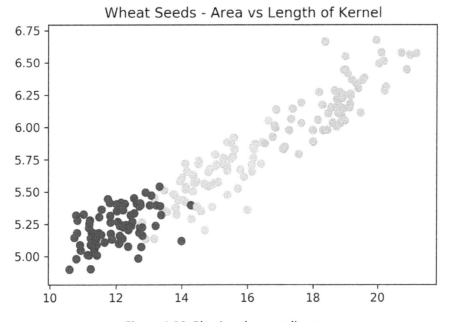

Figure 1.28: Plotting the coordinates

11. Calculate the silhouette score as follows:

    ```
    silhouette_score(X[['A','LK']], labels)
    ```

    The output is as follows:

    ```
    0.5875704550892767
    ```

By completing this activity, you have gained hands-on experience of tuning a k-means clustering algorithm for a real-world dataset. The seeds dataset is seen as a classic "hello world"-type problem in the data science space and is helpful for testing foundational techniques.

> **NOTE**
>
> To access the source code for this specific section, please refer to https://packt.live/2JPZ4M8.
>
> You can also run this example online at https://packt.live/2Ocncuh.

CHAPTER 02: HIERARCHICAL CLUSTERING

ACTIVITY 2.01: COMPARING K-MEANS WITH HIERARCHICAL CLUSTERING

Solution:

1. Import the necessary packages from scikit-learn (**KMeans**, **AgglomerativeClustering**, and **silhouette_score**), as follows:

   ```
   from sklearn.cluster import KMeans
   from sklearn.cluster import AgglomerativeClustering
   from sklearn.metrics import silhouette_score
   import pandas as pd
   import matplotlib.pyplot as plt
   ```

2. Read the wine dataset into the Pandas DataFrame and print a small sample:

   ```
   wine_df = pd.read_csv("wine_data.csv")
   print(wine_df.head())
   ```

   The output is as follows:

   ```
       OD_read  Proline
   0     3.92   1065.0
   1     3.40   1050.0
   2     3.17   1185.0
   3     3.45   1480.0
   4     2.93    735.0
   ```

   Figure 2.25: The output of the wine dataset

3. Visualize the wine dataset to understand the data structure:

   ```
   plt.scatter(wine_df.values[:,0], wine_df.values[:,1])
   plt.title("Wine Dataset")
   plt.xlabel("OD Reading")
   plt.ylabel("Proline")
   plt.show()
   ```

The output is as follows:

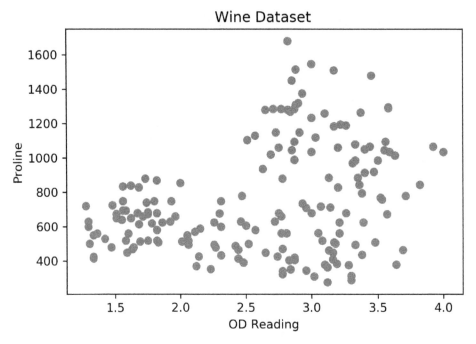

Figure 2.26: A plot of raw wine data

4. Use the **sklearn** implementation of k-means on the wine dataset, knowing that there are three wine types:

```
km = KMeans(3)
km_clusters = km.fit_predict(wine_df)
```

5. Use the **sklearn** implementation of hierarchical clustering on the wine dataset:

```
ac = AgglomerativeClustering(3, linkage='average')
ac_clusters = ac.fit_predict(wine_df)
```

6. Plot the predicted clusters from k-means as follows:

```
plt.scatter(wine_df.values[:,0], \
            wine_df.values[:,1], c=km_clusters)
plt.title("Wine Clusters from K-Means Clustering")
plt.xlabel("OD Reading")
plt.ylabel("Proline")
plt.show()
```

The output is as follows

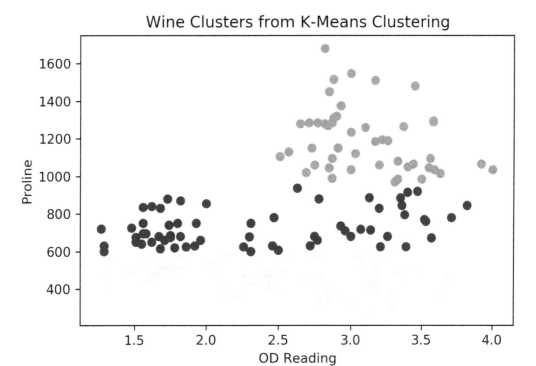

Figure 2.27: A plot of clusters from k-means clustering

7. Plot the predicted clusters from hierarchical clustering as follows:

```
plt.scatter(wine_df.values[:,0], \
            wine_df.values[:,1], c=ac_clusters)
plt.title("Wine Clusters from Agglomerative Clustering")
plt.xlabel("OD Reading")
plt.ylabel("Proline")
plt.show()
```

The output is as follows:

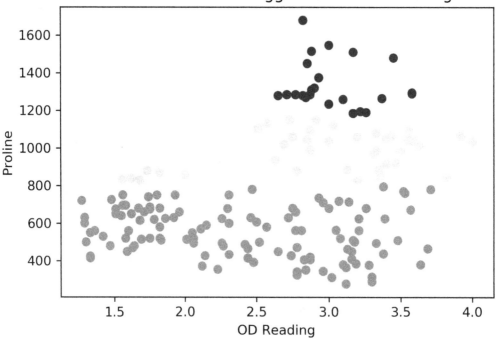

Figure 2.28: A plot of clusters from agglomerative clustering

> **NOTE**
>
> In *Figure 2.23* and *Figure 2.24*, each color represents a single cluster. The colors of the clusters will change every time the code is executed.

8. Compare the silhouette score of each clustering method:

```
print("Silhouette Scores for Wine Dataset:\n")
print("K-Means Clustering: ", silhouette_score\
      (wine_df, km_clusters))
print("Agg Clustering: ", silhouette_score(wine_df, ac_clusters))
```

The output will be as follows:

```
Silhouette Scores for Wine Dataset:

K-Means Clustering:  0.5809421087616886
Agg Clustering:  0.5988495817462
```

As you can see from the preceding silhouette metric, agglomerative clustering narrowly beats k-means clustering when it comes to separating the clusters by mean intra-cluster distance. This is not the case for every version of agglomerative clustering, however. Instead, try different linkage types and examine how the silhouette score and clustering changes between each.

> **NOTE**
>
> To access the source code for this specific section, please refer to https://packt.live/2AFA60Z.
>
> You can also run this example online at https://packt.live/3fe2lTi.

CHAPTER 03: NEIGHBORHOOD APPROACHES AND DBSCAN

ACTIVITY 3.01: IMPLEMENTING DBSCAN FROM SCRATCH

Solution:

1. Generate a random cluster dataset as follows:

```
from sklearn.cluster import DBSCAN
from sklearn.datasets import make_blobs
import matplotlib.pyplot as plt
import numpy as np
%matplotlib inline
X_blob, y_blob = make_blobs(n_samples=500, \
                            centers=4, n_features=2, \
                            random_state=800)
```

2. Visualize the generated data:

```
plt.scatter(X_blob[:,0], X_blob[:,1])
plt.show()
```

The output is as follows:

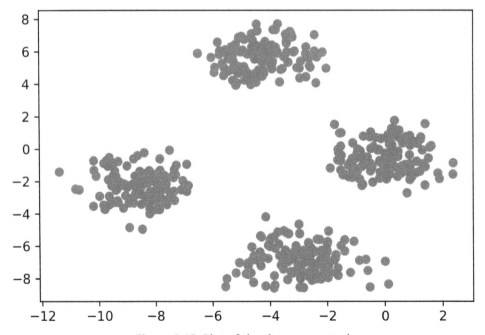

Figure 3.15: Plot of the data generated

3. Create functions from scratch that allow you to call DBSCAN on a dataset:

Activity3.01.ipynb

```
def scratch_DBSCAN(x, eps, min_pts):
    """
    param x (list of vectors): your dataset to be clustered
    param eps (float): neighborhood radius threshold
    param min_pts (int): minimum number of points threshold for
    a neighborhood to be a cluster
    """

    # Build a label holder that is comprised of all 0s
    labels = [0]* x.shape[0]

    # Arbitrary starting "current cluster" ID
    C = 0
```

The complete code for this step can be found at https://packt.live/3c1rONO.

4. Use your created DBSCAN implementation to find clusters in the generated dataset. Feel free to use hyperparameters as you see fit, tuning them based on their performance in *Step 5*:

```
labels = scratch_DBSCAN(X_blob, 0.6, 5)
```

5. Visualize the clustering performance of your DBSCAN implementation:

```
plt.scatter(X_blob[:,0], X_blob[:,1], c=labels)
plt.title("DBSCAN from Scratch Performance")
plt.show()
```

The output is as follows:

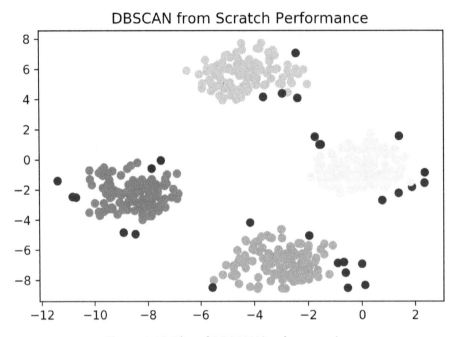

Figure 3.16: Plot of DBSCAN implementation

In the preceding output, you can see that there are four clearly defined clusters in our generated data. The non-highlighted points fall out of neighborhood range and thus are considered noise. While it may not be ideal since not every point is accounted for, for most business cases, this noise is acceptable. If it is not acceptable in your scenario, you can tune the supplied hyperparameters to be more forgiving of distance.

As you may have noticed, it takes quite some time for a custom implementation to run. This is because we explored the non-vectorized version of this algorithm for the sake of clarity. In most cases, you should aim to use the DBSCAN implementation provided by scikit-learn, as it is highly optimized.

> **NOTE**
>
> To access the source code for this specific section, please refer to https://packt.live/3c1rONO.
>
> You can also run this example online at https://packt.live/2ZVoFuO.

ACTIVITY 3.02: COMPARING DBSCAN WITH K-MEANS AND HIERARCHICAL CLUSTERING

Solution:

1. Import the necessary packages:

   ```
   from sklearn.cluster \
   import KMeans, AgglomerativeClustering, DBSCAN
   from sklearn.metrics import silhouette_score
   import pandas as pd
   import matplotlib.pyplot as plt
   %matplotlib inline
   ```

2. Load the wine dataset from *Chapter 2, Hierarchical Clustering*, and familiarize yourself again with what the data looks like:

   ```
   # Load Wine data set
   wine_df = pd.read_csv("wine_data.csv")
   # Show sample of data set
   print(wine_df.head())
   ```

   The output is as follows:

	OD_read	Proline
0	3.92	1065.0
1	3.40	1050.0
2	3.17	1185.0
3	3.45	1480.0
4	2.93	735.0

   Figure 3.17: First five rows of the wine dataset

3. Visualize the data:

```
plt.scatter(wine_df.values[:,0], wine_df.values[:,1])
plt.title("Wine Dataset")
plt.xlabel("OD Reading")
plt.ylabel("Proline")
plt.show()
```

The output is as follows:

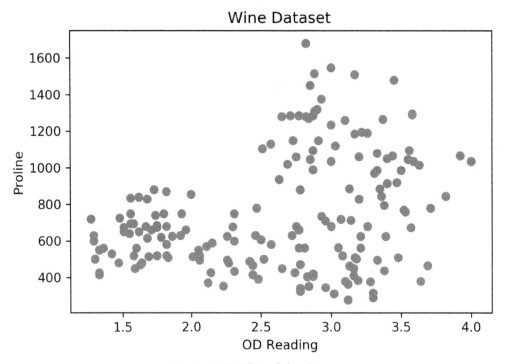

Figure 3.18: Plot of the data

4. Generate clusters using k-means, agglomerative clustering, and DBSCAN:

```
# Generate clusters from K-Means
km = KMeans(3)
km_clusters = km.fit_predict(wine_df)
# Generate clusters using Agglomerative Hierarchical Clustering
ac = AgglomerativeClustering(3, linkage='average')
ac_clusters = ac.fit_predict(wine_df)
```

5. Evaluate a few different options for DSBSCAN hyperparameters and their effect on the silhouette score:

```
db_param_options = [[20,5],[25,5],[30,5],[25,7],[35,7],[40,5]]
for ep,min_sample in db_param_options:
    # Generate clusters using DBSCAN
    db = DBSCAN(eps=ep, min_samples = min_sample)
    db_clusters = db.fit_predict(wine_df)
    print("Eps: ", ep, "Min Samples: ", min_sample)
    print("DBSCAN Clustering: ", \
        silhouette_score(wine_df, db_clusters))
```

The output is as follows:

```
Eps:  20 Min Samples:  5
DBSCAN Clustering:  0.3997987919957757
Eps:  25 Min Samples:  5
DBSCAN Clustering:  0.35258611037074095
Eps:  30 Min Samples:  5
DBSCAN Clustering:  0.43763797761597306
Eps:  25 Min Samples:  7
DBSCAN Clustering:  0.2711660466706248
Eps:  35 Min Samples:  7
DBSCAN Clustering:  0.46000630149335495
Eps:  40 Min Samples:  5
DBSCAN Clustering:  0.5739675293567901
```

Figure 3.19: Printing the silhouette score for clusters

6. Generate the final clusters based on the highest silhouette score (**eps**: 35, **min_samples**: 3):

```
# Generate clusters using DBSCAN
db = DBSCAN(eps=40, min_samples = 5)
db_clusters = db.fit_predict(wine_df)
```

7. Visualize clusters generated using each of the three methods:

```
plt.title("Wine Clusters from K-Means")
plt.scatter(wine_df['OD_read'], wine_df['Proline'], \
            c=km_clusters,s=50, cmap='tab20b')
plt.show()
plt.title("Wine Clusters from Agglomerative Clustering")
plt.scatter(wine_df['OD_read'], wine_df['Proline'], \
            c=ac_clusters,s=50, cmap='tab20b')
plt.show()
plt.title("Wine Clusters from DBSCAN")
plt.scatter(wine_df['OD_read'], wine_df['Proline'], \
            c=db_clusters,s=50, cmap='tab20b')
plt.show()
```

The output is as follows:

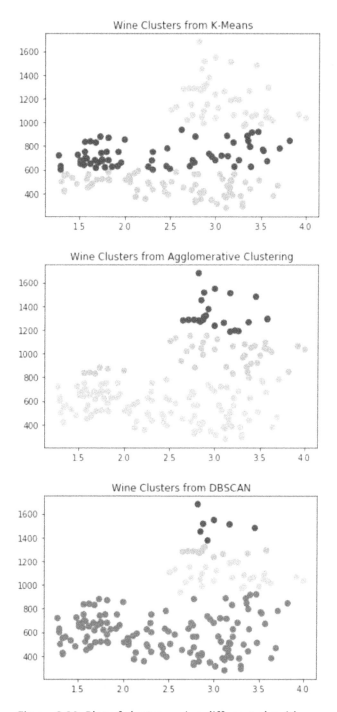

Figure 3.20: Plot of clusters using different algorithms

8. Evaluate the silhouette score of each approach:

```
# Calculate Silhouette Scores
print("Silhouette Scores for Wine Dataset:\n")
print("K-Means Clustering: ", \
      silhouette_score(wine_df, km_clusters))
print("Agg Clustering: ", \
      silhouette_score(wine_df, ac_clusters))
print("DBSCAN Clustering: ", \
      silhouette_score(wine_df, db_clusters))
```

The output is as follows:

```
Silhouette Scores for Wine Dataset:

K-Means Clustering:   0.5809421087616886
Agg Clustering:   0.5988495817462
DBSCAN Clustering:   0.5739675293567901
```

As you can see, DBSCAN isn't automatically the best choice for your clustering needs. One key trait that makes it different from other algorithms is the use of noise as a potential clustering. In some cases, this is great, as it removes outliers; however, there may be situations where it is not tuned well enough and classifies too many points as noise. You can improve the silhouette score further by tuning the hyperparameters while fitting your clustering algorithms – try a few different combinations and see how they affect your score.

> **NOTE**
>
> To access the source code for this specific section, please refer to https://packt.live/2BNSQvC.
>
> You can also run this example online at https://packt.live/3iElboS.

CHAPTER 04: DIMENSIONALITY REDUCTION TECHNIQUES AND PCA

ACTIVITY 4.01: MANUAL PCA VERSUS SCIKIT-LEARN

Solution:

1. Import the **pandas**, **numpy**, and **matplotlib** plotting libraries and the scikit-learn **PCA** model:

   ```
   import pandas as pd
   import numpy as np
   import matplotlib.pyplot as plt
   from sklearn.decomposition import PCA
   ```

2. Load the dataset and select only the sepal features as per the previous exercises. Display the first five rows of the data:

   ```
   df = pd.read_csv('../Seed_Data.csv')
   df = df[['A', 'LK']]
   df.head()
   ```

   The output is as follows:

	A	LK
0	15.26	5.763
1	14.88	5.554
2	14.29	5.291
3	13.84	5.324
4	16.14	5.658

   Figure 4.36: The first five rows of the data

3. Compute the **covariance** matrix for the data:

   ```
   cov = np.cov(df.values.T)
   cov
   ```

   The output is as follows:

   ```
   array([[8.46635078, 1.22470367],
          [1.22470367, 0.19630525]])
   ```

4. Transform the data using the scikit-learn API and only the first principal component. Store the transformed data in the **sklearn_pca** variable:

   ```
   model = PCA(n_components=1)
   sklearn_pca = model.fit_transform(df.values)
   ```

5. Transform the data using the manual PCA and only the first principal component. Store the transformed data in the **manual_pca** variable:

   ```
   eigenvectors, eigenvalues, _ = \
   np.linalg.svd(cov, full_matrices=False)
   P = eigenvectors[0]
   manual_pca = P.dot(df.values.T)
   ```

6. Plot the **sklearn_pca** and **manual_pca** values on the same plot to visualize the difference:

   ```
   plt.figure(figsize=(10, 7))
   plt.plot(sklearn_pca, label='Scikit-learn PCA')
   plt.plot(manual_pca, label='Manual PCA', linestyle='--')
   plt.xlabel('Sample')
   plt.ylabel('Transformed Value')
   plt.legend()
   plt.show()
   ```

The output is as follows:

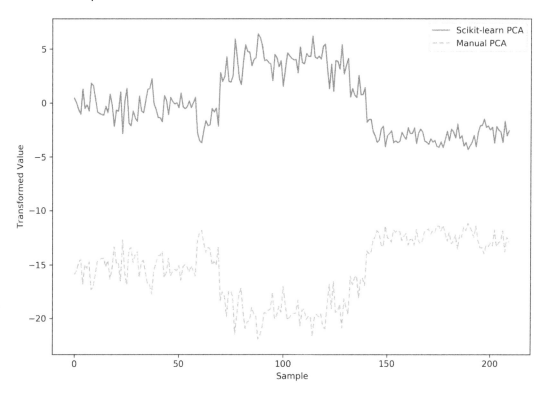

Figure 4.37: A plot of the data

7. Notice that the two plots look almost identical, except that one is a mirror image of the other and there is an offset between the two. Display the components of the **sklearn_pca** and **manual_pca** models:

   ```
   model.components_
   ```

   The output is as follows:

   ```
   array([[0.98965371, 0.14347657]])
   ```

8. Now, print **P**:

   ```
   P
   ```

   The output is as follows:

   ```
   array([-0.98965371, -0.14347657])
   ```

Notice the difference in the signs; the values are identical, but the signs are different, producing the mirror image result. This is just a difference in convention, and nothing meaningful.

9. Multiply the **manual_pca** models by **-1** and replot:

```
manual_pca *= -1
plt.figure(figsize=(10, 7))
plt.plot(sklearn_pca, label='Scikit-learn PCA')
plt.plot(manual_pca, label='Manual PCA', linestyle='--')
plt.xlabel('Sample')
plt.ylabel('Transformed Value')
plt.legend()
plt.show()
```

The output is as follows:

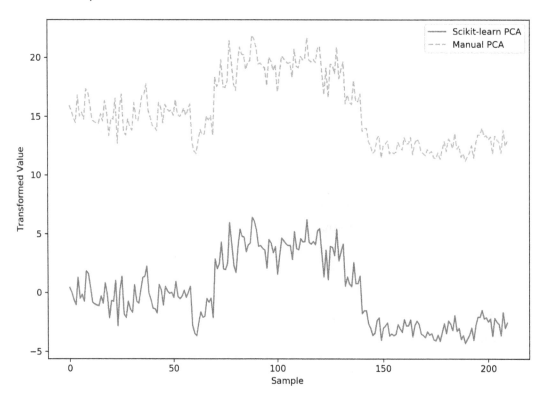

Figure 4.38: Replotted data

10. Now, all we need to do is deal with the offset between the two. The scikit-learn API subtracts the mean of the data prior to the transform. Subtract the mean of each column from the dataset before completing the transform with manual PCA:

    ```
    mean_vals = np.mean(df.values, axis=0)
    offset_vals = df.values - mean_vals
    manual_pca = P.dot(offset_vals.T)
    ```

11. Multiply the result by **-1**:

    ```
    manual_pca *= -1
    ```

12. Replot the individual **sklearn_pca** and **manual_pca** values:

    ```
    plt.figure(figsize=(10, 7))
    plt.plot(sklearn_pca, label='Scikit-learn PCA')
    plt.plot(manual_pca, label='Manual PCA', linestyle='--')
    plt.xlabel('Sample')
    plt.ylabel('Transformed Value')
    plt.legend()
    plt.show()
    ```

The output is as follows:

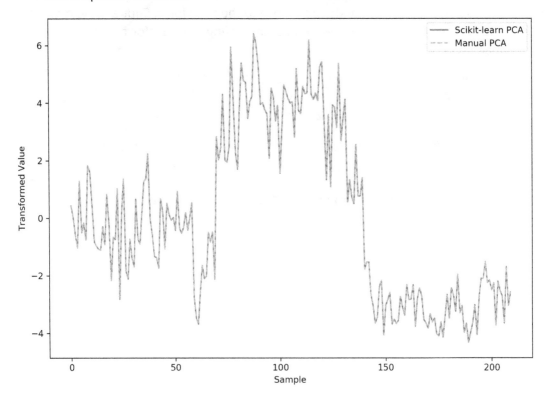

Figure 4.39: Replotting the data

The final plot will demonstrate that the dimensionality reduction completed by the two methods is, in fact, the same. The differences lie in the differences in the signs of the **covariance** matrices, as the two methods simply use a different feature as the baseline for comparison. Finally, there is also an offset between the two datasets, which is attributed to the mean samples being subtracted before executing the transform in the scikit-learn PCA.

> **NOTE**
>
> To access the source code for this specific section, please refer to https://packt.live/2O9MEk4.
>
> You can also run this example online at https://packt.live/3gBntTU.

ACTIVITY 4.02: PCA USING THE EXPANDED SEEDS DATASET

Solution:

1. Import **pandas** and **matplotlib**. To enable 3D plotting, you will also need to import **Axes3D**:

```
import pandas as pd
import numpy as np
import matplotlib.pyplot as plt
from sklearn.decomposition import PCA
from mpl_toolkits.mplot3d import Axes3D #Required for 3D plotting
```

2. Read in the dataset and select the **A**, **LK**, and **C** columns:

```
df = pd.read_csv('../Seed_Data.csv')[['A', 'LK', 'C']]
df.head()
```

The output is as follows:

	A	LK	C
0	15.26	5.763	0.8710
1	14.88	5.554	0.8811
2	14.29	5.291	0.9050
3	13.84	5.324	0.8955
4	16.14	5.658	0.9034

Figure 4.40: Area, length, and compactness of the kernel

3. Plot the data in three dimensions:

```
fig = plt.figure(figsize=(10, 7))
# Where Axes3D is required
ax = fig.add_subplot(111, projection='3d')
ax.scatter(df['A'], df['LK'], df['C'])
ax.set_xlabel('Area of Kernel')
ax.set_ylabel('Length of Kernel')
ax.set_zlabel('Compactness of Kernel')
ax.set_title('Expanded Seeds Dataset')
plt.show()
```

The output is as follows:

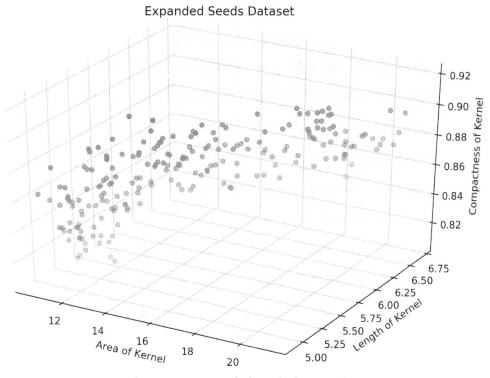

Figure 4.41: Expanded Seeds dataset plot

4. Create a **PCA** model without specifying the number of components:

   ```
   model = PCA()
   ```

5. Fit the model to the dataset:

   ```
   model.fit(df.values)
   ```

   The output is as follows:

   ```
   PCA(copy=True, iterated_power='auto', n_components=None,
       random_state=None,
       svd_solver='auto', tol=0.0, whiten=False)
   ```

6. Display the eigenvalues or **explained_variance_ratio_**:

   ```
   model.explained_variance_ratio_
   ```

   The output is as follows:

   ```
   array([9.97794495e-01, 2.19418709e-03, 1.13183333e-05])
   ```

We want to reduce the dimensionality of the dataset, but still keep at least 90% of the variance. What is the minimum number of components required to keep 90% of the variance?

Only the first component is required for at least a 90% variance. The first component provides 99.7% of the variance within the dataset.

7. Create a new **PCA** model, this time specifying the number of components required to retain at least 90% of the variance:

```
model = PCA(n_components=1)
```

8. Transform the data using the new model:

```
data_transformed = model.fit_transform(df.values)
```

9. Restore the transformed data to the original dataspace:

```
data_restored = model.inverse_transform(data_transformed)
```

10. Plot the restored data in three dimensions in one subplot and the original data in a second subplot to visualize the effect of removing some of the variance:

```
fig = plt.figure(figsize=(10, 14))

# Original Data
ax = fig.add_subplot(211, projection='3d')
ax.scatter(df['A'], df['LK'], df['C'], label='Original Data');
ax.set_xlabel('Area of Kernel');
ax.set_ylabel('Length of Kernel');
ax.set_zlabel('Compactness of Kernel');
ax.set_title('Expanded Seeds Dataset');

# Transformed Data
ax = fig.add_subplot(212, projection='3d')
ax.scatter(data_restored[:,0], data_restored[:,1], \
           data_restored[:,2], label='Restored Data');
ax.set_xlabel('Area of Kernel');
ax.set_ylabel('Length of Kernel');
ax.set_zlabel('Compactness of Kernel');
ax.set_title('Restored Seeds Dataset');
```

The output is as follows:

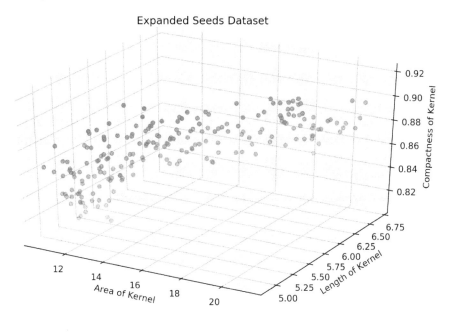

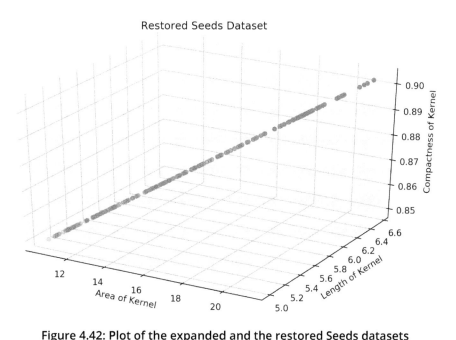

Figure 4.42: Plot of the expanded and the restored Seeds datasets

Looking at the preceding diagram, we can see that we have removed much of the noise within the data, but retained the most important information regarding the trends within the data. You can see that, in general, the compactness of the wheat kernel increases as the area increases.

> **NOTE**
>
> When applying PCA, it is important to keep in mind the size of the data being modeled, along with the available system memory. The singular value decomposition process involves separating the data into the eigenvalues and eigenvectors and can be quite memory-intensive. If the dataset is too large, you may either be unable to complete the process, suffer significant performance loss, or lock up your system.
>
> To access the source code for this specific section, please refer to https://packt.live/2ZVpaFc.
>
> You can also run this example online at https://packt.live/3glrR3D.

CHAPTER 05: AUTOENCODERS

ACTIVITY 5.01: THE MNIST NEURAL NETWORK

Solution:

In this activity, you will train a neural network to identify images in the MNIST dataset and reinforce your skills in training neural networks:

1. Import **pickle**, **numpy**, **matplotlib**, and the **Sequential** and **Dense** classes from Keras:

   ```
   import pickle
   import numpy as np
   import matplotlib.pyplot as plt
   from keras.models import Sequential
   from keras.layers import Dense

   import tensorflow.python.util.deprecation as deprecation
   deprecation._PRINT_DEPRECATION_WARNINGS = False
   ```

2. Load the **mnist.pkl** file, which contains the first 10,000 images and corresponding labels from the MNIST dataset that are available in the accompanying source code. The MNIST dataset is a series of 28 x 28 grayscale images of handwritten digits 0 through 9. Extract the images and labels:

   ```
   with open('mnist.pkl', 'rb') as f:
       data = pickle.load(f)
   images = data['images']
   labels = data['labels']
   ```

3. Plot the first 10 samples along with the corresponding labels:

   ```
   plt.figure(figsize=(10, 7))
   for i in range(10):
       plt.subplot(2, 5, i + 1)
       plt.imshow(images[i], cmap='gray')
       plt.title(labels[i])
       plt.axis('off')
   ```

The output is as follows:

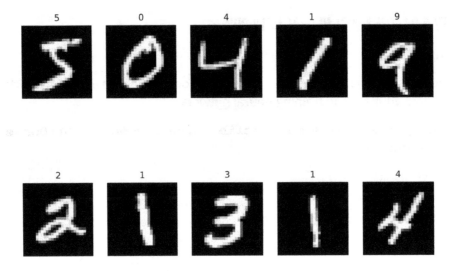

Figure 5.40: First 10 samples

4. Encode the labels using one-hot encoding:

```
one_hot_labels = np.zeros((images.shape[0], 10))
for idx, label in enumerate(labels):
    one_hot_labels[idx, label] = 1
one_hot_labels
```

The output is as follows:

```
array([[0., 0., 0., ..., 0., 0., 0.],
       [1., 0., 0., ..., 0., 0., 0.],
       [0., 0., 0., ..., 0., 0., 0.],
       ...,
       [0., 0., 0., ..., 0., 0., 0.],
       [0., 0., 0., ..., 0., 0., 1.],
       [0., 0., 0., ..., 1., 0., 0.]])
```

5. Prepare the images for input into a neural network. As a hint, there are two separate steps in this process:

```
images = images.reshape((-1, 28 ** 2))
images = images / 255.
```

6. Construct a neural network model in Keras that accepts the prepared images, has a hidden layer of 600 units with a ReLU activation function, and an output of the same number of units as classes. The output layer uses a **softmax** activation function:

   ```
   model = Sequential([Dense(600, input_shape=(784,), \
                       activation='relu'), \
                       Dense(10, activation='softmax'),])
   ```

7. Compile the model using multiclass cross-entropy, stochastic gradient descent, and an accuracy performance metric:

   ```
   model.compile(loss='categorical_crossentropy', \
                 optimizer='sgd', metrics=['accuracy'])
   ```

8. Train the model. How many epochs are required to achieve at least 95% classification accuracy on the training data? Let's have a look:

   ```
   model.fit(images, one_hot_labels, epochs=20)
   ```

   The output is as follows:

   ```
   Epoch 15/20
   10000/10000 [==============================] - 3s 334us/step - loss: 0.2385 - acc: 0.9359
   Epoch 16/20
   10000/10000 [==============================] - 4s 359us/step - loss: 0.2315 - acc: 0.9388 1s - loss:
   Epoch 17/20
   10000/10000 [==============================] - 3s 345us/step - loss: 0.2257 - acc: 0.9388
   Epoch 18/20
   10000/10000 [==============================] - 4s 359us/step - loss: 0.2195 - acc: 0.9407
   Epoch 19/20
   10000/10000 [==============================] - 4s 373us/step - loss: 0.2141 - acc: 0.9431
   Epoch 20/20
   10000/10000 [==============================] - 3s 347us/step - loss: 0.2087 - acc: 0.9429
   <keras.callbacks.History at 0x19abd2fbd08>
   ```

   Figure 5.41: Training the model

   15 epochs are required to achieve at least 95% classification accuracy on the training set.

In this example, we have measured the performance of the neural network classifier using the data that the classifier was trained with. In general, this method should not be used as it typically reports a higher level of accuracy than you should expect from the model. In supervised learning problems, there are a number of **cross-validation** techniques that should be used instead. As this is a course on unsupervised learning, cross-validation lies outside the scope of this book.

> **NOTE**
>
> To access the source code for this specific section, please refer to https://packt.live/2VZpLnZ.
>
> You can also run this example online at https://packt.live/2Z9ueGz.

ACTIVITY 5.02: SIMPLE MNIST AUTOENCODER

Solution:

1. Import **pickle**, **numpy**, and **matplotlib**, as well as the **Model**, **Input**, and **Dense** classes, from Keras:

    ```
    import pickle
    import numpy as np
    import matplotlib.pyplot as plt
    from keras.models import Model
    from keras.layers import Input, Dense

    import tensorflow.python.util.deprecation as deprecation
    deprecation._PRINT_DEPRECATION_WARNINGS = False
    ```

2. Load the images from the supplied sample of the MNIST dataset that is provided with the accompanying source code (**mnist.pkl**):

    ```
    with open('mnist.pkl', 'rb') as f:
        images = pickle.load(f)['images']
    ```

3. Prepare the images for input into a neural network. As a hint, there are **two** separate steps in this process:

    ```
    images = images.reshape((-1, 28 ** 2))
    images = images / 255.
    ```

4. Construct a simple autoencoder network that reduces the image size to 10 x 10 after the encoding stage:

```
input_stage = Input(shape=(784,))
encoding_stage = Dense(100, activation='relu')(input_stage)
decoding_stage = Dense(784, activation='sigmoid')(encoding_stage)
autoencoder = Model(input_stage, decoding_stage)
```

5. Compile the autoencoder using a binary cross-entropy loss function and **adadelta** gradient descent:

```
autoencoder.compile(loss='binary_crossentropy', \
                    optimizer='adadelta')
```

6. Fit the encoder model:

```
autoencoder.fit(images, images, epochs=100)
```

The output is as follows:

```
Epoch 95/100
10000/10000 [==============================] - 2s 223us/step - loss: 0.0756
Epoch 96/100
10000/10000 [==============================] - 2s 225us/step - loss: 0.0755
Epoch 97/100
10000/10000 [==============================] - 2s 226us/step - loss: 0.0754
Epoch 98/100
10000/10000 [==============================] - 2s 225us/step - loss: 0.0754
Epoch 99/100
10000/10000 [==============================] - 2s 225us/step - loss: 0.0753
Epoch 100/100
10000/10000 [==============================] - 2s 225us/step - loss: 0.0752
<keras.callbacks.History at 0x24da05b2d48>
```

Figure 5.42: Training the model

7. Calculate and store the output of the encoding stage for the first five samples:

```
encoder_output = Model(input_stage, encoding_stage)\
                 .predict(images[:5])
```

8. Reshape the encoder output to 10 x 10 (10 x 10 = 100) pixels and multiply by 255:

```
encoder_output = encoder_output.reshape((-1, 10, 10)) * 255
```

9. Calculate and store the output of the decoding stage for the first five samples:

```
decoder_output = autoencoder.predict(images[:5])
```

10. Reshape the output of the decoder to 28 x 28 and multiply by 255:

    ```
    decoder_output = decoder_output.reshape((-1, 28, 28)) * 255
    ```

11. Plot the original image, the encoder output, and the decoder:

    ```
    images = images.reshape((-1, 28, 28))
    plt.figure(figsize=(10, 7))
    for i in range(5):
        plt.subplot(3, 5, i + 1)
        plt.imshow(images[i], cmap='gray')
        plt.axis('off')
        plt.subplot(3, 5, i + 6)
        plt.imshow(encoder_output[i], cmap='gray')
        plt.axis('off')
        plt.subplot(3, 5, i + 11)
        plt.imshow(decoder_output[i], cmap='gray')
        plt.axis('off')
    ```

    The output is as follows:

Figure 5.43: The original image, the encoder output, and the decoder

So far, we have shown how a simple single hidden layer in both the encoding and decoding stage can be used to reduce the data to a lower dimension space. We can also make this model more complicated by adding additional layers to both the encoding and the decoding stages.

> **NOTE**
>
> To access the source code for this specific section, please refer to https://packt.live/3f5ZSdH.
>
> You can also run this example online at https://packt.live/2W0ZkhP.

ACTIVITY 5.03: MNIST CONVOLUTIONAL AUTOENCODER

Solution:

1. Import `pickle`, `numpy`, `matplotlib`, and the `Model` class from `keras.models`, and import `Input`, `Conv2D`, `MaxPooling2D`, and `UpSampling2D` from `keras.layers`:

   ```
   import pickle
   import numpy as np
   import matplotlib.pyplot as plt
   from keras.models import Model
   from keras.layers \
   import Input, Conv2D, MaxPooling2D, UpSampling2D

   import tensorflow.python.util.deprecation as deprecation
   deprecation._PRINT_DEPRECATION_WARNINGS = False
   ```

2. Load the data:

   ```
   with open('mnist.pkl', 'rb') as f:
       images = pickle.load(f)['images']
   ```

3. Rescale the images to have values between 0 and 1:

   ```
   images = images / 255.
   ```

4. We need to reshape the images to add a single depth channel for use with convolutional stages. Reshape the images to have a shape of 28 x 28 x 1:

   ```
   images = images.reshape((-1, 28, 28, 1))
   ```

5. Define an input layer. We will use the same shape input as an image:

   ```
   input_layer = Input(shape=(28, 28, 1,))
   ```

6. Add a convolutional stage with 16 layers or filters, a 3 x 3 weight matrix, a ReLU activation function, and using the same padding, which means the output has the same length as the input image:

   ```
   hidden_encoding = \
   Conv2D(16, # Number of layers or filters in the weight matrix \
           (3, 3), # Shape of the weight matrix \
           activation='relu', \
           padding='same', # How to apply the weights to the images \
           )(input_layer)
   ```

7. Add a max pooling layer to the encoder with a 2 x 2 kernel:

   ```
   encoded = MaxPooling2D((2, 2))(hidden_encoding)
   ```

8. Add a decoding convolutional layer:

   ```
   hidden_decoding = \
   Conv2D(16, # Number of layers or filters in the weight matrix \
           (3, 3), # Shape of the weight matrix \
           activation='relu', \
           padding='same', # How to apply the weights to the images \
           )(encoded)
   ```

9. Add an upsampling layer:

   ```
   upsample_decoding = UpSampling2D((2, 2))(hidden_decoding)
   ```

10. Add the final convolutional stage, using one layer as per the initial image depth:

    ```
    decoded = \
    Conv2D(1, # Number of layers or filters in the weight matrix \
            (3, 3), # Shape of the weight matrix \
            activation='sigmoid', \
            padding='same', # How to apply the weights to the images \
            )(upsample_decoding)
    ```

11. Construct the model by passing the first and last layers of the network to the **Model** class:

    ```
    autoencoder = Model(input_layer, decoded)
    ```

12. Display the structure of the model:

    ```
    autoencoder.summary()
    ```

    The output is as follows:

    ```
    Layer (type)                 Output Shape              Param #
    =================================================================
    input_1 (InputLayer)         (None, 28, 28, 1)         0
    _____
    conv2d_1 (Conv2D)            (None, 28, 28, 16)        160
    _____
    max_pooling2d_1 (MaxPooling2 (None, 14, 14, 16)        0
    _____
    conv2d_2 (Conv2D)            (None, 14, 14, 16)        2320
    _____
    up_sampling2d_1 (UpSampling2 (None, 28, 28, 16)        0
    _____
    conv2d_3 (Conv2D)            (None, 28, 28, 1)         145
    =================================================================
    Total params: 2,625
    Trainable params: 2,625
    Non-trainable params: 0
    ```

    Figure 5.44: Structure of the model

13. Compile the autoencoder using a binary cross-entropy loss function and **adadelta** gradient descent:

    ```
    autoencoder.compile(loss='binary_crossentropy', \
                        optimizer='adadelta')
    ```

14. Now, let's fit the model; again, we pass the images as the training data and as the desired output. Train for 20 epochs as convolutional networks take a lot longer to compute:

    ```
    autoencoder.fit(images, images, epochs=20)
    ```

The output is as follows:

```
Epoch 15/20
10000/10000 [==============================] - 25s 2ms/step - loss: 0.0638
Epoch 16/20
10000/10000 [==============================] - 25s 3ms/step - loss: 0.0638
Epoch 17/20
10000/10000 [==============================] - 24s 2ms/step - loss: 0.0636
Epoch 18/20
10000/10000 [==============================] - 25s 2ms/step - loss: 0.0635
Epoch 19/20
10000/10000 [==============================] - 25s 3ms/step - loss: 0.0634
Epoch 20/20
10000/10000 [==============================] - 25s 2ms/step - loss: 0.0633
<keras.callbacks.History at 0x2acfe717a08>
```

Figure 5.45: Training the model

15. Calculate and store the output of the encoding stage for the first five samples:

    ```
    encoder_output = Model(input_layer, encoded).predict(images[:5])
    ```

16. Reshape the encoder output for visualization, where each image is X*Y in size:

    ```
    encoder_output = encoder_output.reshape((-1, 14 * 14, 16))
    ```

17. Get the output of the decoder for the first five images:

    ```
    decoder_output = autoencoder.predict(images[:5])
    ```

18. Reshape the decoder output to 28 x 28 in size:

    ```
    decoder_output = decoder_output.reshape((-1, 28, 28))
    ```

19. Reshape the original images back to 28 x 28 in size:

    ```
    images = images.reshape((-1, 28, 28))
    ```

20. Plot the original image, the mean encoder output, and the decoder:

```
plt.figure(figsize=(10, 7))
for i in range(5):
    # Plot the original digit images
    plt.subplot(3, 5, i + 1)
    plt.imshow(images[i], cmap='gray')
    plt.axis('off')
    # Plot the encoder output
    plt.subplot(3, 5, i + 6)
    plt.imshow(encoder_output[i], cmap='gray')
    plt.axis('off')
    # Plot the decoder output
    plt.subplot(3, 5, i + 11)
    plt.imshow(decoder_output[i], cmap='gray')
    plt.axis('off')
```

The output is as follows:

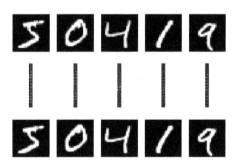

Figure 5.46: The original image, the encoder output, and the decoder

At the end of this activity, you will have developed an autoencoder comprising convolutional layers within the neural network. Note the improvements made in the decoder representations. This architecture has a significant performance benefit over fully connected neural network layers and is extremely useful in working with image-based datasets and generating artificial data samples.

> **NOTE**
>
> To access the source code for this specific section, please refer to https://packt.live/2CdpIxY.
>
> You can also run this example online at https://packt.live/3iKz8l2.

CHAPTER 06: T-DISTRIBUTED STOCHASTIC NEIGHBOR EMBEDDING

ACTIVITY 6.01: WINE T-SNE

Solution:

1. Import **pandas**, **numpy**, and **matplotlib**, as well as the **t-SNE** and **PCA** models from scikit-learn:

   ```
   import pandas as pd
   import numpy as np
   import matplotlib.pyplot as plt
   from sklearn.decomposition import PCA
   from sklearn.manifold import TSNE
   ```

2. Load the Wine dataset using the **wine.data** file included in the accompanying source code and display the first five rows of data:

   ```
   df = pd.read_csv('wine.data', header=None)
   df.head()
   ```

   The output is as follows:

	0	1	2	3	4	5	6	7	8	9	10	11	12	13
0	1	14.23	1.71	2.43	15.6	127	2.80	3.06	0.28	2.29	5.64	1.04	3.92	1065
1	1	13.20	1.78	2.14	11.2	100	2.65	2.76	0.26	1.28	4.38	1.05	3.40	1050
2	1	13.16	2.36	2.67	18.6	101	2.80	3.24	0.30	2.81	5.68	1.03	3.17	1185
3	1	14.37	1.95	2.50	16.8	113	3.85	3.49	0.24	2.18	7.80	0.86	3.45	1480
4	1	13.24	2.59	2.87	21.0	118	2.80	2.69	0.39	1.82	4.32	1.04	2.93	735

   Figure 6.25: The first five rows of the Wine dataset

3. The first column contains the labels; extract this column and remove it from the dataset:

   ```
   labels = df[0]
   del df[0]
   ```

4. Execute PCA to reduce the dataset to the first six components:

   ```
   model_pca = PCA(n_components=6)
   wine_pca = model_pca.fit_transform(df)
   ```

5. Determine the amount of variance within the data described by these six components:

   ```
   np.sum(model_pca.explained_variance_ratio_)
   ```

   The output is as follows:

   ```
   0.99999314824536
   ```

6. Create a t-SNE model using a specified random state and a **verbose** value of 1:

   ```
   tsne_model = TSNE(random_state=0, verbose=1)
   tsne_model
   ```

   The output is as follows:

   ```
   TSNE(angle=0.5, early_exaggeration=12.0, init='random', learning_rate=200.0,
        method='barnes_hut', metric='euclidean', min_grad_norm=1e-07,
        n_components=2, n_iter=1000, n_iter_without_progress=300, n_jobs=None,
        perplexity=30.0, random_state=0, verbose=1)
   ```

   Figure 6.26: Creating a t-SNE model

7. Fit the PCA data to the t-SNE model:

   ```
   wine_tsne = tsne_model.fit_transform\
               (wine_pca.reshape((len(wine_pca), -1)))
   ```

   The output is as follows:

   ```
   [t-SNE] Computing 91 nearest neighbors...
   [t-SNE] Indexed 178 samples in 0.000s...
   [t-SNE] Computed neighbors for 178 samples in 0.304s...
   [t-SNE] Computed conditional probabilities for sample 178 / 178
   [t-SNE] Mean sigma: 9.207049
   [t-SNE] KL divergence after 250 iterations with early exaggeration: 51.192078
   [t-SNE] KL divergence after 750 iterations: 0.128432
   ```

   Figure 6.27: Fitting the PCA data to the t-SNE model

8. Confirm that the shape of the t-SNE fitted data is two-dimensional:

   ```
   wine_tsne.shape
   ```

   The output is as follows:

   ```
   (178, 2)
   ```

9. Create a scatter plot of the two-dimensional data:

```
plt.figure(figsize=(10, 7))
plt.scatter(wine_tsne[:,0], wine_tsne[:,1])
plt.title('Low Dimensional Representation of Wine')
plt.show()
```

The output is as follows:

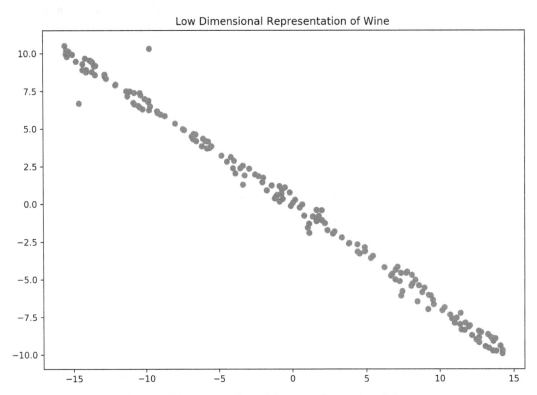

Figure 6.28: Scatter plot of the two-dimensional data

10. Create a secondary scatter plot of the two-dimensional data with the class labels applied to visualize any clustering that may be present:

```
MARKER = ['o', 'v', '^',]
plt.figure(figsize=(10, 7))
plt.title('Low Dimensional Representation of Wine')
for i in range(1, 4):
    selections = wine_tsne[labels == i]
    plt.scatter(selections[:,0], selections[:,1], \
                marker=MARKER[i-1], label=f'Wine {i}', s=30)
```

```
    plt.legend()
plt.show()
```

The output is as follows:

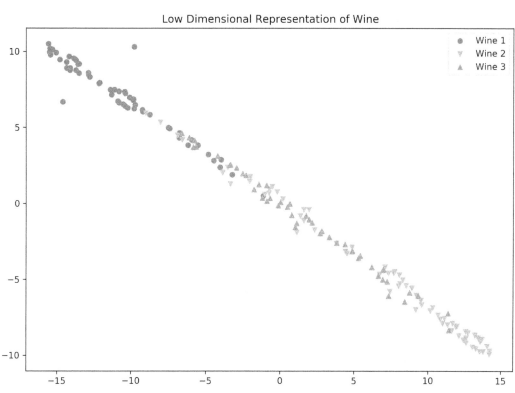

Figure 6.29: Secondary plot of the two-dimensional data

Note that while there is an overlap between the wine classes, it can also be seen that there is some clustering within the data. The first wine class is predominantly positioned in the top left-hand corner of the plot, the second wine class in the bottom-right, and the third wine class is between the first two. This representation certainly couldn't be used to classify individual wine samples with great confidence, but it shows an overall trend and the series of clusters contained within the high-dimensional data that we were unable to see earlier.

In this section, we covered the basics of generating SNE plots. The ability to represent high-dimensional data in a low-dimensional space is critical, especially for developing a thorough understanding of the data at hand. Occasionally, these plots can be tricky to interpret as the exact relationships are sometimes contradictory, sometimes leading to misleading structures.

ACTIVITY 6.02: T-SNE WINE AND PERPLEXITY

Solution:

1. Import **pandas**, **numpy**, and **matplotlib**, as well as the **t-SNE** and **PCA** models from scikit-learn:

   ```
   import pandas as pd
   import numpy as np
   import matplotlib.pyplot as plt
   from sklearn.decomposition import PCA
   from sklearn.manifold import TSNE
   ```

2. Load the Wine dataset and inspect the first five rows:

   ```
   df = pd.read_csv('wine.data', header=None)
   df.head()
   ```

   The output is as follows:

	0	1	2	3	4	5	6	7	8	9	10	11	12	13
0	1	14.23	1.71	2.43	15.6	127	2.80	3.06	0.28	2.29	5.64	1.04	3.92	1065
1	1	13.20	1.78	2.14	11.2	100	2.65	2.76	0.26	1.28	4.38	1.05	3.40	1050
2	1	13.16	2.36	2.67	18.6	101	2.80	3.24	0.30	2.81	5.68	1.03	3.17	1185
3	1	14.37	1.95	2.50	16.8	113	3.85	3.49	0.24	2.18	7.80	0.86	3.45	1480
4	1	13.24	2.59	2.87	21.0	118	2.80	2.69	0.39	1.82	4.32	1.04	2.93	735

   Figure 6.30: The first five rows of the Wine dataset

3. The first column provides the labels; extract them from the DataFrame and store them in a separate variable. Ensure that the column is removed from the DataFrame:

```
labels = df[0]
del df[0]
```

4. Execute PCA on the dataset and extract the first six components:

```
model_pca = PCA(n_components=6)
wine_pca = model_pca.fit_transform(df)
wine_pca = wine_pca.reshape((len(wine_pca), -1))
```

5. Construct a loop that iterates through the perplexity values (1, 5, 20, 30, 80, 160, 320). For each loop, generate a t-SNE model with the corresponding perplexity and print a scatter plot of the labeled wine classes. Note the effect of different perplexity values:

```
MARKER = ['o', 'v', '^',]
for perp in [1, 5, 20, 30, 80, 160, 320]:
    tsne_model = TSNE(random_state=0, verbose=1, perplexity=perp)
    wine_tsne = tsne_model.fit_transform(wine_pca)
    plt.figure(figsize=(10, 7))
    plt.title(f'Low Dimensional Representation of Wine. \
              Perplexity {perp}');
    for i in range(1, 4):
        selections = wine_tsne[labels == i]
        plt.scatter(selections[:,0], selections[:,1], \
                    marker=MARKER[i-1], label=f'Wine {i}', s=30)
        plt.legend()
plt.show()
```

A perplexity value of 1 fails to separate the data into any particular structure:

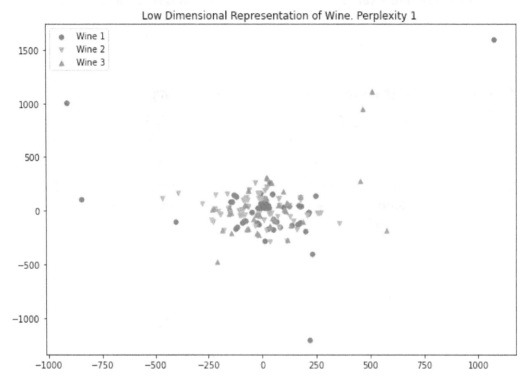

Figure 6.31: Plot for perplexity of 1

Increasing the perplexity to 5 leads to a very non-linear structure that is difficult to separate, and it's hard to identify any clusters or patterns:

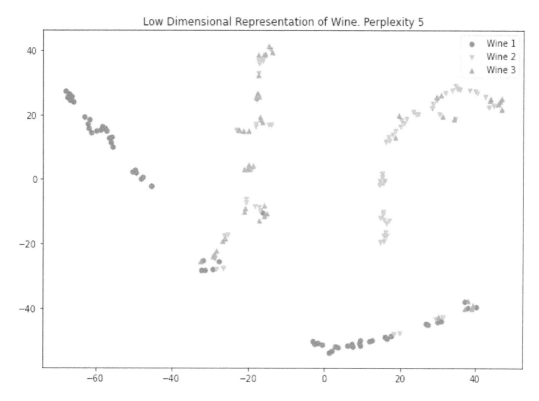

Figure 6.32: Plot for perplexity of 5

A perplexity of 20 finally starts to show some sort of horse-shoe structure. While visually obvious, this can still be tricky to implement:

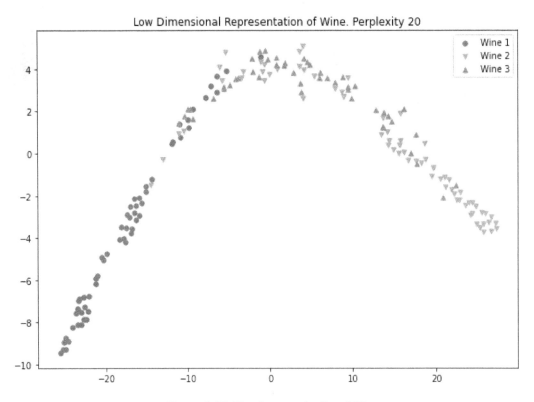

Figure 6.33: Plot for perplexity of 20

A perplexity value of 30 demonstrates quite good results. There is a linear relationship between the projected structure with some separation between the types of wine:

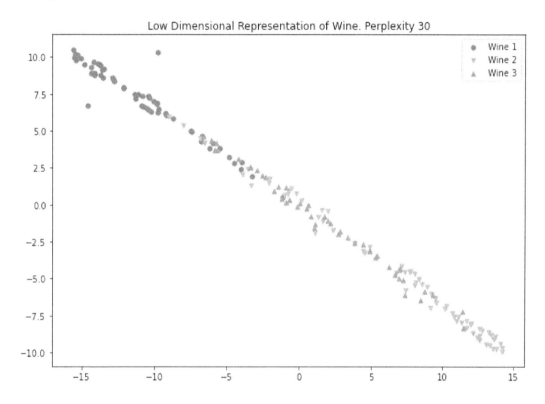

Figure 6.34: Plot for perplexity of 30

470 | Appendix

Finally, the last two images in the activity show the extent to which the plots can become increasingly complex and non-linear with increasing perplexity:

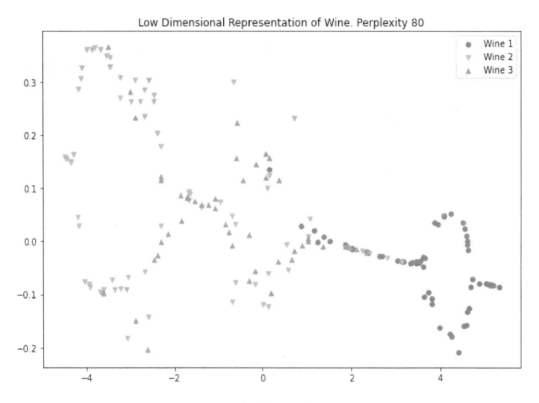

Figure 6.35: Plot for perplexity of 80

Here's the plot for a perplexity value of 160:

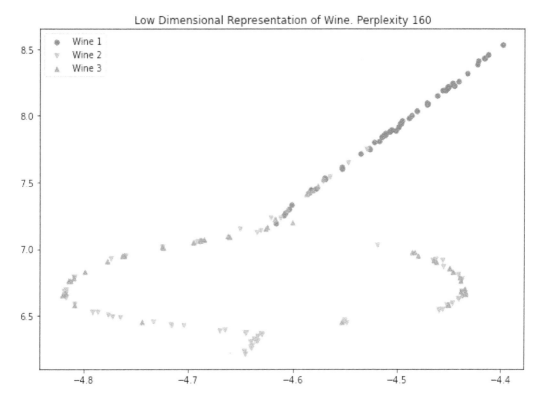

Figure 6.36: Plot for perplexity of 160

Finally, here's the plot for a perplexity value of 320:

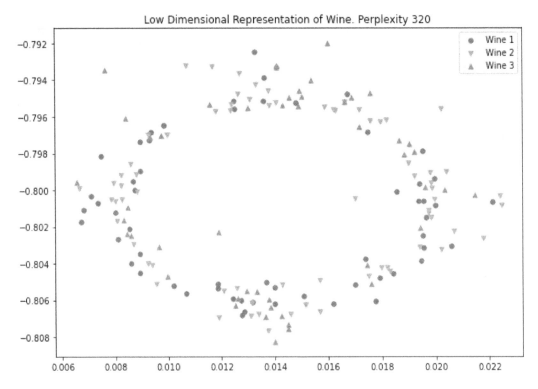

Figure 6.37: Plot for perplexity of 320

By looking at the individual plots for each of the perplexity values, the effect perplexity has on the visualization of data is immediately obvious. Very small or very large perplexity values produce a range of unusual shapes that don't indicate the presence of any persistent pattern. The most plausible value seems to be 30 (*Figure 6.35*), which produced the most linear plot.

In this activity, we demonstrated the need to be careful when selecting the perplexity and that some iteration may be required to determine the correct value.

> **NOTE**
>
> To access the source code for this specific section, please refer to https://packt.live/3faqESn.
>
> You can also run this example online at https://packt.live/2AF12Oi.

ACTIVITY 6.03: T-SNE WINE AND ITERATIONS

Solution:

1. Import **pandas**, **numpy**, and **matplotlib**, as well as the **t-SNE** and **PCA** models from scikit-learn:

   ```
   import pandas as pd
   import numpy as np
   import matplotlib.pyplot as plt
   from sklearn.decomposition import PCA
   from sklearn.manifold import TSNE
   ```

2. Load the Wine dataset and inspect the first five rows:

   ```
   df = pd.read_csv('wine.data', header=None)
   df.head()
   ```

   The output is as follows:

	0	1	2	3	4	5	6	7	8	9	10	11	12	13
0	1	14.23	1.71	2.43	15.6	127	2.80	3.06	0.28	2.29	5.64	1.04	3.92	1065
1	1	13.20	1.78	2.14	11.2	100	2.65	2.76	0.26	1.28	4.38	1.05	3.40	1050
2	1	13.16	2.36	2.67	18.6	101	2.80	3.24	0.30	2.81	5.68	1.03	3.17	1185
3	1	14.37	1.95	2.50	16.8	113	3.85	3.49	0.24	2.18	7.80	0.86	3.45	1480
4	1	13.24	2.59	2.87	21.0	118	2.80	2.69	0.39	1.82	4.32	1.04	2.93	735

   Figure 6.38: The first five rows of the Wine dataset

3. The first column provides the labels; extract these from the DataFrame and store them in a separate variable. Ensure that the column is removed from the DataFrame:

```
labels = df[0]
del df[0]
```

4. Execute PCA on the dataset and extract the first six components:

```
model_pca = PCA(n_components=6)
wine_pca = model_pca.fit_transform(df)
wine_pca = wine_pca.reshape((len(wine_pca), -1))
```

5. Construct a loop that iterates through the iteration values (**250, 500, 1000**). For each loop, generate a t-SNE model with the corresponding number of iterations and identical number of iterations without progress values:

```
MARKER = ['o', 'v', '1', 'p' ,'*', '+', 'x', 'd', '4', '.']
for iterations in [250, 500, 1000]:
    model_tsne = TSNE(random_state=0, verbose=1, \
                      n_iter=iterations, \
                      n_iter_without_progress=iterations)
    wine_tsne = model_tsne.fit_transform(wine_pca)
```

6. Construct a scatter plot of the labeled wine classes. Note the effect of different iteration values:

```
    plt.figure(figsize=(10, 7))
    plt.title(f'Low Dimensional Representation of Wine \
(iterations = {iterations})')
    for i in range(10):
        selections = wine_tsne[labels == i]
        plt.scatter(selections[:,0], selections[:,1], \
                    alpha=0.7, marker=MARKER[i], s=10);
        x, y = selections.mean(axis=0)
        plt.text(x, y, str(i), \
                 fontdict={'weight': 'bold', 'size': 30})
    plt.show()
```

The output is as follows:

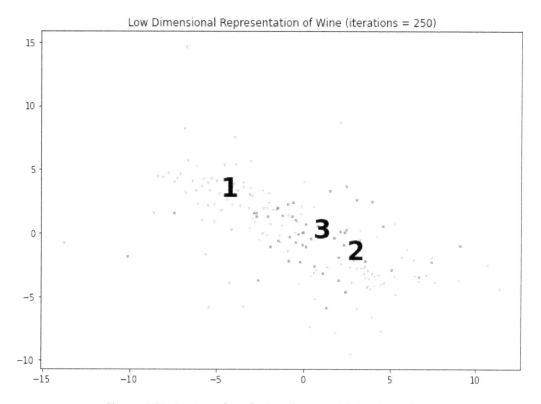

Figure 6.39: Scatter plot of wine classes with 250 iterations

Here's the plot for 500 iterations:

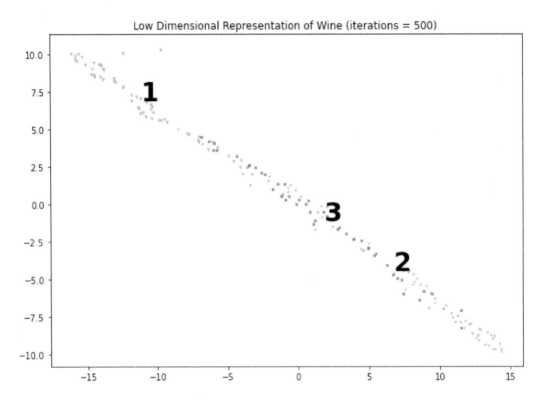

Figure 6.40: Scatter plot of wine classes with 500 iterations

Here's the plot for 1,000 iterations:

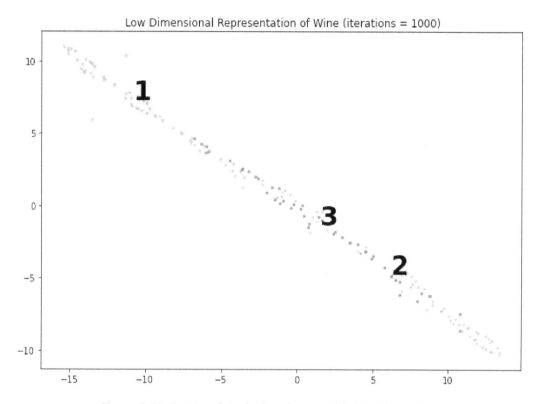

Figure 6.41: Scatter plot of wine classes with 1,000 iterations

Again, we can see an improvement in the structure of the data as the number of iterations increases. Even in a relatively simple dataset such as this, 250 iterations is not sufficient to project any structure of data into the lower-dimensional space.

As we observed in this activity, there is a balance to find in setting the iteration parameter. In this example, 250 iterations were insufficient, and at least 1,000 iterations were required to stabilize the data.

> **NOTE**
>
> To access the source code for this specific section, please refer to https://packt.live/2ZOJuYv.
>
> You can also run this example online at https://packt.live/2Z8wEoP.

CHAPTER 07: TOPIC MODELING

ACTIVITY 7.01: LOADING AND CLEANING TWITTER DATA

Solution:

1. Import the necessary libraries:

    ```
    import warnings
    warnings.filterwarnings('ignore')
    import langdetect
    import matplotlib.pyplot
    import nltk
    nltk.download('wordnet')
    nltk.download('stopwords')
    import numpy
    import pandas
    import pyLDAvis
    import pyLDAvis.sklearn
    import regex
    import sklearn
    ```

2. Load the LA Times health Twitter data (**latimeshealth.txt**) from https://packt.live/2Xje5xF.

    > **NOTE**
    >
    > Pay close attention to the delimiter (it is neither a comma nor a tab) and double-check the header status.

    The code looks as follows:

    ```
    path = 'latimeshealth.txt'
    df = pandas.read_csv(path, sep="|", header=None)
    df.columns = ["id", "datetime", "tweettext"]
    ```

3. Run a quick exploratory analysis to ascertain the data size and structure:

   ```
   def dataframe_quick_look(df, nrows):
       print("SHAPE:\n{shape}\n".format(shape=df.shape))
       print("COLUMN NAMES:\n{names}\n".format(names=df.columns))
       print("HEAD:\n{head}\n".format(head=df.head(nrows)))
   dataframe_quick_look(df, nrows=2)
   ```

   The output is as follows:

   ```
   SHAPE:
   (4171, 3)

   COLUMN NAMES:
   Index(['id', 'datetime', 'tweettext'], dtype='object')

   HEAD:
                     id                         datetime  \
   0  576760256031682561   Sat Mar 14 15:02:15 +0000 2015
   1  576715414811471872   Sat Mar 14 12:04:04 +0000 2015

                                               tweettext
   0  Five new running shoes that aim to go the extr...
   1  Gym Rat: Disq class at Crunch is intense worko...
   ```

   Figure 7.49: Shape, column names, and head of data

4. Extract the tweet text and convert it to a list object:

   ```
   raw = df['tweettext'].tolist()
   print("HEADLINES:\n{lines}\n".format(lines=raw[:5]))
   print("LENGTH:\n{length}\n".format(length=len(raw)))
   ```

   The output is as follows:

   ```
   HEADLINES:
   ['Five new running shoes that aim to go the extra mile http://lat.ms/1ELp3wU', 'Gym Rat: Disq class
   at Crunch is intense workout on pulley system http://lat.ms/1EKOFdr', 'Noshing through thousands of
   ideas at Natural Products Expo West http://lat.ms/1EHqywg', 'Natural Products Expo also explores bea
   uty, supplements and more http://lat.ms/1EHqyfE', 'Free Fitness Weekends in South Bay beach cities a
   im to spark activity http://lat.ms/1EH3SMC']

   LENGTH:
   4171
   ```

   Figure 7.50: Headlines and their length

5. Write a function to perform language detection and tokenization on white spaces, and then replace the screen names and URLs with **SCREENNAME** and **URL**, respectively. The function should also remove punctuation, numbers, and the **SCREENNAME** and **URL** replacements. Convert everything to lowercase, except **SCREENNAME** and **URL**. It should remove all stop words, perform lemmatization, and keep words with five or more letters only.

> **NOTE**
>
> Screen names start with the @ symbol.

The code is as follows:

`Activity7.01-Activity7.03.ipynb`

```
def do_language_identifying(txt):
    try:
        the_language = langdetect.detect(txt)
    except:
        the_language = 'none'
    return the_language
def do_lemmatizing(wrd):
    out = nltk.corpus.wordnet.morphy(wrd)
    return (wrd if out is None else out)
```

The complete code for this step can be found at https://packt.live/3e3VifV.

6. Apply the function defined in *Step 5* to every tweet:

```
clean = list(map(do_tweet_cleaning, raw))
```

7. Remove elements of the output list equal to **None**:

```
clean = list(filter(None.__ne__, clean))
print("HEADLINES:\n{lines}\n".format(lines=clean[:5]))
print("LENGTH:\n{length}\n".format(length=len(clean)))
```

The output is as follows:

```
HEADLINES:
[['running', 'shoes', 'extra'], ['class', 'crunch', 'intense', 'workout', 'pulley', 'system'], ['tho
usand', 'natural', 'product'], ['natural', 'product', 'explore', 'beauty', 'supplement'], ['fitnes
s', 'weekend', 'south', 'beach', 'spark', 'activity']]

LENGTH:
4095
```

Figure 7.51: Headlines and length after removing None

8. Turn the elements of each tweet back into a string. Concatenate using white space:

```
clean_sentences = [" ".join(i) for i in clean]
print(clean_sentences[0:10])
```

The first 10 elements of the output list should resemble the following:

```
['running shoes extra', 'class crunch intense workout pulley system', 'thousand natural product', 'n
atural product explore beauty supplement', 'fitness weekend south beach spark activity', 'kayla harr
ison sacrifice', 'sonic treatment alzheimers disease', 'ultrasound brain restore memory alzheimers n
eedle onlyso farin mouse', 'apple researchkit really medical research', 'warning chantix drink takin
g might remember']
```

Figure 7.52: Tweets cleaned for modeling

Keep the notebook open for future activities. By completing this activity, you should now be fairly comfortable working with textual data and preparing it for topic modeling. An important callout would be to recognize the subtle differences in data cleaning needs between the exercises and the activity. Modeling is not a one-size-fits-all process, which is obvious if you spend sufficient time exploring the data before starting any modeling work.

> **NOTE**
>
> To access the source code for this specific section, please refer to https://packt.live/3e3VifV.
>
> You can also run this example online at https://packt.live/3fegXlU. You must execute the entire Notebook in order to get the desired result.

ACTIVITY 7.02: LDA AND HEALTH TWEETS

Solution:

1. Specify the **number_words**, **number_docs**, and **number_features** variables:

    ```
    number_words = 10
    number_docs = 10
    number_features = 1000
    ```

2. Create a bag-of-words model and assign the feature names to another variable for use later on:

    ```
    vectorizer1 = sklearn.feature_extraction.text\
                  .CountVectorizer(analyzer="word", \
                                   max_df=0.95, \
                                   min_df=10, \
                                   max_features=number_features)
    clean_vec1 = vectorizer1.fit_transform(clean_sentences)
    print(clean_vec1[0])
    feature_names_vec1 = vectorizer1.get_feature_names()
    ```

    The output is as follows:

    ```
    (0, 320)    1
    ```

3. Identify the optimal number of topics:

    Activity7.01-Activity7.03.ipynb

    ```
    def perplexity_by_ntopic(data, ntopics):
        output_dict = {"Number Of Topics": [], \
                       "Perplexity Score": []}
        for t in ntopics:
            lda = sklearn.decomposition\
                  .LatentDirichletAllocation(n_components=t, \
                                             learning_method="online", \
                                             random_state=0)
            lda.fit(data)
            output_dict["Number Of Topics"].append(t)
            output_dict["Perplexity Score"]\
                .append(lda.perplexity(data))
    ```

    The complete code for this step can be found at https://packt.live/3e3VifV.

The output is as follows:

	Number Of Topics	Perplexity Score
0	2	350.307315
1	4	401.997523
2	6	425.870398
3	8	461.893848
4	10	475.041804
5	12	481.215539
6	14	505.080188
7	16	509.406520
8	18	527.267805
9	20	533.020490

Figure 7.53: Number of topics versus the perplexity score DataFrame

4. Fit the LDA model using the optimal number of topics:

```
lda = sklearn.decomposition.LatentDirichletAllocation\
      (n_components=optimal_num_topics, \
       learning_method="online", \
       random_state=0)
lda.fit(clean_vec1)
```

The output is as follows:

```
LatentDirichletAllocation(batch_size=128, doc_topic_prior=None,
                  evaluate_every=-1, learning_decay=0.7,
                  learning_method='online', learning_offset=10.0,
                  max_doc_update_iter=100, max_iter=10,
                  mean_change_tol=0.001, n_components=2, n_jobs=None,
                  perp_tol=0.1, random_state=0, topic_word_prior=None,
                  total_samples=1000000.0, verbose=0)
```

Figure 7.54: The LDA model

5. Create and print the word-topic table:

`Activity7.01-Activity7.03.ipynb`

```
def get_topics(mod, vec, names, docs, ndocs, nwords):
    # word to topic matrix
    W = mod.components_
    W_norm = W / W.sum(axis=1)[:, numpy.newaxis]
    # topic to document matrix
    H = mod.transform(vec)
    W_dict = {}
    H_dict = {}
```

The complete code for this step can be found at https://packt.live/3e3VifV.

The output is as follows:

	Topic0	Topic1
Word0	(0.0784, study)	(0.0414, latfit)
Word1	(0.0334, health)	(0.0308, cancer)
Word2	(0.024, people)	(0.0211, patient)
Word3	(0.0192, brain)	(0.0178, could)
Word4	(0.018, researcher)	(0.0168, doctor)
Word5	(0.0176, woman)	(0.0165, heart)
Word6	(0.0159, report)	(0.0147, disease)
Word7	(0.0154, obesity)	(0.0146, healthcare)
Word8	(0.0145, scientist)	(0.0145, weight)
Word9	(0.014, death)	(0.0105, expert)

Figure 7.55: Word-topic table for the health tweet data

> **NOTE**
>
> The results can differ slightly from what is shown because of the optimization algorithms that support both LDA and NMF. Many of the functions do not have a seed setting capability.

6. Print the document-topic table:

```
print(H_df)
```

The output is as follows:

```
                                                         Topic0
Doc0   (0.943, RT @LATgreatreads: Black babies are le...
Doc1   (0.9373, Finally, here's some NON-anecdotal ev...
Doc2   (0.9373, MT @DebbieGoffa ICYMI: Pioneering AID...
Doc3   (0.9373, MT @LATerynbrown: Obamacare stories t...
Doc4   (0.9373, This trial in monkeys is the stronges...
Doc5   (0.9373, Side crunches can make love handles v...
Doc6   (0.937, Once again, medical experts confirm th...
Doc7   (0.9349, RT @LATerynbrown: "Finding the gene i...
Doc8   (0.9346, Parents, give your kids advice about ...
Doc9   (0.9333, RT @bfisher_qsi: #LATFit 17,500 steps...

                                                         Topic1
Doc0   (0.9438, RT @LATgreatreads: Many U.S. troops w...
Doc1   (0.9406, Supplements to boost "low T" increase...
Doc2   (0.938, "Creating a contest with somebody's li...
Doc3   (0.9374, Want to live for a long time? Prepare...
Doc4   (0.9373, Scientists show how math can help hea...
Doc5   (0.9339, Doctors often delay vaccines for youn...
Doc6   (0.9286, Southern California hospital executiv...
Doc7   (0.9284, Bisphenol levels that changed brain d...
Doc8   (0.9283, What do folks in the US, Canada, Fran...
Doc9   (0.9283, Warning: TVs, cars &amp; computers ma...
```

Figure 7.56: Document topic table

7. Create a biplot visualization:

```
lda_plot = pyLDAvis.sklearn.prepare(lda, clean_vec1, \
                                    vectorizer1, R=10)
pyLDAvis.display(lda_plot)
```

The output is as follows:

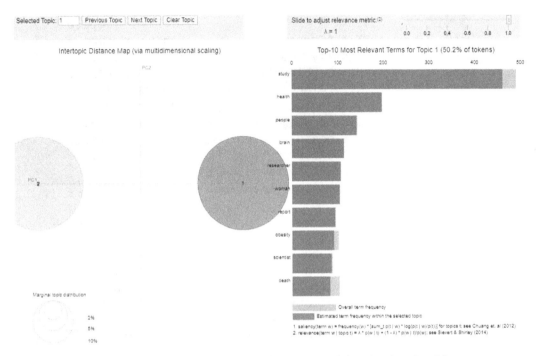

Figure 7.57: A histogram and biplot for the LDA model trained on health tweets

8. Keep the notebook open for future modeling.

Before discussing the next topic modeling methodology, non-negative matrix factorization, let's work through another bag-of-words modeling approach. You'll recall that the **CountVectorizer** algorithm returned the simple count of the number of times each word appeared in each document. In this new approach, called TF-IDF or Term Frequency – Inverse Document Frequency, weights representing the importance of each word to each document are returned instead of the raw counts.

Both the **CountVectorizer** and **TfidfVectorizer** approaches are equally valid. When and how each is used depends on the corpus, the topic modeling methodology being used, and the amount of noise in the documents. For this next exercise, we will use **TfidfVectorizer** and use the output to build the NMF models appearing later on in the chapter.

> **NOTE**
>
> To access the source code for this specific section, please refer to https://packt.live/3e3VifV.
>
> You can also run this example online at https://packt.live/3fegXlU. You must execute the entire Notebook in order to get the desired result.

ACTIVITY 7.03: NON-NEGATIVE MATRIX FACTORIZATION

Solution:

1. Create the appropriate bag-of-words model and output the feature names as another variable:

```
vectorizer2 = sklearn.feature_extraction.text.TfidfVectorizer\
              (analyzer="word", \
               max_df=0.5,\
               min_df=20,\
               max_features=number_features,\
               smooth_idf=False)
clean_vec2 = vectorizer2.fit_transform(clean_sentences)
print(clean_vec2[0])
feature_names_vec2 = vectorizer2.get_feature_names()
```

2. Define and fit the NMF algorithm using the number of topics (**n_components**) value from *Activity 7.02, LDA and Health Tweets*:

```
nmf = sklearn.decomposition.NMF(n_components=optimal_num_topics, \
                                init="nndsvda", \
                                solver="mu", \
                                beta_loss="frobenius", \
                                random_state=0, \
                                alpha=0.1, \
                                l1_ratio=0.5)
nmf.fit(clean_vec2)
```

The output is as follows:

```
NMF(alpha=0.1, beta_loss='frobenius', init='nndsvda', l1_ratio=0.5,
    max_iter=200, n_components=2, random_state=0, shuffle=False, solver='mu',
    tol=0.0001, verbose=0)
```

Figure 7.58: Defining the NMF model

3. Get the topic-document and word-topic tables. Take a few minutes to explore the word groupings and try to define the abstract topics:

```
W_df, H_df = get_topics(mod=nmf, vec=clean_vec2, \
                        names=feature_names_vec2, \
                        docs=raw, \
                        ndocs=number_docs, \
                        nwords=number_words)
print(W_df)
```

The output is as follows:

	Topic0	Topic1
Word0	(0.3764, study)	(0.5933, latfit)
Word1	(0.0258, cancer)	(0.049, steps)
Word2	(0.0208, people)	(0.0449, today)
Word3	(0.0184, obesity)	(0.0405, exercise)
Word4	(0.0184, health)	(0.0274, healthtips)
Word5	(0.0181, brain)	(0.0258, workout)
Word6	(0.0173, suggest)	(0.0204, getting)
Word7	(0.0168, weight)	(0.0193, fitness)
Word8	(0.0153, woman)	(0.0143, great)
Word9	(0.0131, death)	(0.0132, morning)

Figure 7.59: Word-topic table

Print the document-topic table as follows:

```
print(H_df)
```

The output is as follows:

```
                                                      Topic0  \
Doc0   (0.2032, How did H7N9 get from wild birds to c...
Doc1   (0.2032, ICYMI MT @LATerynbrown @chadterhune L...
Doc2   (0.2032, You can get kids to cut down on scree...
Doc3   (0.2032, Avatars often reflect a reality about...
Doc4   (0.2032, If you've been eating 5 servings of f...
Doc5   (0.2032, Four economical workout machines will...
Doc6   (0.2032, State Dept: Security at Libya consula...
Doc7   (0.2032, California seeks limits on small-busi...
Doc8   (0.2032, King Richard III's body provides some...
Doc9   (0.2032, For seniors -- even seniors with memo...

                                                      Topic1
Doc0   (0.2276, Get too much sun today? An anti-infla...
Doc1   (0.2276, Summer camp with a twist: Physical th...
Doc2   (0.2276, How the month you were conceived can ...
Doc3   (0.2276, Study: Contraceptives could prevent a...
Doc4   (0.2276, At one point in the 1990s, 18% of hig...
Doc5   (0.2276, RT @lauraelizdavis: .@renelynch inspi...
Doc6   (0.2276, 'Immunize, immunize!': Doctors counte...
Doc7   (0.2276, RT @annagorman: USC buys Verdugo Hill...
Doc8   (0.2276, FDA approves a brain wave device to h...
Doc9   (0.2276, RT @julie_cart: Single dose of ADHD d...
```

Figure 7.60: The topic-document tables with probabilities

4. Adjust the model parameters and rerun *Step 3* and *Step 4*.

In this activity, we worked through an example topic modeling scenario using the TF-IDF bag-of-words model and non-negative matrix factorization. What is really important in these examples is trying to understand what the algorithms are doing—not just how to fit them—and to comprehend the results. Working with text data is often complex and it is crucial to realize that not every algorithm will return meaningful results every time. Sometimes, the results are just not useful. That is not a reflection on the algorithm or the practitioner; it is simply an example of the challenge of extracting insights from data.

> **NOTE**
>
> To access the source code for this specific section, please refer to https://packt.live/3e3VifV.
>
> You can also run this example online at https://packt.live/3fegXlU. You must execute the entire Notebook in order to get the desired result.

CHAPTER 08: MARKET BASKET ANALYSIS

ACTIVITY 8.01: LOADING AND PREPARING FULL ONLINE RETAIL DATA

Solution:

1. Import the required libraries:

   ```
   import matplotlib.pyplot as plt
   import mlxtend.frequent_patterns
   import mlxtend.preprocessing
   import numpy
   import pandas
   ```

2. Load the online retail dataset file:

   ```
   online = pandas.read_excel(io="./Online Retail.xlsx", \
                              sheet_name="Online Retail", \
                              header=0)
   ```

3. Clean and prep the data for modeling, including turning the cleaned data into a list of lists:

   ```
   online['IsCPresent'] = (online['InvoiceNo'].astype(str)\
                           .apply(lambda x: 1 \
                                  if x.find('C') != -1 else 0))

   online1 = (online.loc[online["Quantity"] > 0]\
                    .loc[online['IsCPresent'] != 1]\
                    .loc[:, ["InvoiceNo", "Description"]].dropna())

   invoice_item_list = []
   for num in list(set(online1.InvoiceNo.tolist())):
       tmp_df = online1.loc[online1['InvoiceNo'] == num]
       tmp_items = tmp_df.Description.tolist()
       invoice_item_list.append(tmp_items)
   ```

4. Encode the data and recast it as a DataFrame. The data is fairly large, so to ensure that everything executes correctly, use a machine with at least 8 GB of memory:

```
online_encoder = mlxtend.preprocessing.TransactionEncoder()
online_encoder_array = \
online_encoder.fit_transform(invoice_item_list)

online_encoder_df = pandas.DataFrame(\
                online_encoder_array, \
                columns=online_encoder.columns_)

online_encoder_df.loc[20125:20135, \
                online_encoder_df.columns.tolist()\
                [100:110]]
```

The output is as follows:

	6 CHOCOLATE LOVE HEART T-LIGHTS	6 EGG HOUSE PAINTED WOOD	6 GIFT TAGS 50'S CHRISTMAS	6 GIFT TAGS VINTAGE CHRISTMAS	6 RIBBONS ELEGANT CHRISTMAS	6 RIBBONS EMPIRE	6 RIBBONS RUSTIC CHARM	6 RIBBONS SHIMMERING PINKS	6 ROCKET BALLOONS	60 CAKE CASES DOLLY GIRL DESIGN
20125	False	False	False	False	False	False	False	False	False	False
20126	False	False	False	False	False	False	False	False	False	False
20127	False	False	False	False	False	False	False	False	False	False
20128	False	False	False	False	False	False	False	False	False	False
20129	False	False	False	False	False	False	False	False	False	False
20130	False	False	False	False	False	False	False	False	False	False
20131	False	False	False	False	False	False	False	False	False	False
20132	False	False	False	False	False	False	False	False	False	False
20133	False	False	False	False	False	False	False	False	False	False
20134	False	False	False	False	False	False	False	False	False	False
20135	False	False	False	False	False	False	False	False	False	False

Figure 8.34: A subset of the cleaned, encoded, and recast DataFrame built from the complete online retail dataset

> **NOTE**
>
> To access the source code for this specific section, please refer to https://packt.live/2Wf2Rcz.
>
> This section does not currently have an online interactive example and will need to be run locally.

ACTIVITY 8.02: RUNNING THE APRIORI ALGORITHM ON THE COMPLETE ONLINE RETAIL DATASET

Solution:

1. Run the Apriori algorithm on the full data with reasonable parameter settings:

```
mod_colnames_minsupport = mlxtend.frequent_patterns\
                    .apriori(online_encoder_df, \
                             min_support=0.01, \
                             use_colnames=True)
mod_colnames_minsupport.loc[0:6]
```

The output is as follows:

	support	itemsets
0	0.013359	(SET 2 TEA TOWELS I LOVE LONDON)
1	0.015793	(10 COLOUR SPACEBOY PEN)
2	0.012465	(12 MESSAGE CARDS WITH ENVELOPES)
3	0.017630	(12 PENCIL SMALL TUBE WOODLAND)
4	0.017978	(12 PENCILS SMALL TUBE RED RETROSPOT)
5	0.017630	(12 PENCILS SMALL TUBE SKULL)
6	0.013309	(12 PENCILS TALL TUBE RED RETROSPOT)

Figure 8.35: The Apriori algorithm results using the complete online retail dataset

2. Filter the results down to the item set containing **10 COLOUR SPACEBOY PEN**. Compare the support value with that under *Exercise 8.06, Executing the Apriori Algorithm*:

```
mod_colnames_minsupport[mod_colnames_minsupport['itemsets'] \
    == frozenset({'10 COLOUR SPACEBOY PEN'})]
```

The output is as follows:

Figure 8.36: Result of item set containing 10 COLOUR SPACEBOY PEN

The support value does change. When the dataset is expanded to include all transactions, the support for this item set drops from 0.0178 to 0.015793. That is, in the reduced dataset used for the exercises, this item set appears in 1.78% of the transactions, while in the full dataset, it appears in approximately 1.6% of transactions.

3. Add another column containing the item set length. Then, filter down to those item sets whose length is two and whose support is in the range [0.02, 0.021]. Are the item sets the same as those found in *Exercise 8.06, Executing the Apriori Algorithm*, Step 6?

```
mod_colnames_minsupport['length'] = (mod_colnames_minsupport\
                                     ['itemsets']\
                                     .apply(lambda x: len(x)))

mod_colnames_minsupport[(mod_colnames_minsupport['length'] == 2) \
                        & (mod_colnames_minsupport['support'] \
                          >= 0.02) \
                        & (mod_colnames_minsupport['support'] \
                          < 0.021)]
```

The output is as follows:

	support	itemsets	length
836	0.020759	(ALARM CLOCK BAKELIKE GREEN, ALARM CLOCK BAKEL...	2
887	0.020362	(CHARLOTTE BAG SUKI DESIGN, CHARLOTTE BAG PINK...	2
923	0.020610	(CHARLOTTE BAG SUKI DESIGN, STRAWBERRY CHARLOT...	2
1105	0.020560	(JUMBO BAG PINK POLKADOT, JUMBO BAG BAROQUE B...	2
1114	0.020908	(JUMBO SHOPPER VINTAGE RED PAISLEY, JUMBO BAG...	2
1116	0.020957	(JUMBO BAG BAROQUE BLACK WHITE, JUMBO STORAGE...	2
1129	0.020560	(JUMBO BAG ALPHABET, JUMBO BAG RED RETROSPOT)	2
1137	0.020163	(JUMBO BAG APPLES, JUMBO BAG PEARS)	2
1203	0.020709	(JUMBO BAG PINK VINTAGE PAISLEY, JUMBO SHOPPER...	2
1218	0.020560	(JUMBO BAG RED RETROSPOT, JUMBO STORAGE BAG SK...	2
1236	0.020610	(JUMBO BAG RED RETROSPOT, RECYCLING BAG RETROS...	2
1328	0.020610	(LUNCH BAG BLACK SKULL., LUNCH BAG APPLE DESIGN)	2
1390	0.020610	(LUNCH BAG SUKI DESIGN , LUNCH BAG PINK POLKADOT)	2
1458	0.020610	(NATURAL SLATE HEART CHALKBOARD , WHITE HANGIN...	2
1581	0.020362	(SET OF 3 CAKE TINS PANTRY DESIGN , SET OF 6 S...	2
1607	0.020163	(WOODLAND CHARLOTTE BAG, STRAWBERRY CHARLOTTE ...	2
1615	0.020262	(WHITE HANGING HEART T-LIGHT HOLDER, WOODEN PI...	2

Figure 8.37: The results of filtering based on length and support

The results did change. Before even looking at the particular item sets and their support values, we see that this filtered DataFrame has fewer item sets than the DataFrame in the preceding exercise. When we use the full dataset, there are fewer item sets that match the filtering criteria; that is, only 17 item sets contain 2 items and have a support value greater than or equal to `0.02`, and less than `0.021`. In the previous exercise, 32 item sets met these criteria.

4. Plot the **support** values:

```
mod_colnames_minsupport.hist("support", grid=False, bins=30)
plt.xlabel("Support of item")
plt.ylabel("Number of items")
plt.title("Frequency distribution of Support")
plt.show()
```

The output is as follows:

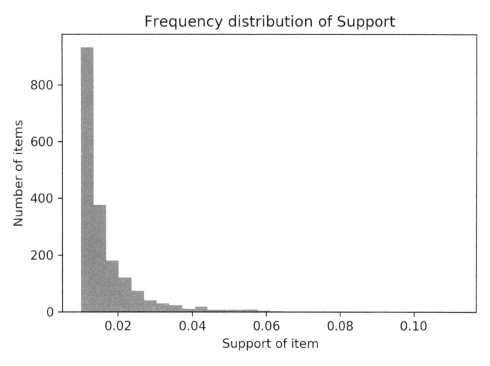

Figure 8.38: The distribution of support values

This plot shows the distribution of support values for the full transaction dataset. As you might have assumed, the distribution is right-skewed; that is, most of the item sets have lower support values and there is a long tail of support values on the higher end of the spectrum. Given how many unique item sets exist, it is not surprising that no single item set appears in a high percentage of the transactions. With this information, we could tell management that even the most prominent item set only appears in approximately 10% of transactions and that the vast majority of item sets appear in fewer than 2% of transactions. These results may not support changes in store layout, but could very well inform pricing and discounting strategies. We would gain more information on how to build these strategies by formalizing a number of association rules.

> **NOTE**
>
> To access the source code for this specific section, please refer to https://packt.live/2Wf2Rcz.
>
> This section does not currently have an online interactive example and will need to be run locally.

ACTIVITY 8.03: FINDING THE ASSOCIATION RULES ON THE COMPLETE ONLINE RETAIL DATASET

Solution:

1. Fit the association rule model on the full dataset. Use the confidence metric and a minimum threshold of **0.6**:

   ```
   rules = mlxtend.frequent_patterns\
           .association_rules(mod_colnames_minsupport, \
                              metric="confidence", \
                              min_threshold=0.6, \
                              support_only=False)
   rules.loc[0:6]
   ```

   The output is as follows:

	antecedents	consequents	antecedent support	consequent support	support	confidence	lift	leverage	conviction
0	(ALARM CLOCK BAKELIKE CHOCOLATE)	(ALARM CLOCK BAKELIKE GREEN)	0.021255	0.048669	0.013756	0.647196	13.297902	0.012722	2.696488
1	(ALARM CLOCK BAKELIKE CHOCOLATE)	(ALARM CLOCK BAKELIKE RED)	0.021255	0.052195	0.014501	0.682243	13.071023	0.013392	2.982798
2	(ALARM CLOCK BAKELIKE ORANGE)	(ALARM CLOCK BAKELIKE GREEN)	0.022100	0.048669	0.013558	0.613483	12.605201	0.012482	2.461292
3	(ALARM CLOCK BAKELIKE GREEN)	(ALARM CLOCK BAKELIKE RED)	0.048669	0.052195	0.031784	0.653061	12.511932	0.029244	2.731908
4	(ALARM CLOCK BAKELIKE RED)	(ALARM CLOCK BAKELIKE GREEN)	0.052195	0.048669	0.031784	0.608944	12.511932	0.029244	2.432722
5	(ALARM CLOCK BAKELIKE IVORY)	(ALARM CLOCK BAKELIKE RED)	0.028308	0.052195	0.018524	0.654386	12.537313	0.017047	2.742380
6	(ALARM CLOCK BAKELIKE ORANGE)	(ALARM CLOCK BAKELIKE RED)	0.022100	0.052195	0.014998	0.678652	13.002217	0.013845	2.949463

   Figure 8.39: The association rules based on the complete online retail dataset

2. Count the number of association rules. Is the number different from that found in *Exercise 8.07, Deriving Association Rules, Step 1*?

   ```
   print("Number of Associations: {}".format(rules.shape[0]))
   ```

   There are **498** association rules. Yes, the count is different.

3. Plot confidence against support:

```
rules.plot.scatter("support", "confidence", \
                   alpha=0.5, marker="*")
plt.xlabel("Support")
plt.ylabel("Confidence")
plt.title("Association Rules")
plt.show()
```

The output is as follows:

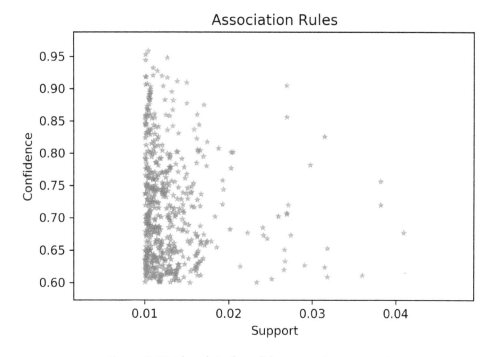

Figure 8.40: The plot of confidence against support

The plot reveals that there are some association rules featuring relatively high support and confidence values for this dataset.

4. Look at the distributions of lift, leverage, and conviction:

```
rules.hist("lift", grid=False, bins=30)
plt.xlabel("Lift of item")
plt.ylabel("Number of items")
plt.title("Frequency distribution of Lift")
plt.show()
```

The output is as follows:

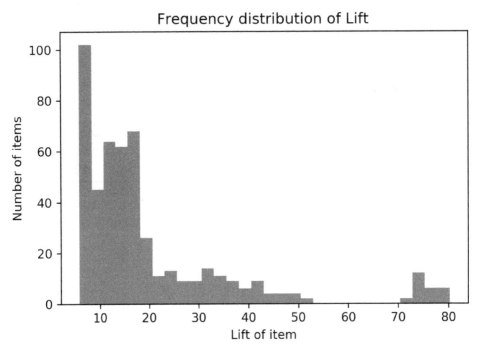

Figure 8.41: The distribution of lift values

Plot the leverage as follows:

```
rules.hist("leverage", grid=False, bins=30)
plt.xlabel("Leverage of item")
plt.ylabel("Number of items")
plt.title("Frequency distribution of Leverage")
plt.show()
```

The output is as follows:

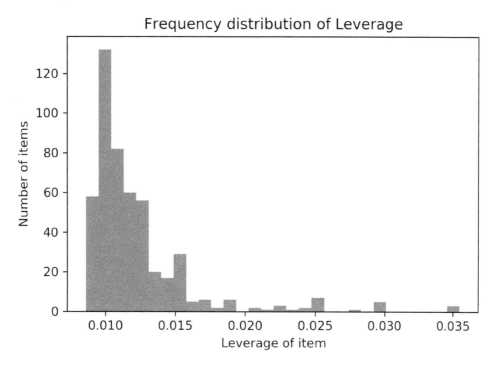

Figure 8.42: The distribution of leverage values

Plot the conviction as follows:

```
plt.hist(rules[numpy.isfinite(rules['conviction'])]\
        .conviction.values, bins = 3)
plt.xlabel("Conviction of item")
plt.ylabel("Number of items")
plt.title("Frequency distribution of Conviction")
plt.show()
```

The output is as follows:

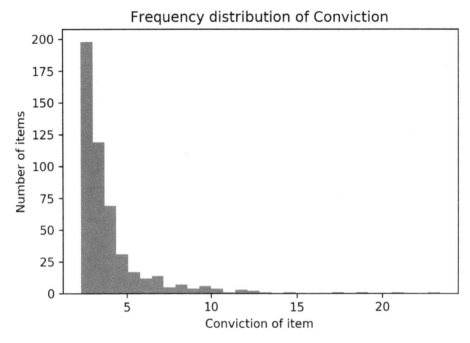

Figure 8.43: The distribution of conviction values

Having derived association rules, we can return to management with additional information, the most important of which would be that there are roughly seven item sets that have reasonably high values for both support and confidence. Look at the scatterplot of confidence against support to see the seven item sets that are separated from all the others. These seven item sets also have high lift values, as can be seen in the lift histogram. It seems that we have identified a number of actionable association rules – rules that we can use to drive business decisions.

> **NOTE**
>
> To access the source code for this specific section, please refer to https://packt.live/2Wf2Rcz.
>
> This section does not currently have an online interactive example and will need to be run locally.

CHAPTER 09: HOTSPOT ANALYSIS

ACTIVITY 9.01: ESTIMATING DENSITY IN ONE DIMENSION

Solution:

1. Open a new notebook and install all the necessary libraries.

   ```
   get_ipython().run_line_magic('matplotlib', 'inline')
   import matplotlib.pyplot as plt
   import numpy
   import pandas
   import seaborn
   import sklearn.model_selection
   import sklearn.neighbors
   seaborn.set()
   ```

2. Sample 1,000 data points from the standard normal distribution. Add 3.5 to each of the last 625 values of the sample (that is, the indices between 375 and 1,000). Set a random state of 100 using **numpy.random.RandomState** to guarantee the same sampled values, and then randomly generate the data points using the **rand.randn(1000)** call:

   ```
   rand = numpy.random.RandomState(100)
   vals = rand.randn(1000)   # standard normal
   vals[375:] += 3.5
   ```

3. Plot the 1,000-point sample data as a histogram and add a scatterplot below it:

   ```
   fig, ax = plt.subplots(figsize=(14, 10))
   ax.hist(vals, bins=50, density=True, label='Sampled Values')
   ax.plot(vals, -0.005 - 0.01 * numpy.random.random(len(vals)), \
           '+k', label='Individual Points')
   ax.legend(loc='upper right')
   plt.show()
   ```

The output is as follows:

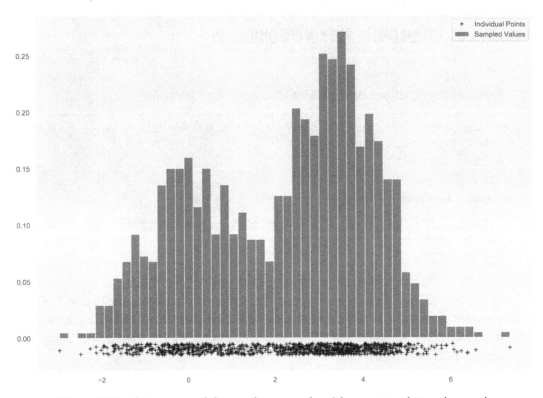

Figure 9.30: A histogram of the random sample with a scatterplot underneath

4. Define a grid of bandwidth values. Then, define and fit a grid search cross-validation algorithm:

```
bandwidths = 10 ** numpy.linspace(-1, 1, 100)
grid = sklearn.model_selection.GridSearchCV\
        (estimator=sklearn.neighbors.KernelDensity(kernel="gaussian"),
        param_grid={"bandwidth": bandwidths}, cv=10)
grid.fit(vals[:, None])
```

The output is as follows:

```
GridSearchCV(cv=10, error_score=nan,
        estimator=KernelDensity(algorithm='auto', atol=0, bandwidth=1.0,
                                breadth_first=True, kernel='gaussian',
                                leaf_size=40, metric='euclidean',
                                metric_params=None, rtol=0),
        iid='deprecated', n_jobs=None,
        param_grid={'bandwidth': array([ 0.1       , 0.10476158, 0.10974
988,  0.1149757 , 0.12045035,
        0.12618569, 0.13219411, 0.13848864, 0.14508288, 0.1519...
        3.27454916, 3.43046929, 3.59381366, 3.76493581, 3.94420606,
        4.1320124 , 4.32876128, 4.53487851, 4.75081016, 4.97702356,
        5.21400829, 5.46227722, 5.72236766, 5.9948425 , 6.28029144,
        6.57933225, 6.8926121 , 7.22080902, 7.56463328, 7.92482898,
        8.30217568, 8.69749003, 9.11162756, 9.54548457, 10.       ])},
        pre_dispatch='2*n_jobs', refit=True, return_train_score=False,
        scoring=None, verbose=0)
```

Figure 9.31: Output of cross-validation model

5. Extract the optimal bandwidth value:

   ```
   best_bandwidth = grid.best_params_["bandwidth"]
   print("Best Bandwidth Value: {}".format(best_bandwidth))
   ```

   The optimal bandwidth value is approximately 0.4.

6. Replot the histogram from *Step 3* and overlay the estimated density:

   ```
   fig, ax = plt.subplots(figsize=(14, 10))
   ax.hist(vals, bins=50, density=True, alpha=0.75, \
           label='Sampled Values')
   x_vec = numpy.linspace(-4, 8, 10000)[:, numpy.newaxis]
   log_density = numpy.exp(grid.best_estimator_.score_samples(x_vec))
   ax.plot(x_vec[:, 0], log_density, \
           '-', linewidth=4, label='Kernel = Gaussian')
   ax.legend(loc='upper right')
   plt.show()
   ```

The output is as follows:

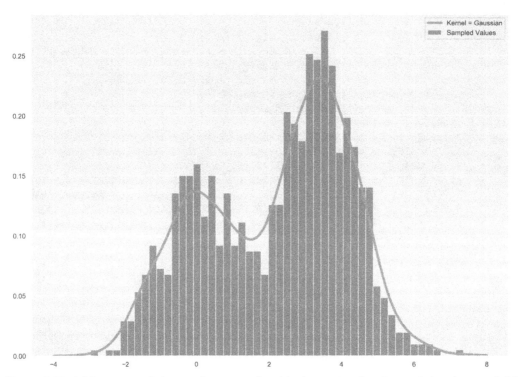

Figure 9.32: A histogram of the random sample with the optimal estimated density overlaid

> **NOTE**
>
> To access the source code for this specific section, please refer to https://packt.live/2wmh5yj.
>
> You can also run this example online at https://packt.live/2W0EAGK.

ACTIVITY 9.02: ANALYZING CRIME IN LONDON

Solution:

1. Load the crime data. Use the path where you saved the downloaded directory, create a list of the year-month tags, use the **read_csv** command to load the individual files iteratively, and then concatenate these files together:

    ```
    # define the file base path
    base_path = "./metro-jul18-dec18/{yr_mon}/{yr_mon}\
    -metropolitan-street.csv"
    print(base_path)
    ```

    The output is as follows:

    ```
    ./metro-jul18-dec18/{yr_mon}/{yr_mon}-metropolitan-street.csv
    ```

2. Define the list of year month combinations as follows:

    ```
    yearmon_list = ["2018-0" + str(i) if i <= 9 else "2018-" + str(i) \
                    for i in range(7, 13)]

    print(yearmon_list)
    ```

    The output is as follows:

    ```
    ['2018-07', '2018-08', '2018-09', \
     '2018-10', '2018-11', '2018-12']
    ```

3. Load the data and print some basic information about it as follows:

    ```
    data_yearmon_list = []

    # read each year month file individually
    #print summary statistics
    for idx, i in enumerate(yearmon_list):
        df = pandas.read_csv(base_path.format(yr_mon=i), \
                             header=0)
        data_yearmon_list.append(df)
        if idx == 0:
            print("Month: {}".format(i))
            print("Dimensions: {}".format(df.shape))
    ```

```
            print("Head:\n{}\n".format(df.head(2)))

    # concatenate the list of year month data frames together
    london = pandas.concat(data_yearmon_list)
```

The output is as follows:

```
Month: 2018-07
Dimensions: (95677, 12)
Head:
                                       Crime ID     Month  \
0  e9fe81ec7a6f5d2a80445f04be3d7e92831dbf3090744e...  2018-07
1  076b796ba1e1ba3f69c9144e2aa7a7bc85b61d51bf7a59...  2018-07

                 Reported by               Falls within  Longitude  \
0  Metropolitan Police Service  Metropolitan Police Service   0.774271
1  Metropolitan Police Service  Metropolitan Police Service  -1.007293

    Latitude                   Location  LSOA code           LSOA name  \
0  51.148147  On or near Bethersden Road  E01024031          Ashford 012B
1  51.893136          On or near Prison  E01017674  Aylesbury Vale 010D

   Crime type      Last outcome category  Context
0  Other theft  Status update unavailable      NaN
1  Other crime     Awaiting court outcome      NaN
```

Figure 9.33: An example of one of the individual crime files

This printed information is just for the first of the loaded files, which will be the criminal information from the Metropolitan Police Service for July 2018. This one file has nearly 100,000 entries. You will notice that there is a great deal of interesting information in this dataset, but we will focus on **Longitude**, **Latitude**, **Month**, and **Crime type**.

4. Print diagnostics of the complete and concatenated dataset:

`Activity9.01-Activity9.02.ipynb`

```
print("Dimensions - Full Data:\n{}\n".format(london.shape))
print("Unique Months - Full Data:\n{}\n".format(london["Month"].unique()))
print("Number of Unique Crime Types - Full Data:\n{}\n"\
    .format(london["Crime type"].nunique()))
```

The complete code for this step can be found at https://packt.live/2wmh5yj.

The output is as follows:

```
Dimensions - Full Data:
(546032, 12)

Unique Months - Full Data:
['2018-07' '2018-08' '2018-09' '2018-10' '2018-11' '2018-12']

Number of Unique Crime Types - Full Data:
14

Unique Crime Types - Full Data:
['Other theft' 'Other crime' 'Violence and sexual offences'
 'Anti-social behaviour' 'Criminal damage and arson' 'Drugs'
 'Possession of weapons' 'Theft from the person' 'Vehicle crime'
 'Burglary' 'Public order' 'Robbery' 'Shoplifting' 'Bicycle theft']

Count Occurrences Of Each Unique Crime Type - Full Type:
Violence and sexual offences    117499
Anti-social behaviour           115448
Other theft                      61833
Vehicle crime                    58857
Burglary                         41145
Criminal damage and arson        28436
Public order                     24655
Theft from the person            22670
Shoplifting                      21296
Drugs                            17292
Robbery                          17060
Bicycle theft                    11362
Other crime                       5223
Possession of weapons             3256
Name: Crime type, dtype: int64
```

Figure 9.34: Descriptors of the full crime dataset

5. Subset the data frame down to four variables (**Longitude**, **Latitude**, **Month**, and **Crime type**):

```
london_subset = london[["Month", "Longitude", "Latitude", \
                        "Crime type"]]
london_subset.head(5)
```

The output is as follows:

	Month	Longitude	Latitude	Crime type
0	2018-07	0.774271	51.148147	Other theft
1	2018-07	-1.007293	51.893136	Other crime
2	2018-07	0.744706	52.038219	Violence and sexual offences
3	2018-07	0.148434	51.595164	Anti-social behaviour
4	2018-07	0.137065	51.583672	Anti-social behaviour

Figure 9.35: Crime data in data frame format

6. Using the `jointplot` function from `seaborn`, fit and visualize three kernel density estimation models for bicycle theft in July, September, and December 2018:

```
crime_bicycle_jul = london_subset\
                [(london_subset["Crime type"] \
                    == "Bicycle theft") \
                 & (london_subset["Month"] == "2018-07")]
seaborn.jointplot("Longitude", "Latitude", \
                crime_bicycle_jul, kind="kde")
```

The output is as follows:

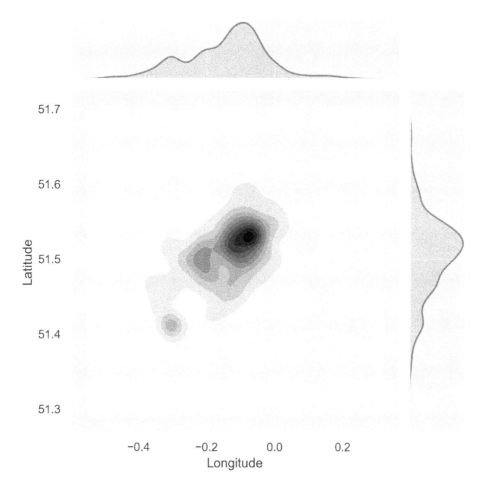

Figure 9.36: The estimated joint and marginal densities for bicycle thefts in July 2018

For the month of September 2018, the code to fit and visualize the kernel density estimation model for bicycle theft is as follows:

```
crime_bicycle_sept = london_subset\
                    [(london_subset["Crime type"]
                     == "Bicycle theft")
                    & (london_subset["Month"] == "2018-09")]
seaborn.jointplot("Longitude", "Latitude", \
                   crime_bicycle_sept, kind="kde")
```

The output is as follows:

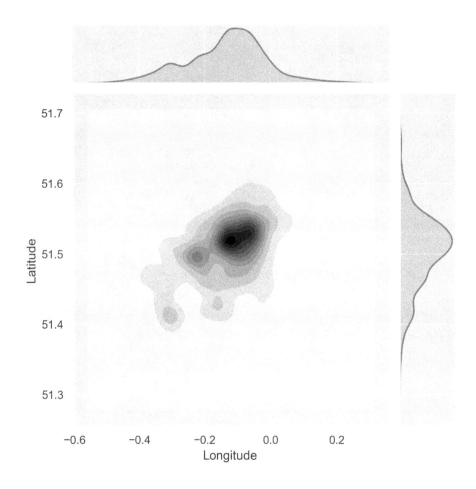

Figure 9.37: The estimated joint and marginal densities for bicycle thefts in September 2018

For the month of December 2018, the code to fit and visualize the kernel density estimation model for bicycle theft is as follows:

```
crime_bicycle_dec = london_subset\
                [(london_subset["Crime type"] \
                == "Bicycle theft")
                & (london_subset["Month"] == "2018-12")]
seaborn.jointplot("Longitude", "Latitude", \
                crime_bicycle_dec, kind="kde")
```

The output is as follows:

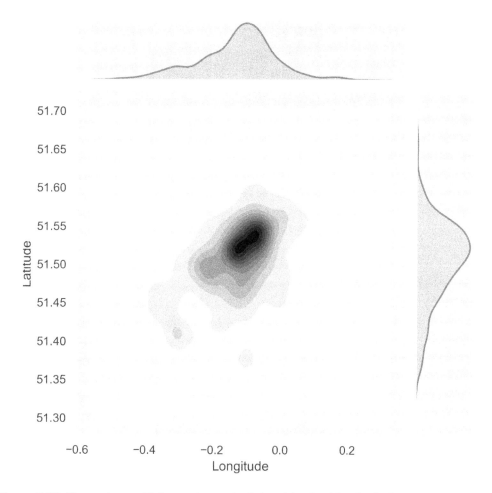

Figure 9.38: The estimated joint and marginal densities for bicycle thefts in December 2018

From month to month, the density of bicycle thefts stays quite constant. There are slight differences between the densities, which is to be expected given that the data that is the foundation of these estimated densities is three one-month samples. Given these results, police or criminologists should be confident in predicting where future bicycle thefts are most likely to occur.

7. Repeat *Step 4*; this time, use shoplifting crimes for the months of August, October, and November 2018:

```
crime_shoplift_aug = london_subset\
                [(london_subset["Crime type"] \
                == "Shoplifting")
                & (london_subset["Month"] == "2018-08")]
seaborn.jointplot("Longitude", "Latitude", \
                crime_shoplift_aug, kind="kde")
```

The output is as follows:

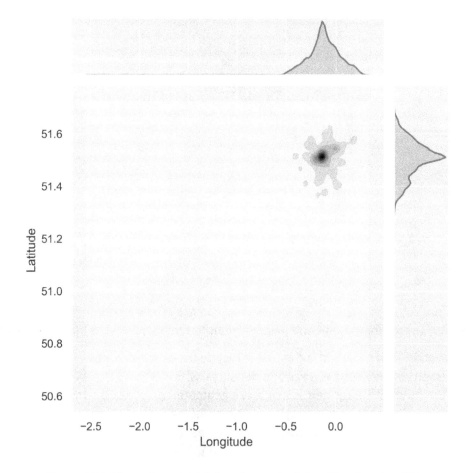

Figure 9.39: The estimated joint and marginal densities for shoplifting incidents in August 2018

For the month of October 2018, the code to fit and visualize the kernel density estimation model for shoplifting crimes is as follows:

```
crime_shoplift_oct = london_subset\
                     [(london_subset["Crime type"] \
                      == "Shoplifting") \
                     & (london_subset["Month"] == "2018-10")]
seaborn.jointplot("Longitude", "Latitude", \
                  crime_shoplift_oct, kind="kde")
```

The output is as follows:

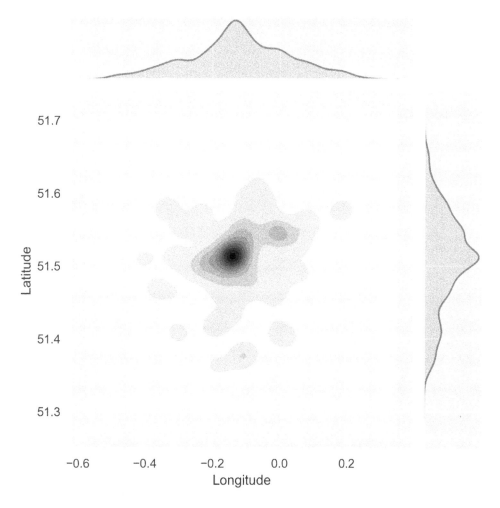

Figure 9.40: The estimated joint and marginal densities for shoplifting incidents in October 2018

For the month of November 2018, the code to fit and visualize the kernel density estimation model for shoplifting crimes is as follows:

```
crime_shoplift_nov = london_subset\
                [(london_subset["Crime type"] \
                 == "Shoplifting") \
                & (london_subset["Month"] == "2018-11")]
seaborn.jointplot("Longitude", "Latitude", \
                crime_shoplift_nov, kind="kde")
```

The output is as follows:

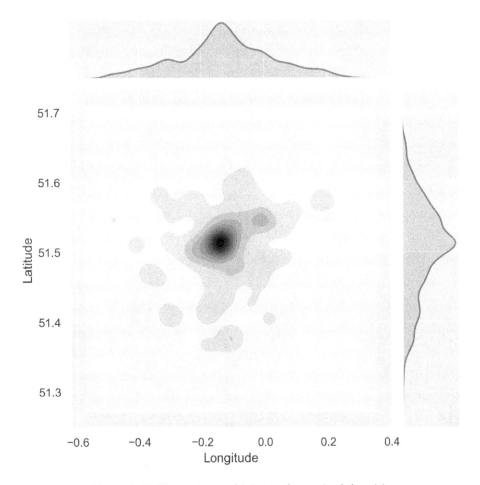

Figure 9.41: The estimated joint and marginal densities for shoplifting incidents in November 2018

Like the bicycle theft results, the shoplifting densities are quite stable across the months. The density from August 2018 looks different from the other two months; however, if you look at the longitude and latitude values, you will notice that the density is very similar – just shifted and scaled. The reason for this is that there were probably a number of outliers forcing the creation of a much larger plotting region.

8. Repeat *Step 5*; this time use burglary crimes for the months of July, October, and December 2018:

```
crime_burglary_jul = london_subset\
                     [(london_subset["Crime type"] == "Burglary") \
                     & (london_subset["Month"] == "2018-07")]
seaborn.jointplot("Longitude", "Latitude", \
                  crime_burglary_jul, kind="kde")
```

516 | Appendix

The output is as follows:

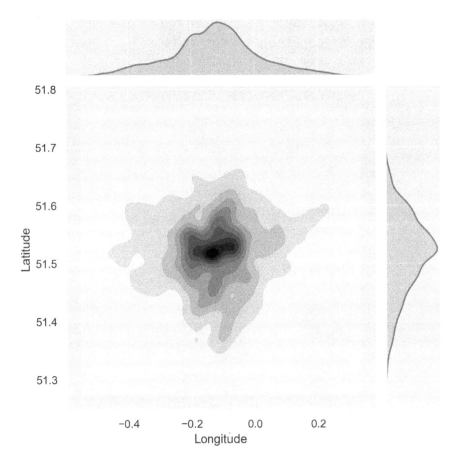

Figure 9.42: The estimated joint and marginal densities for burglaries in July 2018

For the month of October 2018, the code to fit and visualize the kernel density estimation model for burglaries is as follows:

```
crime_burglary_oct = london_subset\
                [(london_subset["Crime type"] == "Burglary")\
                & (london_subset["Month"] == "2018-10")]
seaborn.jointplot("Longitude", "Latitude", \
            crime_burglary_oct, kind="kde")
```

The output is as follows:

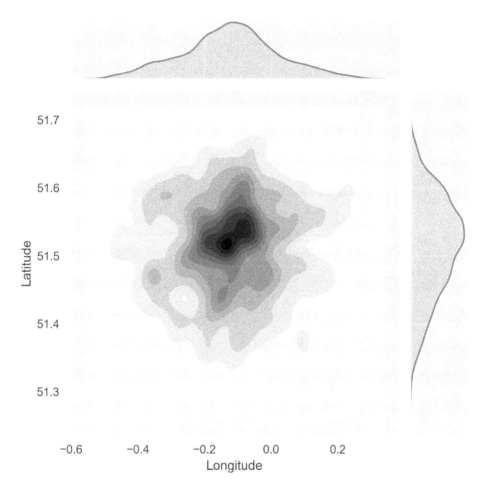

Figure 9.43: The estimated joint and marginal densities for burglaries in October 2018

For the month of December 2018, the code to fit and visualize the kernel density estimation model for burglaries is as follows:

```
crime_burglary_dec = london_subset\
                [(london_subset["Crime type"] == "Burglary")\
                & (london_subset["Month"] == "2018-12")]
seaborn.jointplot("Longitude", "Latitude", \
                crime_burglary_dec, kind="kde")
```

The output is as follows:

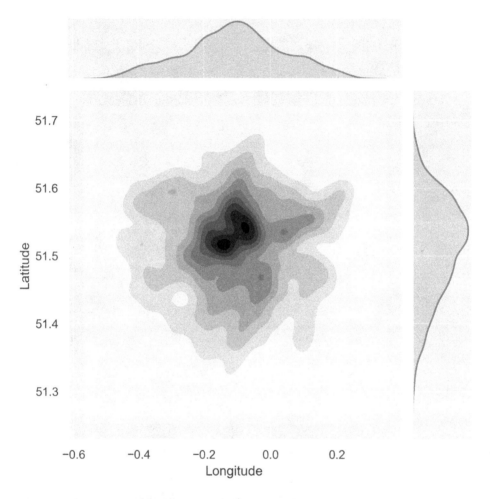

Figure 9.44: The estimated joint and marginal densities for burglaries in December 2018

Once again, we can see that the distributions are quite similar across the months. The only difference is that the densities seem to widen or spread from July to December. As always, the noise and inherent lack of information contained in the sample data is causing small shifts in the estimated densities.

> **NOTE**
>
> To access the source code for this specific section, please refer to https://packt.live/2wmh5yj.
>
> You can also run this example online at https://packt.live/2W0EAGK.

INDEX

A

anaconda: 319, 376, 404
apriori: 313, 318, 325, 327, 343-355, 365, 367
argmax: 184, 289
argmin: 25
asarray: 289
astype: 334

B

bayesian: 274
biplot: 286-287, 291, 295-297
boston: 214

C

cluster: 4, 7, 11, 17-20, 23, 25, 29-30, 33, 39, 42-48, 50-54, 58, 60-61, 66, 69, 72-78, 80, 84, 86, 88-89, 91, 93, 95, 102, 224, 226-227, 241, 243, 316, 395

D

datetime: 331
dbscan: 69, 71, 73, 75-77, 79-82, 84-96
dendrogram: 43, 46-47, 49-51, 54, 60, 69, 72, 74
dtypes: 331

E

eigenvalue: 116-117
epochs: 183, 186, 190, 192, 196, 198, 203, 206, 213, 236

F

framework: 47, 155, 161, 169, 171, 173, 175, 186

G

gaussian: 6, 53, 108, 212-214, 243, 274, 383-384, 386-391, 406, 409

I

imshow: 178, 181, 191, 197-198, 204, 215-216, 226

J

jointplot: 397, 399, 412

L

langdetect: 248-249, 261

M

mlxtend: 319, 329, 339-341, 348-350, 358-359, 367
multiplot: 377

P

pandas: 30-31, 33, 65, 111, 114-115, 118, 121, 123, 128, 133, 136, 143, 147, 150, 228, 235, 242, 249, 256, 279, 319, 330, 340, 342, 377, 396
pyldavis: 249, 286-287, 295
pyplot: 20-21, 31, 47, 54, 60, 80, 89, 111, 123, 128, 136, 143, 147, 161, 165, 175, 188, 194, 200, 214, 230, 237, 249, 289, 319, 377

S

seaborn: 376-377, 380, 396-397, 399, 402, 405, 408, 411-412

T

tensorflow: 169, 171, 175, 188, 195, 200
tokenizers: 262
t-snes: 243

W

wordnet: 250, 265